AF356608

Environmental Odour

Environmental Odour

Editors

Günther Schauberger
Martin Piringer
Chuandong Wu
Jacek Koziel

MDPI • Basel • Beijing • Wuhan • Barcelona • Belgrade • Manchester • Tokyo • Cluj • Tianjin

Editors
Günther Schauberger
University of Veterinary
Medicine
Austria

Martin Piringer
Zentralanstalt für Meteorologie
und Geodynamik
Austria

Chuandong Wu
University of Science and
Technology Beijing
China

Jacek Koziel
Iowa State University
USA

Editorial Office
MDPI
St. Alban-Anlage 66
4052 Basel, Switzerland

This is a reprint of articles from the Special Issue published online in the open access journal *Atmosphere* (ISSN 2073-4433) (available at: https://www.mdpi.com/journal/atmosphere/special_issues/SI_Environmental_Odour).

For citation purposes, cite each article independently as indicated on the article page online and as indicated below:

LastName, A.A.; LastName, B.B.; LastName, C.C. Article Title. *Journal Name* **Year**, *Volume Number*, Page Range.

ISBN 978-3-0365-2363-7 (Hbk)
ISBN 978-3-0365-2364-4 (PDF)

Contents

About the Editors

Günther Schauberger

Nationality: Austria

Date of birth: 8th March 1957

Place of birth: Vienna (Austria)

Professor for Biometeorology at the University of Veterinary Medicine Vienna since 1997 Head of the WG Environmental Health Member of the Climate and Air Quality Commission of the Austrian Academy of Sciences National delegate, working group chair and member of EU-COST Actions, VDI working groups, and the Commission International du Génie Rural (CIGR)

Working areas: environmental odour (emission modelling, meteorological aspects, impact assessement of annoyance, dose-response relationship); climate change and global warming (impact on confined livestock builings, adaptiion measures, econometric aspects); solar and artificial ultraviolet radiation (health impact on humans and animals, personal dosimetry, dose response relationship).

Martin Piringer

Nationality: Austria

Date of birth: 18th February 1956

Place of birth: Vienna (Austria)

Head of Section Environmental Meteorology at ZAMG, the Austrian Weather Service since 2009. National delegate, working group chair and member of EU-COST Actions, Special Issue Editor of COST reports, of the journals *Atmosphere and Urban Climate, editor of Advances in Science and Research for Environmental Meteorology*, Co-Convenor of the Session for Environmental Meteorology at the yearly scientific conference of EMS European Meteorological Society, member of different working groups for odour guidelines in Austria. He has published about 100 peer-reviewed publications and was a long-year referee for scientific journals. Working areas: dispersion modelling including odour, urban meteorology, boundary-layer meteorology conducting vertical soundings with tethersonde systems, sodars, and ceilometers. Retired in March 2021.

Chuandong Wu

Nationality: China

Date of birth: 17th November 1990

Dr. Chuandong Wu is a faculty member at the University of Science and Technology, Beijing, Department of Chemistry, since 2017. He has over 10 years of experience in environmental odour research, including odour analysis by instrumental and sensory methods and odour concentration prediction models. He has published over 30 peer-reviewed journal publications, including *Atmospheric Environment, Chemosphere, Chemical Engineering Journal, Waste Management, Atmosphere,* etc. He is a guest editor for "Environmental Odour" Special Issue in Atmosphere.

Jacek Koziel Prof. Koziel is a faculty member at Iowa State University, Department of Agricultural and Biosystems Engineering. He has over 30 years of experience in conducting engineering research, including the development and assessment of mitigation technologies for odour and gas emissions at and around livestock operations. He published over 190 peer-reviewed journal publications. He is serving as a Senior Editor of *Biosystems Engineering*, associate Editor-in-Chief for *AgriEngineering*, Editorial Board member for *IJERH* and *Atmosphere*. Prof. Koziel is a guest editor for 'Livestock Odor and Air Quality' special issue/collection in *Atmosphere*. He is Fulbright Scholar and member of the American Association for the Advancement of Science.

Preface to "Environmental Odour"

The journal *Atmosphere* is well known for regularly providing Special Issues dealing with topics of high scientific interest and actuality. The current Special Issue "Environmental Odour" is a follow-up of the previous Special Issue "Environmental Odour: Emission, Dispersion, and the Assessment of Annoyance" published in 2020 by two of the current four guest editors. This Special Issue was even more successful than its predecessor: besides the editorial, it is composed of nine original scientific papers and one review paper, convincing the journal to publish this Special Issue as an e-book. This Special Issue deals with the entire spectrum, from the estimation or measurement of odour emissions and the dispersion of odorous substances in the atmosphere to the determination of setback or separation distances and an estimation of odour annoyance levels in a neighbourhood. Each research paper had a specific focus; most considered one element of this chain, while some aimed to cover the entire chain. In particular, this Special Issue encouraged contributions dealing with field trials and dispersion modelling to assess the degree of annoyance and the quantitative success of abatement measures.

The review summarises odour legislation in selected European, North- and South-American, Oceanic and Asian countries. Odour-related assessment criteria tend to be highly variable between countries, individual states, provinces and even counties and towns. The discussion of odour in legislation is a broad topic, relevant and extensively treated in this review. The paper ends with a list of questions that may be used to discuss the formulation of odour regulation.

Several papers of the Special Issue deal with the identification of odour emissions, often from wastewater treatment plants. The reliable determination of their odourant compounds is still challenging. Diverse strategies are in use, starting with gas chromatography and mass spectrometry to identify the spectrum of odorous volatile organic compounds, or using the Odour Profile Method with an odour patrol program to define odour nuisance changes over time. Uncertainties in the quantification of odour measurements are also addressed, caused by, among others, the selection of a panel, the kind of sampling and the stability of the samples.

The assessment of annoyance in the surroundings of an odour source includes the modelling of the dilution of odorous substances in the atmosphere and an evaluation using odour impact criteria (OIC). Odour nuisance can be characterised using trained assessors and questionnaires; atmospheric dispersion models such as CALPUFF or LASAT are used to calculate ambient odour concentrations, and OIC are used to determine separation distances. The latter show a large variation from author to author, depending mainly on the OIC used, which are issued on a national level and vary from country to country. The international harmonisation of OICs is seen as an urgent undertaking for the scientific and the regulator community to ensure analogous separation distances for an equivalent level of protection in the future. The Editorial of this Special Issue summarises the content of the original contributions in more detail and also tackles open scientific questions.

Günther Schauberger, Martin Piringer, Chuandong Wu, Jacek Koziel

Editors

Editorial

Environmental Odour

Günther Schauberger [1],*, Martin Piringer [2],*, Chuandong Wu [3] and Jacek A. Koziel [4]

[1] WG Environmental Health, Department of Biomedical Sciences, University of Veterinary Medicine Vienna, Veterinärplatz 1, 1210 Vienna, Austria

[2] Department of Environmental Meteorology, Central Institute for Meteorology and Geodynamics, HoheWarte 38, 1190 Vienna, Austria

[3] School of Chemistry and Biological Engineering, University of Science and Technology Beijing, Beijing 100083, China; wuchuandong@ustb.edu.cn

[4] Department of Agricultural & Biosystems Engineering, Iowa State University, 4350 Elings Hall, Ames, IA 50011, USA; koziel@iastate.edu

* Correspondence: Gunther.Schauberger@vetmeduni.ac.at (G.S.); Martin.Piringer@zamg.ac.at (M.P.)

Citation: Schauberger, G.; Piringer, M.; Wu, C.; Koziel, J.A. Environmental Odour. *Atmosphere* **2021**, *12*, 1293. https://doi.org/10.3390/atmos12101293

Received: 29 September 2021
Accepted: 2 October 2021
Published: 4 October 2021

Publisher's Note: MDPI stays neutral with regard to jurisdictional claims in published maps and institutional affiliations.

Environmental odour is perceived as a major nuisance by the rural and urban population. The sources of odorous substances are manifold. In urban areas, these include restaurants and services, small manufacturing and other sources that might cause complaints. Wastewater treatment plants, landfill sites and other infrastructures are the expected major odour sources in the suburbs. These problems are often caused by accelerated urban growth. On rural sites, livestock farming and manure spreading on fields, composting plants and biogas reactors are blamed for severe odour annoyance. In fact, environmental odours are a common cause of public complaints by residents to local authorities and regional or national environmental agencies. This Special Issue deals with the entire spectrum, from the estimation or measurement of odour emissions and the dispersion of odorous substances in the atmosphere to the determination of setback or separation distances and an estimation of odour annoyance levels in a neighbourhood. Each research paper had a specific focus; most consider one element of this chain, while some try to cover the entire chain. In particular, this Special Issue encouraged contributions dealing with field trials and dispersion modelling to assess the degree of annoyance and the quantitative success of abatement measures.

This Special Issue, "Environmental Odour", comprises one review and nine original papers. A review by Bokowa et al. [1] summarises odour legislation in selected European countries (France, Germany, Austria, Hungary, the UK, Spain, the Netherlands, Italy and Belgium), North America (the USA and Canada) and South America (Chile and Colombia), as well as Oceania (Australia and New Zealand) and Asia (Japan and China). Many countries have incorporated odour controls into their legislation. However, odour-related assessment criteria tend to be highly variable between countries, individual states, provinces and even counties and towns. The discussion of odour in legislation ranges from no specific mention of odour in the environmental legislation that regulates pollutants known to have an odour impact to extensive details about odour source testing, odour dispersion modelling, ambient odour monitoring, setback distances, process operations and odour control technologies and procedures. The paper ends with a list of questions that may be used to discuss the formulation of odour regulation. As Brancher et al. [2] outlined, the odour impact criteria (OICs) of different jurisdictions do not a priori ensure analogous separation distances for an equivalent level of protection. This must be addressed first, when more homogeneous odour-related assessment criteria among different countries are intended.

Several papers deal mainly with the identification of odour emissions from wastewater treatment plants (WWTPs). The reliable determination of their odourant compounds is still challenging. Gao et al. [3] identified odorous volatile organic compounds (VOCs) from domestic wastewater at different processing units using gas chromatography-ion mobility spectrometry (GC-IMS) and gas chromatography quadrupole-time-of-flight mass spectrometry (GC-QTOF-MS). The results of the latter approach confirmed the odour contribution of organic sulfur compounds in wastewater before primary sedimentation and ruled out the significance of most of the hydrocarbons in wastewater odour. Varied volatile compounds were detected using GC-IMS, mainly oxygen-containing VOCs including alcohols, fatty acids, aldehydes and ketones with low odour threshold values. The GC-IMS technique may provide an efficient profiling method for the changes of inlet water and the performance of the treatment process at WWTPs.

Bian et al. [4] used the Odour Profile Method (OPM) with an odour patrol program; the OPM was based on a seven-level sugar scale for the gustatory sense to calibrate the perception of the intensity of odourants at a school within one mile of the Los Angeles County landfill. A landfill odour wheel was used to identify the odour type. This study shows that an Odour Patrol using the OPM can accurately define odour nuisance changes over time. The OPM not only confirmed the mitigation of a landfill odour problem, but also determined the odour character, the odour intensity, the odour frequency and the odour duration during this study period.

Cipriano et al. [5] discussed uncertainties in the quantification of odour measurements caused by, among others, the selection of a panel (required by dynamic olfactometry), the sampling and the stability of the samples. Proficiency tests (PTs) can help evaluate such contributions. They are, however, often implemented by only using dry gas cylinders containing stable compounds. Consequently, uncertainties related to the sampling activity cannot be assessed. In particular, high odour levels and the presence of water vapour in emission sources can create significant biases due to the sampling techniques used and the chemical reactions that can occur before analysis. Cipriano et al. [5] created an upgraded protocol for implementing PTs for odour determinations in conditions very similar to reality (i.e., high temperatures, high water contents and the presence of chemical interferents).

Hansen et al. [6] compared the Sum of Odour Activity Values (SOAV) method with the odour detection threshold measured using olfactometry and investigated the assumption of additivity. The odour activity value was used for the conversion of chemical concentration values into odour concentrations. Synthetic pig house air with odourants at realistic concentration levels was used in the study (hydrogen sulfide, methanethiol, trimethylamine, butanoic acid and 4-methyl phenol). An olfactometer with only Polytetrafluoroethylene (PTFE) is in contact with the sample air was used to estimate odour threshold values (OTVs) and the odour detection threshold for samples with two to five odourants. The results showed a good correlation (R^2 = 0.88) between the SOAV estimated based on the OTVs for panellists in the present study and values found in the literature. For the majority of the samples, the ratio between the odour detection threshold and the SOAV was not significantly different from one, which indicated that the OAV for individual odourants in a mixture can be considered additive. In conclusion, the assumption of the additivity between odourants measured in pig house air seems reasonable, but the strength of the method is determined by the OTV data used. The SOAV concept was, in the first Special Issue of Environmental Odour used by Park [7] and discussed in detail by Wu et al. [8].

The assessment of annoyance in the surroundings of an odour source is a complex issue that, apart from the estimation of odour emissions, includes the dilution of odorous substances in the atmosphere and an evaluation using OICs. Zarra et al. [9], for an Italian WWTP, and Zhang et al. [10], for a WWTP in Northern China, characterise odour nuisance using trained assessors and questionnaires, applied atmospheric dispersion modelling to calculate ambient odour concentrations and used OICs to determine separation distances. Although both use the Lagrangian dispersion model CALPUFF, the resulting isopleths of separation distances are very different, which is also attributable to the different OICs used. In contrast, Zarra et al. [9] calculated separation distances for hourly average odour concentration threshold values of 1.0 and 1.5 ou_E m^{-3} and the 85th and the 98th percentile, resulting in separation distances of up to a few 1000 m around the source. Zhang et al. [10] applied threshold values from 1 to 5 ou_E m^{-3} and percentiles from 70 to 98. The best predictor of odour exposure was obtained with a threshold value of 4 ou_E m^{-3} at the 99th percentile, resulting in separation distances of only a few hundred metres. However, both groups of authors reported a good agreement of the model-calculated separation distances with the odour nuisance levels obtained from the questionnaires and the trained assessors. An essential contribution of these papers is a dose–response function between the odour exposure and the annoying potential of WWTP odour.

Ravina et al. [11] analysed separation distances around a WWTP in Northern Italy. Odour dispersion modelling was carried out again with the CALPUFF model. For low odour concentration thresholds (C_T = 1 ou_E m^{-3}), the results showed that two different years (2018 and 2019) provided similar patterns of the separation distances. The difference between the two years tended to increase by increasing the concentration threshold value (C_T = 3 ou_E m^{-3} and C_T = 5 ou_E m^{-3}). The second phase of the assessment was the selection of the open field correction method for wind velocity used in the calculation of odour emission rates (OERs). The following three different relationships were considered: the power law, the logarithmic law and the Deaves–Harris (D–H) law. The results showed that OERs and separation distances varied, depending on the selected method. Taking the power law as the reference, the average variability of the separation distances was between −7% (D–H law) and +10% (logarithmic law). Higher variability (up to 25%) was found for single transport distances. The study provides knowledge toward a better alignment of the concept of the odour impact criteria.

Piringer et al. [12] investigated the impact of odour sources as livestock buildings on neighbouring residential areas due to climate change. Separation distances were calculated for two Central European sites with considerable livestock activity influenced by different orographic and climatic conditions. Two climate scenarios were considered, namely, the time period 1981–2010 (present climate) and the period 2036–2065 (predicted future climate). Based on the provided climatic parameters, stability classes were derived as an input for local-scale air pollution modelling. The separation distances were determined using the Lagrangian particle diffusion model LASAT. The main findings comprise the changes of stability classes between the present and the future climate and the resulting changes in the modelled odour impact. The model results based on different schemes for stability classification were compared. With respect to the selected climate scenarios and the variety of the stability schemes, a bandwidth of the affected separation distances resulted. The investigation revealed the extent, to which livestock husbandry will have to adapt to climate change, e.g., with impacts on today's licensing (permitting) processes.

Countries with no specific requirements for managing environmental odour can promote the use of empirical equations as a first-guess or screening tool to estimate possible areas affected by odour annoyance. Brancher et al. [13] compared separation distances obtained from selected empirical equations with those from dispersion models AERMOD and LASAT for sites in Brazil, China and Austria. As the separation distance shape often resembles the wind distribution of a site, wind data should be included in such approaches. Otherwise, the resultant separation distance shape is simply given by an idealised circle around the emission source. The results of this investigation suggested that some empirical equations reach their limitation in the sense that they are not successful in capturing the inherent complexity of dispersion models. However, empirical equations, developed for Germany and Austria, have the potential to deliver reasonable results, especially if used within the conditions for which they were designed. The main advantage of empirical equations lies in the simplification of the meteorological input data and their use in a fast and straightforward approach.

This Special Issue presents a broad perspective of the current status and main aspects of environmental odour as highlighted by the contributing scientific community. Although the results discussed here summarise cutting-edge research on air quality, they also open additional scientific questions, confirming that the topic of environmental odour still presents substantial challenges. While the quantification of odour emissions is, to a great extent, successfully regulated [3–5], OICs, which are necessary to assess annoyance in residential areas around odour sources, are issued on national levels and vary from country to country [2,14]. Some countries such as China, Japan and South Korea use odour standards based on limit values for ambient odour concentration rather than OICs. Therefore, the international harmonisation of OICs is seen as an urgent undertaking for the scientific and the regulator community to ensure analogous separation distances for an equivalent level of protection in the future.

Author Contributions: All four guest editors (G.S., M.P., C.W., and J.A.K.) contributed to this editorial. All authors have read and agreed to the published version of the manuscript.

Funding: This editorial received no external funding.

Acknowledgments: The editors would like to thank the authors from countries all over the world for their valuable contributions, the reviewers for their constructive comments and suggestions that helped to improve the manuscripts and Calvin Li from the editorial office for his excellent support in processing and publishing this issue.

Conflicts of Interest: The authors declare no conflict of interest.

References

1. Bokowa, A.; Diaz, C.; Koziel, J.A.; McGinley, M.; Barclay, J.; Schauberger, G.; Guillot, J.-M.; Sneath, R.; Capelli, L.; Zorich, V.; et al. Summary and overview of the odour regulations worldwide. *Atmosphere* **2021**, *12*, 206. [CrossRef]
2. Brancher, M.; Piringer, M.; Grauer, A.F.; Schauberger, G. Do odour impact criteria of different jurisdictions ensure analogous separation distances for an equivalent level of protection? *J. Environ. Manag.* **2019**, *240*, 394–403. [CrossRef] [PubMed]
3. Gao, W.; Yang, X.; Zhu, X.; Jiao, R.; Zhao, S.; Yu, J.; Wang, D. Limitations of GC-QTOF-MS technique in identification of odorous compounds from wastewater: The application of GC-IMS as supplement for odor profiling. *Atmosphere* **2021**, *12*, 265. [CrossRef]
4. Bian, Y.; Gong, H.; Suffet, I.H. The use of the odor profile method with an "odor patrol" panel to evaluate an odor impacted site near a landfill. *Atmosphere* **2021**, *12*, 472. [CrossRef]
5. Cipriano, D.; Cefalì, A.M.; Allegrini, M. Experimenting with odour proficiency tests implementation using synthetic bench loops. *Atmosphere* **2021**, *12*, 761. [CrossRef]
6. Hansen, M.J.; Adamsen, A.P.S.; Wu, C.; Feilberg, A. Additivity between key odorants in pig house air. *Atmosphere* **2021**, *12*, 1008. [CrossRef]
7. Park, S. Odor characteristics and concentration of malodorous chemical compounds emitted from a combined sewer system in Korea. *Atmosphere* **2020**, *11*, 667. [CrossRef]
8. Wu, C.; Liu, J.; Zhao, P.; Piringer, M.; Schauberger, G. Conversion of the chemical concentration of odorous mixtures into odour concentration and odour intensity: A comparison of methods. *Atmos. Environ.* **2016**, *127*, 283–292. [CrossRef]
9. Zarra, T.; Belgiorno, V.; Naddeo, V. Environmental odour nuisance assessment in urbanized area: Analysis and comparison of different and integrated approaches. *Atmosphere* **2021**, *12*, 690. [CrossRef]
10. Zhang, Y.; Yang, W.; Schauberger, G.; Wang, J.; Geng, J.; Wang, G.; Meng, J. Determination of dose–response relationship to derive odor impact criteria for a wastewater treatment plant. *Atmosphere* **2021**, *12*, 371. [CrossRef]
11. Ravina, M.; Bruzzese, S.; Panepinto, D.; Zanetti, M. Analysis of separation distances under varying odour emission rates and meteorology: A wwtp case study. *Atmosphere* **2020**, *11*, 962. [CrossRef]
12. Piringer, M.; Knauder, W.; Baumann-Stanzer, K.; Anders, I.; Andre, K.; Schauberger, G. Odour impact assessment in a changing climate. *Atmosphere* **2021**, *12*, 1149. [CrossRef]
13. Brancher, M.; Piringer, M.; Knauder, W.; Wu, C.; Griffiths, K.D.; Schauberger, G. Are empirical equations an appropriate tool to assess separation distances to avoid odour annoyance? *Atmosphere* **2020**, *11*, 678. [CrossRef]
14. Sommer-Quabach, E.; Piringer, M.; Petz, E.; Schauberger, G. Comparability of separation distances between odour sources and residential areas determined by various national odour impact criteria. *Atmos. Environ.* **2014**, *95*, 20–28. [CrossRef]

Article

Odour Impact Assessment in a Changing Climate

Martin Piringer [1,*], **Werner Knauder** [1], **Kathrin Baumann-Stanzer** [1], **Ivonne Anders** [2], **Konrad Andre** [3] **and Günther Schauberger**

[1]   Section for Environmental Meteorology, Zentralanstalt für Meteorologie und Geodynamik, Hohe Warte 38, 1190 Vienna, Austria; werner.knauder@zamg.ac.at (W.K.); kathrin.baumann-stanzer@zamg.ac.at (K.B.-S.)
[2]   Deutsches Klimarechenzentrum, Bundesstraße 45a, 20146 Hamburg, Germany; anders@dkrz.de
[3]   Section for Climate Research, Zentralanstalt für Meteorologie und Geodynamik, Hohe Warte 38, 1190 Vienna, Austria; konrad.andre@zamg.ac.at
[4]   WG Environmental Health, Unit for Physiology and Biophysics, University of Veterinary Medicine, Veterinärplatz 1, 1210 Vienna, Austria; gunther.schauberger@vetmeduni.ac.at
*   Correspondence: martin.piringer@zamg.ac.at

Abstract: (1) Background: The impact of odour sources as stock farms on neighbouring residential areas might increase in the future because the relevant climatic parameters will be modified due to climate change. (2) Methodology: Separation distances are calculated for two Central European sites with considerable livestock activity influenced by different orographic and climatic conditions. Furthermore, two climate scenarios are considered, namely, the time period 1981–2010 (present climate) and the period 2036–2065 (future climate). Based on the provided climatic parameters, stability classes are derived as input for local-scale air pollution modelling. The separation distances are determined using the Lagrangian particle diffusion model LASAT. (3) Results: Main findings comprise the changes of stability classes between the present and the future climate and the resulting changes in the modelled odour impact. Model results based on different schemes for stability classification are compared. With respect to the selected climate scenarios and the variety of the stability schemes, a bandwidth of affected separation distances results. (4) Conclusions: The investigation reveals to what extent livestock husbandry will have to adapt to climate change, e.g., with impacts on today's licensing processes.

Keywords: livestock; odour dispersion modelling; separation distance; climate change; stability classification

Citation: Piringer, M.; Knauder, W.; Baumann-Stanzer, K.; Anders, I.; Andre, K.; Schauberger, G. Odour Impact Assessment in a Changing Climate. *Atmosphere* **2021**, *12*, 1149. https://doi.org/10.3390/atmos12091149

Academic Editors: Laura Maria Teresa Capelli and Roberto Bellasio

Received: 28 July 2021
Accepted: 17 August 2021
Published: 6 September 2021

Publisher's Note: MDPI stays neutral with regard to jurisdictional claims in published maps and institutional affiliations.

1. Introduction

The calculation of ambient concentrations of airborne emissions is usually performed by dispersion models [1]. The influence of the climate change signal on the dispersion parameters is analysed in this article. In the current investigation, the emission of odorous substances from confined livestock production is analysed, because the perception of odour is one of the most frequent causes for environmental complaints. The ambient odour concentration is evaluated by the separation distance, which is determined by percentiles between 85% and 98% of the exceedance probability of a preselected threshold [2,3]. Some advantages and disadvantages of the proposed methodology (use of dispersion models with regional climate data) is shown and discussed in this manuscript.

A separation distance is an easy to understand means to quantify the impact of an odour-emitting facility such as a livestock farm on the neighbouring residents. The direction-dependent separation distance divides the area around a source in a zone which is protected from annoyance and a zone closer than the separation distance where annoyance can be expected. The protection level depends on the land use category; the higher the protection level, the farther the separation distance [4]. Separation distances are determined by the use of local-scale atmospheric dispersion models usually based on

5

annual representative time series of the relevant meteorological parameters, namely, wind speed, wind direction and some information on atmospheric stability.

In a changing climate, it is expected that the predicted temperature rise will also affect parameters needed to determine atmospheric stability. In [5], it is investigated how the climatic changes in wind conditions and stability will affect the separation distances around livestock farms dedicated to protecting the neighbourhood from odour annoyance. In the referenced study, atmospheric stability is determined using cloudiness and radiation data combined with wind speed. In the current investigation, stability classes determined based on global radiation during daytime and the vertical temperature gradient during night are used in addition, and the resulting separation distances are compared to those presented in [5].

2. Materials and Methods

The two focus regions, Upper Austria north of the Alpine chain (centred around Wels, 48.16° N, 14.07° E) (Figure 1)and South-Eastern Styria (centred around Feldbach, 46.95° N, 15.88° E), are located within class Cfb (warm temperature, fully humid, warm summers) following the climate classification of Köppen and Geiger [6]. Instead of single sites, regions were chosen for the simulations, representative for livestock farming in Austria. The average meteorological conditions of the two regions are different: whereas the area around Wels north of the main Alpine chain is windy and humid, the focus region Feldbach south of the main Alpine chain is slightly more influenced by the Mediterranean and the Eastern European climate. For this investigation, meteorological parameters were needed which were derived from regional dynamical climate model output. With respect to the observed temperature, a so-called reference year was constructed from model output of a COSMO-CLM [7] simulation for the time period 1981–2010 covering the Alpine Region with a spatial resolution of 9 km forced by ERA-interim reanalysis data [8]. Simulated annual time series of the meteorological parameters air temperature, wind direction, wind speed, radiation balance, global radiation and total cloud cover were provided for the two regions of investigation on an hourly basis representative for the climate period 1981–2010. For the future climatic scenario, the climate change signal was derived from EURO-CORDEX regional climate model simulations for the mean temperature only, which was the most robust signal [9,10]. The climate change scenario was RCP4.5 which assumes an emission maximum in the middle of the century and a stabilisation of the CO_2 concentration at the end of the century. The ensemble mean temperature change was added to the observed temperature, and a second reference year was derived from the regional-scale climate model simulations, which was representative for the future climate period of 2036–2065.

Figure 1. Map of Austria showing the two focus regions (red areas) Wels (in the north) and Feldbach (in the south-east).

The dispersion model LASAT [11], which simulates the dispersion and the transport of a representative sample of tracer particles (in this case odorants) utilising a random walk process (Lagrangian simulation),was used to calculate separation distances. It computed the transport of passive trace substances in the lower atmosphere (up to heights of about 2000 m) on a local and regional scale (up to distances of about 150 km). LASAT is usually run with the Klug–Manier (K-M) stability scheme [12]. K-M classes were numbered from I to V and represented a simplified characterisation of the turbulence situation:

Dispersion categories V and IV comprise very unstable and unstable conditions, meaning good vertical mixing in the boundary layer. They did not occur during night-time. Category V occurred only between May and September.

Dispersion categories III/2 and III/1 are classified as neutral. III/2 occurred predominantly at daytime, III/1 predominantly at night-time and during sunrise and sunset. These categories are typical for cloudy and windy conditions.

Dispersion categories II and I comprise stable and very stable conditions, mostly, but not exclusively, at night.

In [5], cloudiness and radiation data, each in combination with wind speed, were used to determine stability classes. Here, in addition, the stability classification scheme proposed by the U.S. EPA (2000) [13] was applied, based on global radiation during daytime and the vertical temperature gradient at night. The scheme is depicted in Table 1, already transformed into K-M classes as required by the dispersion model LASAT. For completeness, the two other schemes are summarised in Tables 2 and 3, taken from [5]. The six Pasquill stability classes A, B, C, D, E and F were assigned to K-M classes as follows: A . . . V, B . . . IV, C . . . III/2, D . . . III/1, E . . . II, F . . . I.

Table 1. Scheme to determine U.S. EPA stability classes (transformed to K-M classes).

DAYTIME (Global Radiation $\geq$ 20 Wm^{-2})				
	Global Radiation (Wm^{-2})			
Wind Speed (ms^{-1})	$\geq$925	925–675	675–175	175–20
<2	V	V	IV	III/1
2–2.9	V	IV	III/2	III/1
3–4.9	IV	IV	III/2	III/1
5–5.9	III/2	III/2	III/1	III/1
$\geq$6	III/2	III/1	III/1	III/1

NIGHTTIME (Global Radiation < 20 Wm^{-2})		
	Vertical Temperature Gradient (K(100 m)$^{-1}$)	
Wind Speed (ms^{-1})	<0	$\geq$0
<2	II	I
2–2.9	III/1	II
$\geq$3	III/1	III/1

Table 2. VDI scheme to determine Klug/Manier stability classes via cloudiness data (taken from [5]).

Wind Speed υ_{10} at 10 m Height (z_0 = 0.1 m)	**Night-Time**		**Daytime**		
	Total Cloud Cover		**Total Cloud Cover**		
in ms^{-1}	0/8 to 6/8	7/8 to 8/8	0/8 to 2/8	3/8 to 5/8	6/8 to 8/8
$\leq$1.2	I	II	IV	IV	IV
1.3 to 2.3	I	II	IV	IV	III/2
2.4 to 3.3	II	III/1	IV	IV	III/2
3.4 to 4.3	III/1	III/1	IV	III/2	III/2
$\geq$4.4	III/1	III/1	III/2	III/1	III/1

Table 3. KTA scheme to determine stability classes based on the radiation balance and the wind speed (taken from [5]).

Wind Speed v_{10} at 10 m Height	Radiation Balance in Wm^{-2}				
	Limits of Categories				
In ms^{-1}	A/B	B/C	C/D	D/E	E/F
0 to 0.9	214	125	60	−2	−9
1.0 to 1.9	214	126	60	−4	−13
2.0 to 2.9	301	162	60	−6	−21
3.0 to 3.9	400	232	63	−12	−34
4.0 to 4.9	495	305	67	−28	−55
5.0 to 5.9	–	376	84	−55	–
6.0 to 6.9	–	450	108	–	–
7.0 to 7.9	–	–	150	–	–
8.0 to 9.9	–	–	240	–	–
≥10.0	All values category D				

Example:
If the conditions 2.0 ms^{-1} ≤ u$_{10}$ < 3.0 ms^{-1} and 162 Wm^{-2} ≥ radiation balance > 60 Wm^{-2} were fulfilled, then category C was used.

The source term data used for the presented LASAT model runs are summarised in Table 4. The odour emission rate and the odour concentration were given in so-called European odour units (ou$_E$). By definition, the amount of exposure at which 50% of the panellists cannot smell the odour but 50% can, is equal to 1 odour unit per cubic metre. The source is assumed non-buoyant, i.e., the effective stack height is equal to the physical stack height. The source was a livestock building, 3 m high, 120 m long and 30 m wide. Along the main axis of the building, 9 stacks were located, centred at the middle of the roof, with an equal distance of 13.3 m. Each stack had a volume flow rate of 20,000 m^3 h^{-1} and an odour emission rate of 1500ou$_E$s^{-1}. The total emission rate corresponded to 1930 fattening pigs, 113,000 broilers or 940 cattle [14]. Odorous substances were treated such as inert gases without modification during travel time. The odour emission of livestock buildings was composed of several tenths of substances. The main odorous substances causing odour sensation from pig houses are summarised by Zahn et al. [15]. Liu [16] gives the concentration and the odour threshold concentration for chemical substances which can be expected in pig houses. The following chemical substances were identified as potentially important odorant substances in pig odour with growing molar mass: ammonia, hydrogen sulphide, methanethiol, trimethylamine, propanoic acid, 2,3-butanedione, butanoic acid, pentanoic acid, 3-methylbutanoic acid, 4-methylphenol and 3-methyl-1H-indole. The source term data of Table 4 are very typical for Austrian livestock units and assumed constant over time; this assumption is discussed in Section 4.

Table 4. Source term data for dispersion calculations.

Stack height	(m)	5.0
Stack diameter	(m)	1.88
Number of stacks		9
Outlet air velocity	(ms^{-1})	2.0
Volume flow rate	(m^3 h^{-1})	180,000
Temperature	(°C)	0
Odour emission rate	(ou$_E$s^{-1})	13,500
Concentration	(ou$_E$m^{-3})	270

Separation distances were calculated for the odour impact criteria defined in [17], namely, exceedance probabilities of 10% for pure residential and 15% for commercial/ industrial and agriculturally dominated areas, each in combination with a preselected odour concentration threshold of 1ou$_E$m^{-3}.

Usually, hourly mean values of concentrations were simulated by a dispersion model. Thus, a transformation of the hourly mean values calculated by the model to short-term concentrations relevant for human odour perception was necessary. The short-term peak concentrations were obtained via a peak-to-mean approach depending on atmospheric stability; the resulting site-independent peak-to-mean attenuation curves are shown in Figure 1 in [5]. For unstable conditions (classes V and IV), the peak-to-mean factors, starting at rather high values of about 10 near the source, rapidly approached 1 at about 100 m from the source. This is in agreement with the premise that vertical turbulent mixing can lead to short periods of local high ground-level concentrations, whereas the ambient mean concentrations were low. For neutral conditions (classes III/2 and III/1), the decrease in the peak-to-mean ratio was more gradual with increasing distance, because vertical mixing was reduced and horizontal diffusion was dominating the dispersion process. The peak-to-mean ratio in 100 m was then between 2 and 4. For stable conditions (classes II and I), the peak-to-mean ratio exceeded 2 only near the source. Stability class III/2 ("neutral") gave the largest peak-to-mean factor for all distances between 100 and 500 m. In [18], the approach was described in detail; in [5], a shortened description was provided.

3. Results

3.1. Stability Classes

Here, only the parameters needed to determine EPA stability classes were presented, apart from the wind data already shown in [5]. The average daily course of global radiation was as expected (Figure 2): daytime values were on average culminating around noon at approx. 400 Wm^{-2} at both sites. The maximum values at Feldbach were slightly higher than at Wels because the south-alpine site was positioned further south and it was less cloudy and windy than the area around Wels. Only at Feldbach, a slight increase in global radiation around noon was calculated for the future climate, whereas no changes were shown for Wels. By definition, global radiation was zero during night-time.

To further confirm that the climate model data were consistent with expectations, seasonal variations in global radiation for Feldbach and Wels in the present climate are depicted in Figure 3. Feldbach showed especially higher noon values than Wels during the winter months. This was expected because the flatland site in Wels experiences more cloudiness and fog during winter than Feldbach. Similar results were obtained for the future climate period (not shown).

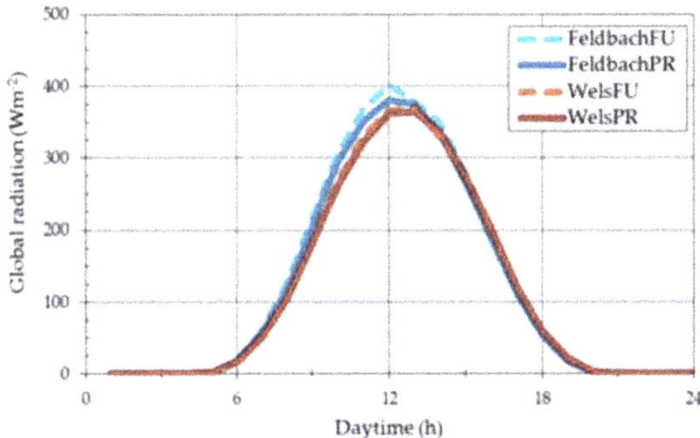

Figure 2. Annual mean of the daily course of global radiation for Feldbach and Wels (PR = present (1981–2010), FU = future (2036–2065)).

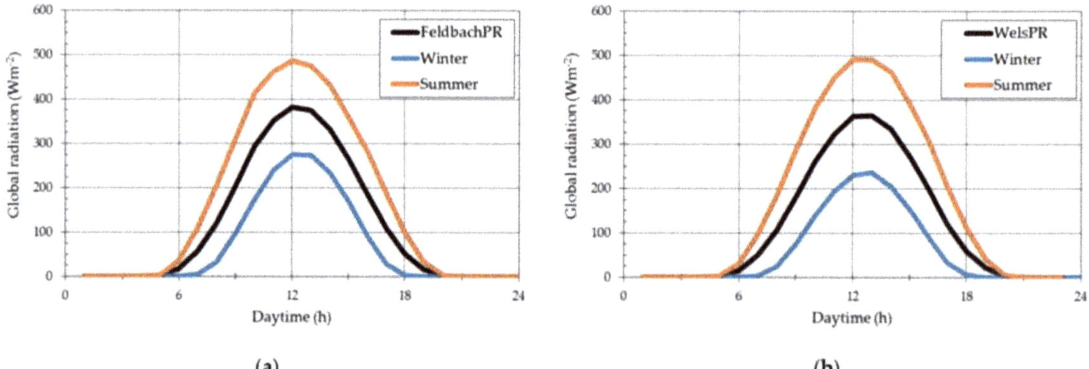

(a)   **(b)**

Figure 3. Winter, summer and annual means of the daily course of global radiation for (**a**) Feldbach and (**b**) Wels; PR—present (1981–2010).

For both climate scenarios, daytime vertical temperature gradients were on average super-adiabatic (Figure 4), which is impossible in the free atmosphere. Day–night differences were more pronounced in Feldbach than in Wels caused by the lower winds and clearer skies at the former. Only at Feldbach, a decrease in the positive and increase in the negative temperature gradients were calculated for the future climate, whereas no changes at Wels occurred. The vertical temperature gradient was used only at night to determine stability classes (Table 1) when the data from the regional dynamical climate model output delivered plausible values.

The consistency of the climate data and the regional differences were further confirmed by Figure 5, which showed the dependence of the vertical temperature gradient on cloud amount at both sites in the present climate. As expected, Feldbach showed a larger variation in nocturnal temperature gradients than Wels, mainly due to the prevalent calm wind conditions at night. Wels, on the contrary, showed less variation during night-time, but a larger variation during daytime. On cloudy days, the daytime course was strongly reduced and lay almost in the expected daily range, with only a slight superadiabatic lapse rate during noon hours. Again, very similar features will occur in the future climate period (not shown).

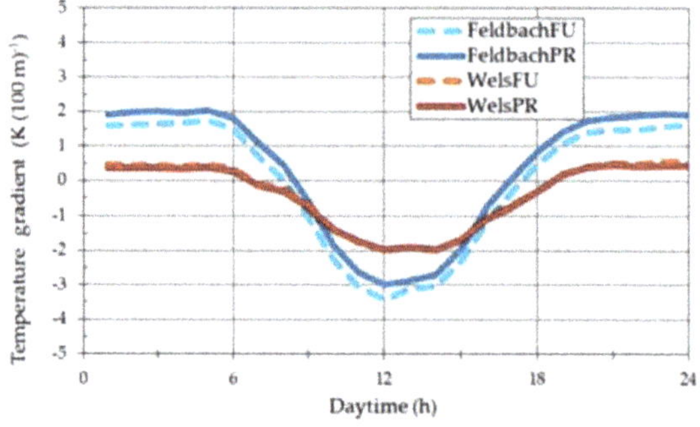

Figure 4. Annual mean of the daily course of the vertical temperature gradient $(K\,(100\,m)^{-1})$ for Feldbach and Wels (PR—present (1981–2010); FU—future (2036–2065)).

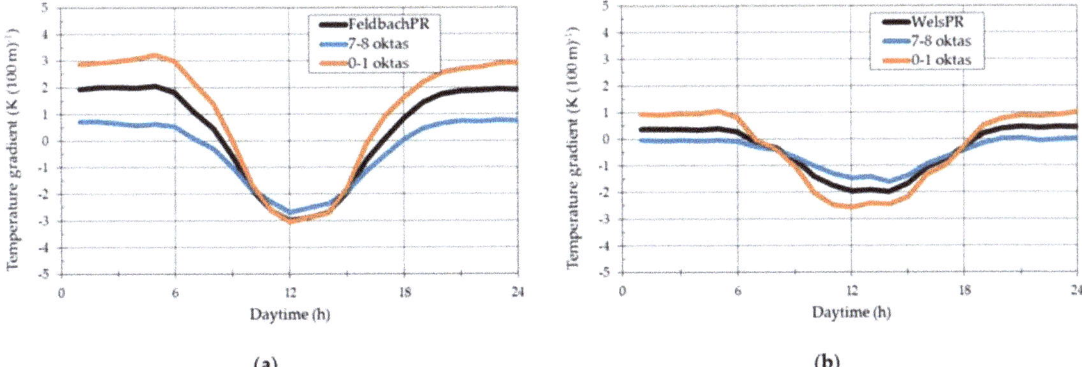

Figure 5. Annual mean of the daily course of the vertical temperature gradient (K $(100\ m)^{-1}$) for low cloud cover (orange line), high cloud cover (blue line) and all data (black line) for (**a**) Feldbach and (**b**) Wels; PR—present (1981–2010).

In Figure 6, the EPA stability classes were grouped from very stable (class I) to very unstable (class V). There were systematic differences in the distribution of stability classes at both sites. At Wels, in the present scenario, class III/1 with more than 50% occurred most frequently (neutral at night). Class I showed a frequency of less than 20%, classes II, III/2 and IV around 10%. The very unstable class V occurred very seldom. In the future, stability class I will occur slightly more frequent, with an increase of about 2%. Class III/1 will occur slightly less frequent, whereas almost no changes were seen for the other classes.

Feldbach was generally characterised by more stable and unstable situations and less neutral conditions, compared to Wels. Classes I and III/1 showed frequencies of 30% and slightly more. For classes II, III/2 and IV, similar frequencies of about 10% were obtained. Comparing the present and the future scenario, the share for stability classes I and IV will be slightly reduced and that for class III/1 slightly increased. Almost no changes occurred for the other stability classes.

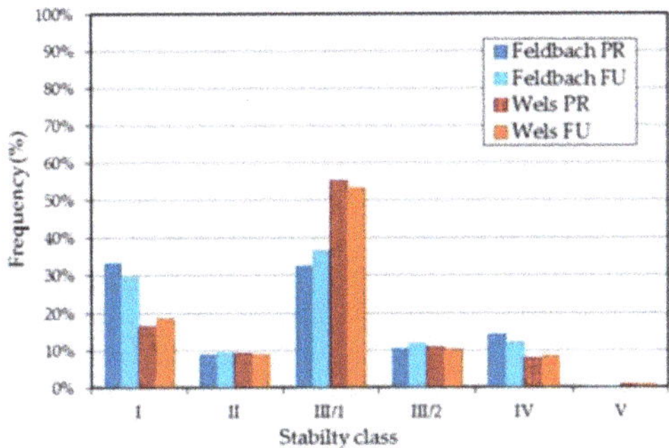

Figure 6. Frequency distribution of EPA stability classes determined according to Table 1 at Feldbach and Wels (PR—present (1981–2010); FU—future (2036–2065)).

For comparison and to facilitate reading, the results for the other two methods (use of cloudiness data and radiation balance) presented in [2] are displayed in Figure 7. Compared to the results for the EPA method (Figure 5), the two other methods showed a much lower occurrence of class III/1, at both sites. Especially when using the KTA method (Figure 7b), a larger abundance for class I was obtained, at both sites. Using different methods to determine atmospheric stability, differences in the frequency distributions of the stability classes hadto be expected [4]. The choice of the method used will mostly depend on the available data; for the current investigation, data were available to apply three different methods to determine atmospheric stability. The resultant effect on the separation distances is shown in Section 3.2.

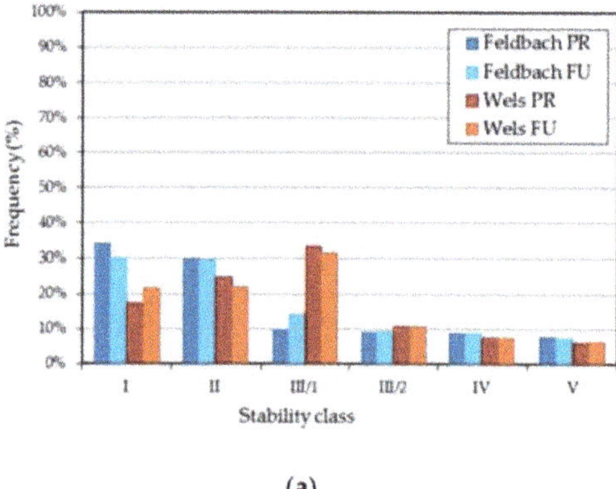
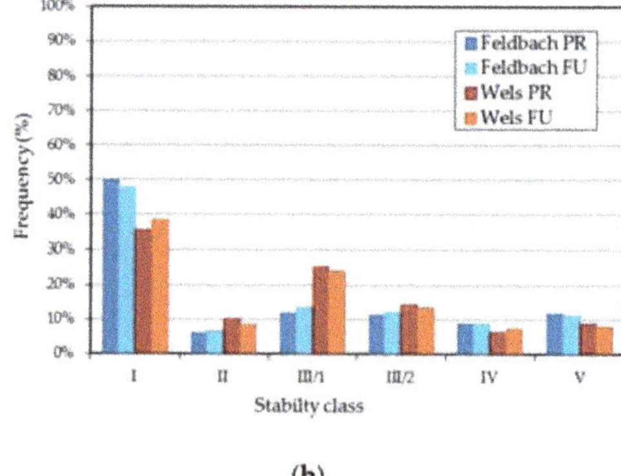

(a)   (b)

Figure 7. Frequency distribution of stability classes determined by (**a**) cloudiness and wind speed (VDI method, Table 2) and (**b**) radiation balance and wind speed (KTA method, Table 3) at Feldbach and Wels (PR—present (1981–2010); FU—future (2036–2065)) (taken from [5]).

3.2. Separation Distances

In Figures 8 and 9, the separation distances for the two selected protection levels determined by the EPA stability method were compared to those delineated from the VDI and the KTA methods [5]. They were calculated for two odour impact criteria according to [15], namely, a threshold of $1ou_E$ m^{-3} and exceedance probabilities of 10% for pure residential areas (Figure 8) and 15% for commercial/industrial and agriculturally dominated areas (Figure 9).The separation distances were shown as isopleths, encompassing the area of exceedance of the given thresholds. An increase in (tolerated) exceedance probability reduced the affected area; a limit value of 15% (Figure 9 was, thus, more unfavourable for residents than a limit value of 10% (a higher level of protection, Figure 8). The black rectangle in the middle of all charts is the livestock building. The area displayed was 1000×1000 m^2, with the odour source in the centre. The black lines depict separation distances obtained by the EPA method, the blue lines those of the VDI method and the red lines those of the KTA method. The affected area in the region centred around Wels is shown in red, the affected area in the region centred around Feldbach in blue. The spatial resolution of the separation distances was 5 m.

The addition of the EPA separation distances did not change the basic findings discussed in [5], i.e., the generally larger separation distances at Wels compared to Feldbach. This was due to the larger wind speeds and the more frequent neutral stability conditions found at Wels. For both the present and the future climate scenario, for both protection levels and at both sites, the EPA method delivered larger separation distances than the other methods. The increase was largest in Feldbach, in the present climate towards the north, in the future climate towards the south-east, and amounted to about 40 m. At Wels, the increase was larger for the lower protection level (Figure 9). At both sites, separation distances determined by the EPA method for the main wind directions will slightly increase in the future. This increase in separation distances means that a larger area might be affected by odour annoyance due to climate change, even if the emissions from the respective site were assumed unchanged. Compared to the other two stability classification schemes, the EPA method delivered the largest area of neighbourhood protection.

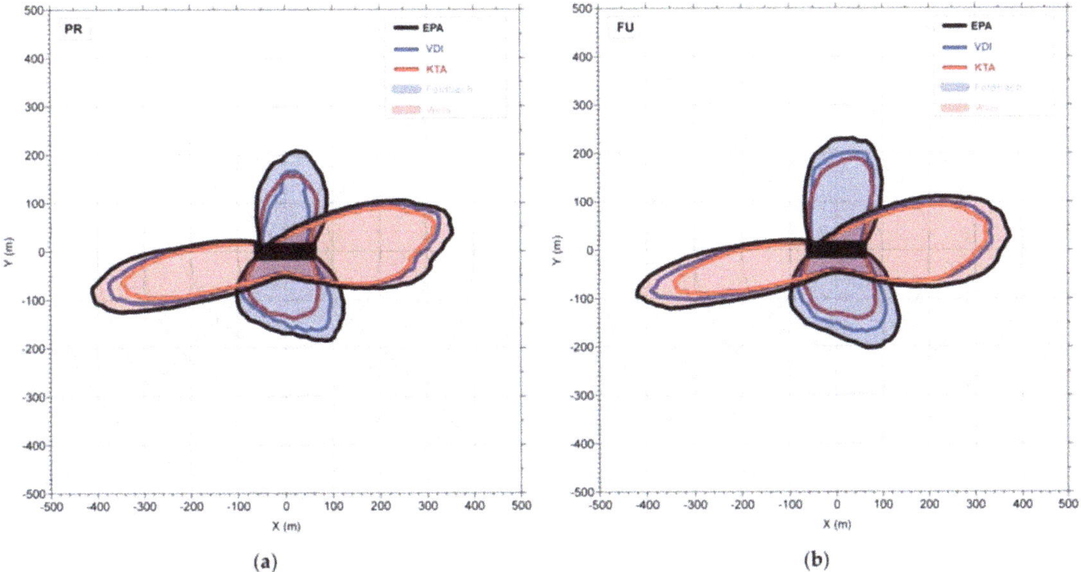

Figure 8. Direction-dependent separation distances (m) for 10% exceedance probability (pure residential areas) at Feldbach and Wels, (**a**) for PR (present climate (1981–2010)) and (**b**) FU (future climate (2036–2065)); based on a combination of global radiation and the vertical temperature gradient (EPA), cloudiness (VDI) and radiation balance (KTA), each in combination with wind speed, to determine stability classes; the black rectangle is the livestock building.

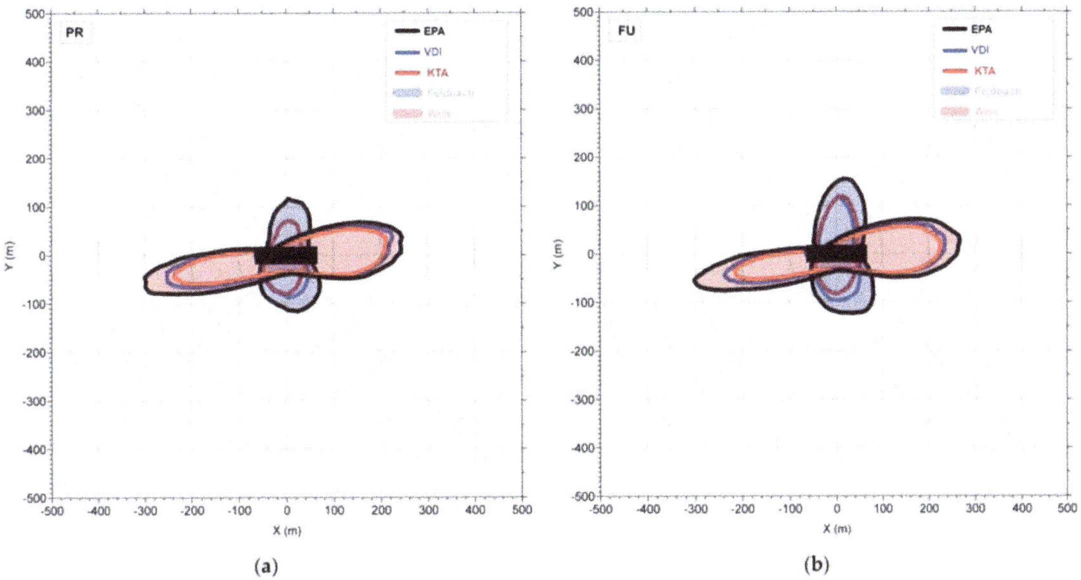

Figure 9. Direction-dependent separation distances (m) for 15% exceedance probability (commercial/industrial and agriculturally dominated areas) at Feldbach and Wels, (**a**) for PR (present climate (1981–2010)) and (**b**) FU (future climate (2036–2065)); based on a combination of global radiation and the vertical temperature gradient (EPA), cloudiness (VDI) and radiation balance (KTA), each in combination with wind speed, to determine stability classes; the black rectangle is the livestock building.

4. Discussion and Conclusions

As demonstrated in Figures 8 and 9, separation distances showed variations, in this case depending on the used meteorological input data. In the presented scenarios, the differences resulted primarily from variations in the frequency distributions of stability classes (Figures 6 and 7). The EPA method (Table 1) generally delivered the largest separation distances because the abundance of neutral stability classes was largest, associated with the highest peak-to-mean factors [5]. This result was valid at both sites despite the very different orographic and meteorological conditions and for both climate scenarios. At the flatland site in Wels, separation distances were generally larger than at Feldbach, due to the on average higher wind speeds and the more frequent neutral atmospheric stratification. At Feldbach, in addition, a very special result showed up, already discussed in [5]: If a frequency of exceedancesof 15% was tolerated, no separation distance north of the livestock unit was determined in the present climate with the cloudiness method (Figure 9a). This was explained by the combination of relatively low wind speeds and low frequencies of neutral stability (Figure 7a) which were associated with the highest peak-to-mean factor.

Earlier investigations revealed a similarly large variety of areas affected by odour annoyance under different orographic and climatic conditions. In [4], it could be demonstrated that separation distances for a site in a narrow valley can be larger than those in the flatlands. Although the flatland site was characterised by higher wind speeds, lower turbulence and somewhat larger peak-to-mean factors than the valley site, the larger separation distances there resulted from a frequent combination of the along-valley channelling of the flow in combination with frequent stable conditions causing higher odour concentrations and leading to the enhanced separation distances. The specific on-site meteorological conditions, thus can, exert a profound and sometimes surprising influence on resulting separation distances. An investigation in [18], carried out at an Austrian flatland site, compared separation distances obtained with a Gaussian and a Lagrangian model using site-specific peak-to-mean ratios and factor four used independent of the distance from the source and the meteorological conditions [12]. A continuous increase in separation distances from the Gaussian over the Lagrangian model to using the factor four (for both models) was found. The separation distances resulting from the factor four were judged unrealistically large, especially in the main wind directions and at sites where low wind speeds and/or stable dispersion categories are prevalent.

In the present investigation, the regional dynamical climate model output instead of observations at meteorological stations was used to determine separation distances (Section 2). The extracted data allowed the use of three different methods to determine stability classes (Tables 1–3). The frequency distributions of stability classes varied from method to method, but generally met the expected range and tendency. In particular, the dominance of neutral stability classes at flatland sites versus the large abundance of stable and unstable classes at valley or basin sites was confirmed also by the use of the model data. The fact that the climate model data delivered unrealistic superadiabatic vertical temperature gradients during daytime at both sites (Figures 4 and 5) was not relevant for the determination of EPA stability classes, as the vertical temperature gradient was used only at night-time (Table 1) when it showed a realistic range of values. The range of separation distances also agreed well with those derived from meteorological data (compare the range of distances in [4,18] with Figures 8 and 9 in this work).

Separation distances in this investigation were calculated assuming constant emissions over time (annual mean value), although odour emissions were often characterised by temporal variability [19,20]. Overall, the results in these papers indicate that the common practice of assuming a constant value for the odour emission can underestimate the separation distances to avoid annoyance, as compared to more realistic scenarios with hourly varying emission rates. Such an underestimation was observed to be more pronounced in the prevailing wind directions. The results also showed that the lower the selected exceedance probability of a specific odour impact criterion, the larger the separation distance underestimation. We selected odour impact criteria with exceedance probabilities of 10 and 15%, which meant that the impact of the assumption of a constant odour emission was small.

The presented results as well as previous investigations revealed the dependency of local-scale dispersion simulations on the meteorological input as a major source of model uncertainty. From model evaluation studies undertaken in the framework of EU-COST (see, e.g., [21–24]), the so-called fitness-for-purpose concept evolved, meaning that the type of model and source of data most apt for the purpose of an investigation should have been used. In the case of a complex terrain, for example, the use of a Lagrangian model is to be preferred over a Gaussian one. With data from three-axis ultrasonic anemometers, site-specific peak-to-mean attenuation curves can be obtained, and wind and stability information are obtained from one instrument [4,18]. If available, such data are preferred over data from conventional (mostly semi-automatic) meteorological stations. Even then, a range of results are obtained, whenever different models or datasets are used.

This investigation showed that, due by climate change, current protection areas might not be sufficient in the future. The Feldbach case revealed that separation distances would have to be increased by approx. 50 m, especially in the main wind directions. In addition, climate change will likely also lead to an increase in emissions, caused, e.g., by the increased animal activity due to heat stress [25], probably further increasing separation distances. In contrast to Feldbach, however, the calculations for Wels showed only marginal changes of the affected area in the future. The influence of climate change on odour dispersion has to, therefore, be investigated on a case-by-case basis, best using the methodology presented in the current work.

If the focus of an investigation lies on neighbourhood protection, a stricter odour impact criterion should be used, leading to larger separation distances in terms of a worst case scenario. For the current investigation, the EPA method was preferred over the VDI and KTA schemes, as it delivered generally larger separation distances. As the climate conditions at the two sites were very different, this outcome can be generalized.

Author Contributions: Conceptualization, writing—original draft preparation, M.P.; methodology, M.P., K.B.-S., I.A. and G.S.; software analysis and validation, W.K. and K.A.; writing—review and editing, M.P., G.S., K.B.-S. and I.A.; visualization, W.K. All authors have read and agreed to the published version of the manuscript.

Funding: The project PiPoCooL "Climate change and future pig and poultry production: implications for animal health, welfare, performance, environment and economic consequences" was funded by the Austrian Climate and Energy Fund in the framework of the Austrian Climate Research Program(ACRP8–PiPoCooL–KR15AC8K12646).

Data Availability Statement: The CORDEX (Coordinated Regional Climate Downscaling Experiment)data are available from the Earth System Grid Federation (ESGF), https://cordex.org/data-access/esgf/, latest access date: 31 August 2016. The data presented in this study are available on request from the corresponding author.

Acknowledgments: We acknowledge the World Climate Research Programme's Working Group on Regional Climate, and the Working Group on Coupled Modelling, former coordinating body of CORDEX, a responsible panel for CMIP5. We also thank the climate modelling groups for producing and making available their model output. We also acknowledge the Earth System Grid Federation infrastructure, an international effort led by the U.S. Department of Energy's Program for Climate Model Diagnosis and Intercomparison, the European Network for Earth System Modelling and other partners in the Global Organisation for Earth System Science Portals (GO-ESSP).

Conflicts of Interest: The authors declare no conflict of interest.

References

1. Leelőssy, Á.; Molnár, F., Jr.; Izsák, F.; Havasi, Á.; Lagzi, I.; Mészáros, R. Dispersion modeling of air pollutants in the atmosphere: A review. *Cent. Euro. J. Geosci.* **2014**, *6*, 257–278. [CrossRef]
2. Brancher, M.; Griffiths, K.D.; Franco, D.; De MeloLisboa, H. A review of odour impact criteria in selected countries around the world. *Chemosphere* **2017**, *168*, 1531–1570. [CrossRef] [PubMed]
3. Sommer-Quabach, E.; Piringer, M.; Petz, E.; Schauberger, G. Comparability of separation distances between odour sources and residential areas determined by various national odour impact criteria. *Atmos. Environ.* **2014**, *95*, 20–28. [CrossRef]
4. Piringer, M.; Knauder, W.; Petz, E.; Schauberger, G. Factors influencing separation distances against odour annoyance calculated by Gaussian and Langrangian dispersion models. *Atmos. Environ.* **2016**, *140*, 69–83. [CrossRef]
5. Piringer, M.; Knauder, W.; Anders, I.; Andre, K.; Zollitsch, W.; Hörtenhuber, S.J.; Baumgartner, J.; Niebuhr, K.; Hennig-Pauka, I.; Schönhart, M.; et al. Climate change impact on the dispersion of airborne emissions and the resulting separation distances to avoid odour annoyance. *Atmos. Environ. X* **2019**, *2*, 100021. [CrossRef]
6. Kottek, M.; Grieser, J.; Beck, C.; Rudolf, B.; Rubel, F. World map of the Köppen-Geiger climate classification updated. *Meteorol. Z.* **2006**, *15*, 259–263. [CrossRef]
7. Rockel, B.; Will, A.; Hense, A. The Regional Climate Model COSMO-CLM (CCLM). *Meteorol. Z.* **2008**, *17*, 347–348. [CrossRef]
8. Dee, D.P.; Uppala, S.M.; Simmons, A.J.; Berrisford, P.; Poli, P.; Kobayashi, S.; Andrae, U.; Balmaseda, M.A.; Balsamo, G.; Bauer, P.; et al. The ERA-Interim reanalysis: Configuration and performance of the data assimilation system. *Q. J. R. Meteorol. Soc.* **2011**, *137*, 553–597. [CrossRef]
9. Smiatek, G.; Kunstmann, H.; Senatore, A. EURO-CORDEX regional climate model analysis for the Greater Alpine Region: Performance and expected future change. *J. Geophys. Res. Atmos.* **2016**, *121*, 7710–7728. [CrossRef]
10. Jacob, D.; Petersen, J.; Eggert, B.; Alias, A.; Bøssing Christensen, O.; Bouwer, L.M.; Braun, A.; Colette, A.; Déqué, M.; Georgievski, G.; et al. EURO-CORDEX: New high-resolution climate change projections for European impact research. *Reg. Environ. Chang.* **2014**, *14*, 563. [CrossRef]
11. Janicke Consulting. *Dispersion Model LASAT Version 3.4, Reference Book*; Janicke Consulting Environmental Physics: Überlingen, Germany, 2019.
12. Bundesministerium für Umwelt, Naturschutz und Reaktorsicherheit. *Erste Allgemeine Verwaltungsvorschrift zum Bundes–Immissionsschutzgesetz (Technische Anleitung zur Reinhaltung der Luft—TA Luft)*; Bundesministerium für Umwelt, Naturschutz und Reaktorsicherheit: Berlin, Germany, 2002.
13. EPA. *Meteorological Monitoring Guidance for Regulatory Modeling Applications*; United States Environmental Protection Agency (EPA): Washington, DC, USA, 2000.
14. VDI. *Emissions and Immissions from Animal Husbandry—Housing Systems and Emissions—Pigs, Cattle, Poultry, Horses*; VDI 3894 Part 1; Verein Deutscher Ingenieure, VDI Verlag: Düsseldorf, Germany, 2011.
15. Zahn, J.A.; DiSpirito, A.A.; Do, Y.S.; Brooks, B.E.; Cooper, E.E.; Hatfield, J.L. Correlation of human olfactory responses to airborne concentrations of malodorous volatile organic compounds emitted from swine effluent. *J. Environ. Qual.* **2001**, *30*, 624–634. [CrossRef] [PubMed]
16. Liu, D. *Application of PTR-MS for Optimization of Odour Removal from Intensive Pig Production with Emphasis on Biofiltration*; Department of Engineering—Biological and Chemical Engineering, Aarhus University: Aarhus, Denmark, 2013.
17. GOAA. Guideline on Odour in Ambient Air. In *Detection and Assessment of Odour in Ambient Air*, 2nd ed.; GOAA: Berlin, Germany, 2008.
18. Piringer, M.; Knauder, W.; Petz, E.; Schauberger, G. A comparison of separation distances against odour annoyance calculated with two models. *Atmos. Environ.* **2015**, *116*, 22–35. [CrossRef]
19. Brancher, M.; Knauder, W.; Piringer, M.; Schauberger, G. Temporal variability in odour emissions: To what extent this matters for the assessment of annoyance using dispersion modelling. *Atmos. Environ. X* **2020**, *5*, 100054. [CrossRef]
20. Schauberger, G.; Piringer, M.; Heber, A.J. Odour emission scenarios for fattening pigs as input for dispersion models: A step from an annual mean value to time series. *Agric. Ecosyst. Environ.* **2014**, *93*, 108–116. [CrossRef]
21. Baumann-Stanzer, K.; Piringer, M.; Polreich, E.; Hirtl, M.; Petz, E.; Bügelmayer, M. User experience with model validation exercises. *Croat. Meteorol. J.* **2008**, *43*, 52–56.
22. Piringer, M.; Baumann-Stanzer, K. Selected results of a model validation exercise. *Adv. Sci. Res.* **2009**, *3*, 13–16. [CrossRef]
23. Baumann-Stanzer, K.; Piringer, M. Validation of regulatory micro-scale air quality models: Modeling odour dispersion and built-up areas. *World Rev. Sci. Technol. Sustain. Dev.* **2011**, *8*, 203–213. [CrossRef]
24. Di Sabatino, S.; Buccolieri, R.; Olesen, H.R.; Ketzel, M.; Berkowicz, R.; Franke, J.; Schatzmann, M.; Schlünzen, K.H.; Leitl, B.; Britter, R.; et al. COST 732 in practice: The MUST model evaluation exercise. *Int. J. Environ. Pollut.* **2011**, *44*, 403–418. [CrossRef]
25. Schauberger, G.; Hennig-Pauka, I.; Zollitsch, W.; Hörtenhuber, S.J.; Baumgartner, J.; Niebuhr, K.; Piringer, M.; Knauder, W.; Anders, I.; Andre, K.; et al. Efficacy of adaptation measures to alleviate heat stress in confined livestock buildings in temperate climate zones. *Biosyst. Eng.* **2020**, *200*, 157–175. [CrossRef]

Article

Additivity between Key Odorants in Pig House Air

Michael Jørgen Hansen [1,*], Anders Peter S. Adamsen [1,2], Chuandong Wu [3] and Anders Feilberg [1]

[1] Department of Biological and Chemical Engineering, Aarhus University, 8000 Aarhus, Denmark; apa@bce.au.dk (A.P.S.A.); af@bce.au.dk (A.F.)

[2] SEGES, Danish Pig Research Centre, 1609 Copenhagen, Denmark

[3] School of Chemistry and Biological Engineering, University of Science and Technology, Beijing 100083, China; wuchuandong@ustb.edu.cn

* Correspondence: michaelj.hansen@bce.au.dk; Tel.: +45-21622710

Abstract: The verification of odor abatement technologies for livestock production based on chemical odorants requires a method for conversion into an odor value that reflects the significance of the individual odorants. The aim of the present study was to compare the SOAV method (Sum of Odor Activity Values) with the odor detection threshold measured by olfactometry and to investigate the assumption of additivity. Synthetic pig house air with odorants at realistic concentration levels was used in the study (hydrogen sulfide, methanethiol, trimethylamine, butanoic acid, and 4-methylphenol). An olfactometer with only PTFE in contact with sample air was used to estimate odor threshold values (OTVs) and the odor detection threshold for samples with two to five odorants. The results show a good correlation ($R^2 = 0.88$) between SOAV estimated based on the OTVs for panelists in the present study and values found in the literature. For the majority of the samples, the ratio between the odor detection threshold and SOAV was not significantly different from one, which indicates that the OAV for individual odorants in a mixture can be considered additive. In conclusion, the assumption of additivity between odorants measured in pig house air seems reasonable, but the strength of the method is determined by the OTV data used.

Keywords: odor; odorants; SOAV; OTV

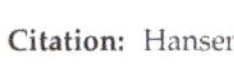

Citation: Hansen, M.J.; Adamsen, A.P.S.; Wu, C.; Feilberg, A. Additivity between Key Odorants in Pig House Air. *Atmosphere* **2021**, *12*, 1008. https://doi.org/10.3390/atmos12081008

Academic Editors: Laura Maria Teresa Capelli and Kimitaka Kawamura

Received: 31 May 2021
Accepted: 26 July 2021
Published: 5 August 2021

Publisher's Note: MDPI stays neutral with regard to jurisdictional claims in published maps and institutional affiliations.

1. Introduction

The verification of odor abatement technologies for animal houses is normally based on dynamic olfactometry, where samples are collected in bags and transported to a laboratory for threshold measurements by human panelists within 30 h [1]. This method has some drawbacks, such as the impaired storage stability of odorants in bags [2–5], the low recovery of odorants in olfactometers [6,7] and high variability between panelists [8,9]. The chemical measurement of odorants is an alternative method with a lower variability and less influence on the odorants from the sampling equipment, and if on-line methods are applied, a high time resolution is achieved [10–12]. However, the chemical measurement of odorants requires a conversion into a theoretical odor concentration that accounts for the contributions of the individual odorants in a mixture. The sum of odor activity values (SOAV), defined as the summation of concentration/odor threshold ratios, is often used to convert concentrations of odorants into a theoretical odor concentration. This approach assumes that the OAV for the individual odorants in a mixture is additive. Some previous studies have compared the SOAV method with the odor detection threshold and have demonstrated additivity between odorants [13–15]. Odor threshold values (OTVs) have been applied in a few studies related to livestock production to evaluate the significance of odorants found in the air matrix [16–18], but the assumption of additivity has to our knowledge not been investigated for this type of air matrix. There are also more advanced conversion methods such as the sums of odor intensity (SOI) or the equivalent odor concentration (EOC), where both the odor threshold and the sensitivity of the odor perception defined as the slope of the Weber–Fechner law are included [19]. However, Weber–Fechner law data required for these methods are relatively scarce. OTV data are available in the literature, and although there is variation between measured OTVs [20–26], the SOAV is a simple method that can be applied in the short term, and it is therefore relevant to investigate the assumption of additivity.

Air from livestock houses contains a large number of odorants, but only a limited number of odorants may account for the majority of the odor perception [10]. Odorants in air from pig houses were used as the case in the present study. Based on previous studies concerned with concentration levels of odorants [10–12] and a recent evaluation of OTVs for pig house air [23], the five most dominant odorants were selected: hydrogen sulfide (H2S), methanethiol (MT), trimethylamine (TMA), butanoic acid (BA) and 4-methylphenol (4MP). A synthetic odor mixture containing these five odorants at realistic concentration levels was used in the study. The aim of the present study was to compare the SOAV method (Sum of Odor Activity Values) with the odor detection threshold measured by olfactometry and to investigate the assumption of additivity. It was concluded that the assumption of additivity for the five odorants included in the present study is reasonable, but the strength of the method is dependent on the OTV input data.

2. Materials and Methods

2.1. Gas Dilution System

H2S, MT, and TMA were introduced to the gas dilution system from pressurized gas cylinders (AGA, Copenhagen, Denmark). BA was generated from a liquid calibration unit (Ionicon Analytik GmbH, Innsbruck, Austria) and 4MP was generated from a permeation tube (VICI Metronics, Inc., Houston, TX, USA) using a permeation oven (Dynacalibrator model 150, VICI Metronics Inc.), see Figure 1. The odorant mixtures were diluted with atmospheric air purified by a Supelpure HC filter (Supelco, Bellefonte, PA, USA). Mass flow controllers (Bronkhorst, The Netherlands) controlled the flow of the dilution air and odorants. The gas dilution system (reduction valve/permeation oven, mass flow controllers, and tubing) was allowed to equilibrate for at least two hours before the measurements were carried out. All tubes and fittings in the gas dilution system were made of PTFE.

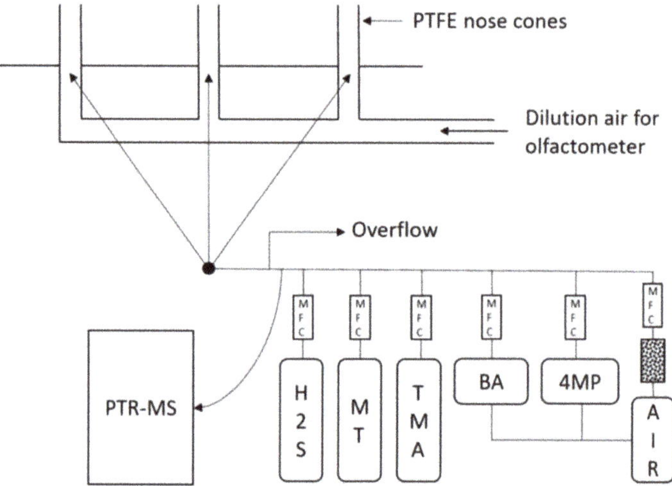

Figure 1. Schematic drawing of gas dilution system and connection to the olfactometer. H2S: Hydrogen sulfide; MT: methanethiol; TMA: trimethylamine; BA: butanoic acid; 4MP: 4-methylphenol; PTR-MS: Proton-Transfer-Reaction Mass Spectrometry; MFC: mass flow controller.

2.2. Analytical Methods

A custom-made glass olfactometer (Eurofins Denmark, Galten, Denmark) was used to estimate the dilution to threshold. The olfactometer was based on the two-alternative forced-choice method (2AFC) and was designed for three panelists at a time. The olfactometer provided a constant flow of ca. 20 L min^{-1} in each nose cone and the presentation time at each dilution step was set at maximum 15 s. The nose cones in the olfactometer were made of PTFE, since it has previously been shown that PTFE only has a very limited effect on the recovery of odorants in an olfactometer [7]. Each nose cone was connected to the gas dilution system with 1/8″ PTFE tubes. A needle valve made of PTFE (Bohlender GmbH, Grünsfeld, Germany) was used to adjust the required flow of the gas mixture to the nose cones and the flow was measured prior to each presentation using a calibrated flowmeter (Agilent Technologies, Ballerup, Denmark). The odorant mixtures were diluted in descending order from ca. 18.000 to 40 times dilution with a step factor of 1.5.

High sensitivity Proton-Transfer-Reaction Mass Spectrometry (HS-PTR-MS Ionicon Analytik GmbH, Innsbruck, Austria) was used to measure concentrations of odorants in the gas dilution system. PTR-MS is based on chemical ionization of compounds with protonated water (H_3O^+) and detection in a quadrupole mass spectrometer. The principle of PTR-MS has been described in detail in a previous study [27]. Standard drift tube conditions were applied with a pressure between 2.1 and 2.2 mbar, a voltage of 600 V and a temperature of 75 °C. The inlet temperature was set at 75 °C. PTR-MS was operated in single ion mode with a dwell time at one second. The mass-specific transmission factors were checked before the measurements with a mixture of eight compounds between m/z 79–237 (AGA, Copenhagen, Denmark). Reaction rate constants between odorants and protonated water were applied for MT, TMA, BA, and 4MP [28] and the humidity dependency of H2S was corrected with a calibration gas (AGA, Copenhagen, Denmark) according to a previously described method [10].

2.3. Experimental Setup

A group of eight panelists from the odor laboratory at Eurofins Denmark was used in the study. All panelists were selected based on the criteria of the European standard for olfactometry [1]. Based on previous studies with odorants in air from pig houses [10–12], a gas mixture was designed with similar concentration levels (see Table 1). The odor detection threshold was measured for individual odorants and for all combinations, in total 26 combinations containing two to five odorants. Each sample was measured twice for all panelists. During each measurement, the actual odorant concentrations were measured in the gas mixture prior to dilution in the olfactometer. Four of the samples including BA were not successfully measured due to technical problems with the liquid calibration unit. Furthermore, the threshold measurements for the sample with only 4MP and the mixture with all five odorants were repeated twice to confirm the results.

Table 1. Average odorant concentration in gas mixture, mean ± standard deviation (SD).

Compound [1]	H2S [2]	MT	TMA	BA	4MP
m/z	35	49	60	89 + 71	109
Reaction rate constant [28], cm^3 molecule^{-1} s^{-1}	-	1.9×10^{-9}	1.58×10^{-9}	2.11×10^{-9}	2.32×10^{-9}
Concentration, ppb$_v$	319 ± 21	14 ± 1	33 ± 1	88 ± 9	11 ± 2

[1] H2S: Hydrogen sulfide; MT: methanethiol; TMA: trimethylamine; BA: butanoic acid; 4MP: 4-methylphenol.
[2] Humidity dependency of H2S was corrected with a calibration gas.

2.4. Data Analysis

The OTVs for individual odorants were estimated as the geometric mean of the individual responses by the panelists. SOAV was estimated as the summation of odorant concentration relative to the odor threshold value (OTV) for individual odorants (Equation (1)).

$$SOAV = \sum [\text{odorant}]/OTV \tag{1}$$

The test for additivity was based on the ratio between the measured odor detection threshold for the individual panelists and SOAV (Equation (2)). A ratio at one indicates that the interaction is additive, a value below one indicates that there is a smaller degree of cooperation between odorants, and a value above 1 indicates that there is a higher degree of cooperation. A t-test was used to investigate if the average ratio for the panelists was significantly different from one. The level of significance was defined as a p-value below 0.05. A qq-plot was used to check the assumption of a normal distribution, and for all measurements a normal distribution was applicable.

$$\text{Ratio} = \text{odor detection threshold}/SOAV \tag{2}$$

The odorant threshold level for samples with two to five odorants was estimated as the odorant concentration at the odor detection threshold level divided by the OTV (Equation (3)). A value below one means that the odorant concentration is at a subthreshold level, and a value above one means that it is at a suprathreshold level.

$$\text{Odorant threshold level} = ([\text{odorant}]/\text{odor detection threshold})/OTV \tag{3}$$

3. Results

3.1. Odor Threshold Values for Odorants

The measured OTVs for the individual odorants included in the present study and the values from the literature are shown in Table 2. The measured OTVs for H2S, TMA, BA and 4MP were two to four times lower than the reported literature values, whereas the OTV for MT was slightly higher. The measured OTV for 4MP was only based on four of the panelists since the other panelists were unable to detect the odorant within the available dilution steps. The threshold estimate for 4MP was repeated at two different days with the same result. In Figure 2, the correlation between SOAV based on measured OTVs and literature values shows a good correlation ($R^2 = 0.88$).

Table 2. Measured and literature values for odor threshold values (OTVs; ppb_v) [min; max].

Compound [1]	H2S	MT	TMA	BA	4MP
Measured OTV	0.2 [0.06; 1]	0.04 [0.01; 0.1]	0.02 [0.01; 0.07]	0.1 [0.02; 0.4]	0.006 [0.003; 0.03]
Literature [23] OTV	0.8 [0.4; 3]	0.03 [0.02; 0.07]	0.08 [0.03; 0.2]	0.2 [0.1; 0.8]	0.02 [0.005; 0.05]

[1] H2S: Hydrogen sulfide; MT: methanethiol; TMA: trimethylamine; BA: butanoic acid; 4MP: 4-methylphenol.

3.2. Additivity between Odorants

The assumption of the SOAV method is that the OAVs for the individual odorants are additive. In order to test this assumption, the ratio between the odor detection threshold measured by olfactometry and SOAV was calculated. In Figure 3, the ratios are shown for the samples with two to five odorants. The results show that, for all samples, the ratio was within 0.6 to 2.3 and with a mean of 1.2. In 17 out of 23 samples, the ratio was not significantly ($p > 0.05$) different from one, which indicates that the effect was additive. For five of the samples, the ratio was significantly ($p < 0.05$) higher than one, which indicates that there was a higher degree of cooperation between the odorants than expected based on SOAV. For one of the samples, the ratio was significantly ($p < 0.05$) lower than one, which indicates that there was a lower degree of cooperation between the odorants than expected based on SOAV. In Figure 4, the odorant threshold level is shown for samples with two to five odorants. Except for 4MP in samples with two and three odorants, the odorant threshold level was below one in all other cases, meaning that the odorants were at subthreshold level at odor detection. Furthermore, the odorant threshold level and the variation in data decreased as the number of odorants increased.

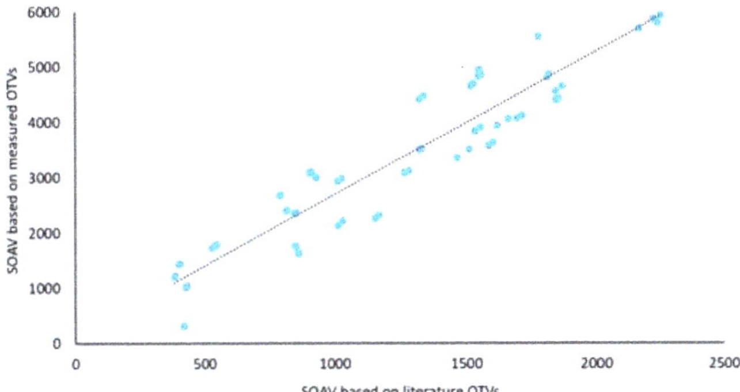

Figure 2. Correlation between sums of odor activity values (SOAV) based on measured and literature odor threshold values (OTVs) [23]. $SOAV_{measured} = 2.59\ SOAV_{literature} + 107$, $R^2 = 0.88$.

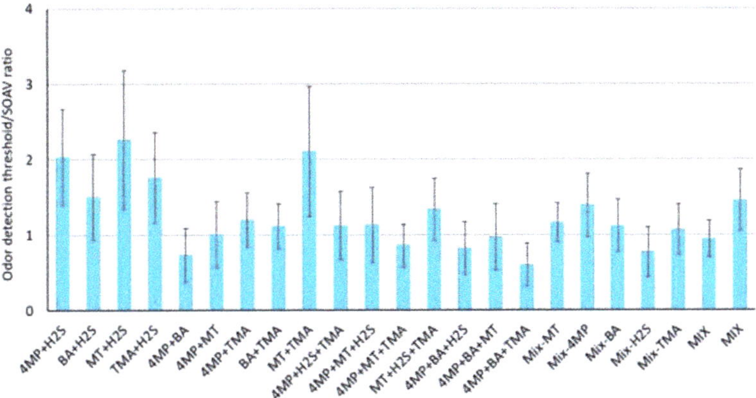

Figure 3. Odor detection threshold/SOAV ratio with 95% confidence intervals. H2S: Hydrogen sulfide; MT: methanethiol; TMA: trimethylamine; BA: butanoic acid; 4MP: 4-methylphenol.

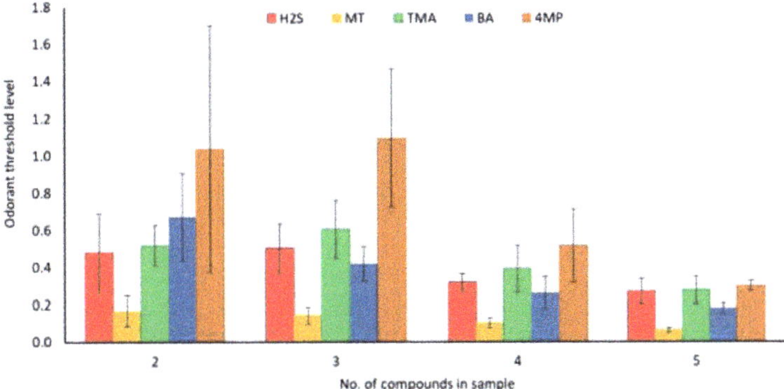

Figure 4. Odorant threshold level of odorants in samples with two to five odorants, mean ± standard deviation (SD). H2S: Hydrogen sulfide; MT: methanethiol; TMA: trimethylamine; BA: butanoic acid; 4MP: 4-methylphenol.

4. Discussion

The aim of the present study was to compare the SOAV method (Sum of Odor Activity Values) with the odor detection threshold measured by olfactometry and to investigate the assumption of additivity. The SOAV method is often applied for evaluating the significance of odorants found in air from livestock houses [16–18], and for estimating the efficiency of odor abatement technologies. The SOAV method is highly dependent on the OTV for the included odorants. Although the present study shows a high correlation between SOAV based on the OTV from measured and literature values, the differences in the applied threshold values will affect the relative effect of an odor abatement technology. However, assuming that significant odorant concentrations are measured with a relatively high precision and reproducibility, the SOAV method will still be a significant improvement for the evaluation of odor abatement technologies. The advantage of the SOAV method is that all odorants in the air matrix are included in the evaluation and with the same proportions as found at the odor source. For olfactometry used in relation to livestock production, it is merely the effect on volatile sulfur compounds that is evaluated since most of the other odorants are lost during storage [2–5] and analysis in the olfactometer [6,7]. It has previously been shown that olfactometry underestimates the effect of air cleaners that have an effect on most of the odorants in pig house air except for volatile sulfur compounds [29], whereas the effect of slurry acidification is overestimated because the concentration of odorants such as volatile fatty acids is increased and volatile sulfur compounds are decreased. Although the SOAV method is an improvement of the evaluation of odor abatement technologies, the discrepancy between OTVs found in the literature and OTVs measured in the present study underlines that the strength of the SOAV method depends on the OTV data quality. OTVs for odorants found in pig house air have been measured in several studies [20–26], but mostly with a rather limited number of panelists and with different methods. In order to improve the strength of the SOAV method, a larger population study would be preferable to obtain a better estimate of the population mean for different odorants and the effect of factors such as age and gender.

An air matrix from pig houses or other odor sources will be much more complex than the five odorants included in the present study. However, the five odorants that were included in the present study normally account for more than 80% of the odor nuisance from pig houses based on the SOAV method [10–12] and are representative of different types of odorants found in pig house air (e.g., volatile sulfur compounds, volatile fatty acids, amines and phenols). The ratio between the odor detection threshold and SOAV was close to one for most of the samples, which indicates that the OAV for odorants in pig house air can be treated as additive. In one sample, the ratio was below one, which indicates a lower degree of cooperation between the odorants, and in some samples, the ratio was above one, indicating some degree of cooperation between odorants. However, it has to be taken into account that although the odorant stimuli were quite stable, there will be a variation related to the odor detection threshold estimated by the panelists that will influence the result. The odorant threshold level (odorant concentration at threshold level) revealed that for most of the odorants, the concentration was as expected at subthreshold level and that the odorant threshold level decreased as the number of odorants increased, which also indicates that the effect of the odorants is additive. For 4MP, the odorant threshold level was above one in samples with two and three odorants. However, only four of the panelists were able to detect 4MP within the available dilution steps in the present study, which means that the OTV for 4MP is underestimated and the odorant threshold level is overestimated. The additive effect of odorants at subthreshold levels is in accordance with other studies about detection probability for binary mixtures [24,26,30], where it was shown that at subthreshold level the interaction between odorants seems to be additive, whereas at levels close to or above the threshold level the interaction is hypo additive. A possible explanation for the additive effect at subthreshold level is that there will be less competition between the molecules for receptors in the olfactory bulb, whereas at higher concentrations (closer to the threshold level) the competition will be higher or there might be a blocking of receptors for other odorants [30]. Furthermore, the variation in the odorant threshold level also decreased as the number of odorants in the samples increased, which is in line with previous studies [14,15]. The lower variation in more complex samples also indicates that investigations of interaction between odorants in a given matrix should reflect the complexity of the matrix and not only a simple approach with binary combinations.

In conclusion, comparing the odor detection threshold with the sum of odor activity values (the SOAV method) for five key odorants found in pig house air (hydrogen sulfide, methanethiol, trimethylamine, butanoic acid, and 4-methylphenol) indicates that the assumption of additivity between odorants at subthreshold level is reasonable. The strength of the SOAV method is highly dependent on the applied OTVs and more effort should be put into population studies of OTVs. Although there is opportunity for the improvement of the SOAV method, it will be a significant improvement in relation to reproducibility and precision compared to dynamic olfactometry.

Author Contributions: Methodology, M.J.H. and A.P.S.A.; formal analysis, M.J.H., C.W. and A.P.S.A.; investigation, M.J.H., C.W. and A.P.S.A.; writing—original draft preparation, M.J.H.; project administration, A.F.; funding acquisition, A.F. All authors have read and agreed to the published version of the manuscript.

Funding: This research was funded by Innovation Fund Denmark, grant number 6150-00030B.

Institutional Review Board Statement: Ethical review and approval were waived for this study because the measurements with human panelists were conducted according to the European standard for dynamic olfactometry.

Informed Consent Statement: Informed consent was obtained from all subjects involved in the study.

Data Availability Statement: The data in the study are available upon request to the corresponding author.

Acknowledgments: The authors acknowledge Morten Sieleman at Eurofins Product Testing A/S for his valuable help with the planning and performance of the odor measurements.

Conflicts of Interest: The authors declare no conflict of interest. The funders had no role in the design of the study; in the collection, analyses, or interpretation of data; in the writing of the manuscript, or in the decision to publish the results.

References

1. CEN. *Air Quality—Determination of Odour Concentration by Dynamic Ofactometry*; EN 13725; European Committee for Standardization: Brussels, Belgium, 2003.
2. Hansen, M.J.; Adamsen, A.P.S.; Feilberg, A.; Jonassen, K.E.N. Stability of odorants from pig production in sampling bags for olfactometry. *J. Environ. Qual.* **2011**, *40*, 1096–1102. [CrossRef]
3. Kasper, P.L.; Oxbøl, A.; Hansen, M.J.; Feilberg, A. Mechanisms of Loss of Agricultural Odorous Compounds in Sample Bags of Nalophan, Tedlar, and PTFE. *J. Environ. Qual.* **2018**, *47*, 246–253. [CrossRef] [PubMed]
4. Kim, K.H.; Ju, D.W.; Joo, S.W. The evaluation of recovery rate associated with the use of thermal desorption systems for the analysis of atmospheric reduced sulfur compounds (RSC) using the GC/PFPD method. *Talanta* **2005**, *67*, 955–959. [CrossRef] [PubMed]
5. Trabue, S.L.; Anhalt, J.C.; Zahn, J.A. Bias of Tedlar bags in the measurement of agricultural odorants. *J. Environ. Qual.* **2006**, *35*, 1668–1677. [CrossRef]
6. Hansen, M.J.; Adamsen, A.P.S.; Feilberg, A. Recovery of Odorants from an Olfactometer Measured by Proton-Transfer-Reaction Mass Spectrometry. *Sensors* **2013**, *13*, 7860–7861. [CrossRef]
7. Kasper, P.; Mannebeck, D.; Oxbøl, A.; Nygaard, J.; Hansen, M.J.; Feilberg, A. Effects of Dilution Systems in Olfactometry on the Recovery of Typical Livestock Odorants Determined by PTR-MS. *Sensors* **2017**, *17*, 1859. [CrossRef] [PubMed]
8. Feilberg, A.; Hansen, M.J.; Pontoppidan, O.; Oxbøl, A.; Jonassen, K. Relevance of n-butanol as a reference gas for odorants and complex odors. *Water Sci. Technol.* **2018**, *77*, 1751–1756. [CrossRef] [PubMed]
9. Klarenbeek, J.V.; Ogink, N.W.M.; van der Voet, H. Odor measurements according to EN 13725: A statistical analysis of variance components. *Atmos. Environ.* **2014**, *86*, 9–15. [CrossRef]
10. Feilberg, A.; Liu, D.; Adamsen, A.P.S.; Hansen, M.J.; Jonassen, K.E.N. Odorant Emissions from Intensive Pig Production Measured by Online Proton-Transfer-Reaction Mass Spectrometry. *Environ. Sci. Technol.* **2010**, *47*, 5894–5900. [CrossRef] [PubMed]

11. Hansen, M.J.; Liu, D.; Guldberg, L.B.; Feilberg, A. Application of Proton-Transfer-Reaction Mass Spectrometry to the Assessment of Odorant Removal in a Biological Air Cleaner for Pig Production. *J. Agric. Food Chem.* **2012**, *60*, 2599–2606. [CrossRef]
12. Liu, D.; Feilberg, A.; Adamsen, A.P.S.; Jonassen, K.E.N. The effect of slurry treatment including ozonation on odorant reduction measured by in-situ PTR-MS. *Atmos. Environ.* **2011**, *45*, 3786–3793. [CrossRef]
13. Guadagni, D.G.; Buttery, R.G.; Okano, S.; Burr, H.K. Additive Effect of Sub-Threshold Concentrations of Some Organic Compounds Associated with Food Aromas. *Nature* **1963**, *200*, 1288–1289. [CrossRef]
14. Laska, M.; Hudson, R. A comparison of the detection thresholds of odour mixtures and their components. *Chem. Senses* **1991**, *16*, 651–662. [CrossRef]
15. Patterson, M.Q.; Stevens, J.C.; Cain, W.S.; Cometto-Muñiz, J.E. Detection thresholds for an olfactory mixture and its three constituent compounds. *Chem. Senses* **1993**, *18*, 723–734. [CrossRef]
16. Parker, D.B.; Koziel, J.A.; Cai, L.; Jacobson, L.D.; Akdeniz, N.; Bereznicki, S.D.; Lim, T.T.; Caraway, E.A.; Zhang, S.; Hoff, S.J.; et al. Odor and Odorous Chemical Emissions from Animal Buildings: Part 6. Odor Activity Value. *Trans. ASABE* **2012**, *55*, 2357–2368. [CrossRef]
17. Trabue, S.; Kerr, B.; Bearson, B.; Ziemer, C. Swine Odor Analyzed by Odor Panels and Chemical Techniques. *J. Environ. Qual.* **2011**, *40*, 1510–1520. [CrossRef]
18. Trabue, S.L.; Scoggin, K.D.; Li, H.; Burns, R.; Xin, H.W. Field sampling method for quantifying odorants in humid environments. *Environ. Sci. Technol.* **2008**, *42*, 3745–3750. [CrossRef]
19. Wu, C.; Liu, J.; Zhao, P.; Piringer, M.; Schauberger, G. Conversion of the chemical concentration of odorous mixtures into odour concentration and odour intensity: A comparison of methods. *Atmos. Environ.* **2016**, *127*, 283–292. [CrossRef]
20. Chappuis, C.J.-F.o.; Niclass, Y.; Vuilleumier, C.; Starkenmann, C. Quantitative Headspace Analysis of Selected Odorants from Latrines in Africa and India. *Environ. Sci. Technol.* **2015**, *49*, 6134–6140. [CrossRef]
21. Cometto-Muñiz, J.E.; Abraham, M.H. Structure-activity relationships on the odor detectability of homologous carboxylic acids by humans. *Exp. Brain Res.* **2010**, *207*, 75–84. [CrossRef] [PubMed]
22. Czerny, M.; Brueckner, R.; Kirchhoff, E.; Schmitt, R.; Buettner, A. The Influence of Molecular Structure on Odor Qualities and Odor Detection Thresholds of Volatile Alkylated Phenols. *Chem. Senses* **2011**, *36*, 539–553. [CrossRef]
23. Hansen, M.J.; Kasper, P.; Adamsen, A.; Feilberg, A. Key Odorants from Pig Production Based on Improved Measurements of Odor Threshold Values Combining Olfactometry and Proton-Transfer-Reaction Mass Spectrometry (PTR-MS). *Sensors* **2018**, *18*, 788. [CrossRef] [PubMed]
24. Miyazawa, T.; Gallagher, M.; Preti, G.; Wise, P.M. Odor Detection of Mixtures of Homologous Carboxylic Acids and Coffee Aroma Compounds by Humans. *J. Agric. Food Chem.* **2009**, *57*, 9895–9901. [CrossRef] [PubMed]
25. Nagata, Y. Measurement of odor threshold by triangle odor bag method. In *Odor Measurement Review*; Ministry of the Environment, Goverment of Japan: Kawasaki City, Japan, 2003; pp. 118–127.
26. Wise, P.M.; Miyazawa, T.; Gallagher, M.; Preti, G. Human odor detection of homologous carboxylic acids and their binary mixtures. *Chem. Senses* **2007**, *32*, 475–482. [CrossRef]
27. De Gouw, J.; Warneke, C. Measurements of volatile organic compounds in the earth's atmosphere using proton-transfer-reaction mass spectrometry. *Mass Spectrom. Rev.* **2007**, *26*, 223–257. [CrossRef]
28. Liu, D.; Nyord, T.; Rong, L.; Feilberg, A. Real-time quantification of emissions of volatile organic compounds from land spreading of pig slurry measured by PTR-MS and wind tunnels. *Sci. Total Environ.* **2018**, *639*, 1079–1087. [CrossRef] [PubMed]
29. Hansen, M.J.; Jonassen, K.E.N.; Feilberg, A. Evaluation of abatement technologies for pig houses by dynamic olfactometry and on-site mass spectrometry. *Chem. Eng. Trans.* **2014**, *40*, 253–258.
30. Cometto-Muñiz, J.E.; Cain, W.S.; Abraham, M.H. Odor detection of single chemicals and binary mixtures. *Behav. Brain Res.* **2005**, *156*, 115–123. [CrossRef]

atmosphere

Article

Experimenting with Odour Proficiency Tests Implementation Using Synthetic Bench Loops

Domenico Cipriano [1],*, Amedeo M. Cefalì [1,2] and Marco Allegrini [3]

[1] RSE—Ricerca sul Sistema Energetico, Via Rubattino 54, 20131 Milan, Italy; amedeomanuel.cefali@rse-web.it

[2] Department of Earth and Environmental Sciences, University of Milano-Bicocca, Piazza della Scienza 1, 20126 Milan, Italy

[3] Politecnico di Milano, Piazza Leonardo da Vinci 32, 20133 Milan, Italy; marco.allegrini1993@libero.it

* Correspondence: domenico.cipriano@rse-web.it; Tel.: +39-329-6873394

Abstract: Uncertainty in the quantification of odour measurements is a difficult (but needed) task. Critical aspects include panel selection (required by dynamic olfactometry), sampling, and stability of the samples. Proficiency tests (PTs) can help evaluate such contributions; however, the classical approach to PTs, in which laboratories analyse real samples taken from the field, are not as applicable in this field, and are often implemented by only using dry gas cylinders containing stable compounds. Consequently, uncertainties related to the sampling activity cannot be assessed. In particular, high odour levels and the presence of water vapour in emission sources can create significative biases due to sampling techniques used and chemical reactions that can occur before analysis. In this work, we present experimental notes, developed using the experimental facility 'LOOP', realised at the RSE research centre in Italy, in order to "help" the definition, in an upgraded protocol for implementing PTs for odour determinations. Using this bench loop is advantageous as it involves the possibility of implementing samples in conditions very similar to reality (i.e., high temperatures, high water content, and the presence of chemical interferents).

Keywords: olfactometry; proficiency test; bench loop; n-butanol; sampling uncertainties

check for **updates**

Citation: Cipriano, D.; Cefalì, A.M.; Allegrini, M. Experimenting with Odour Proficiency Tests Implementation Using Synthetic Bench Loops. *Atmosphere* **2021**, *12*, 761. https://doi.org/10.3390/atmos12060761

Academic Editors: Günther Schauberger, Martin Piringer, Chuandong Wu and Jacek Koziel

Received: 21 April 2021
Accepted: 8 June 2021
Published: 12 June 2021

Publisher's Note: MDPI stays neutral with regard to jurisdictional claims in published maps and institutional affiliations.

1. Introduction

In Europe, odour concentration is determined by means of dynamic olfactometry, as standardised in EN 13,725 [1]. As 'odour' is defined as the effect of a gas sample to the human nose, with no relation to the chemical composition of the sample itself, the determination of uncertainty is very delicate and complicated. One major source is related to the human panel, where components are selected on an empirical basis.

Other factors affecting measurement uncertainty are sample collections, transport, and storage [2,3].

Proficiency tests (PTs) are fundamental tools to improve knowledge on odour measurements; generally, PT schemes are based on analyses by participating laboratories of reference materials [4,5], for example, environmental samples [6,7] delivered by proficiency test providers (PTP). In many cases, sampling is not part of the process, so sampling uncertainties cannot be assessed, but are often considered negligible, and are included in the preparation process of the references.

In this particular application, the effects, due to sampling and stability of the sample, cannot be neglected, particularly because of the possible presence of high humidity levels that require specific sampling procedures (i.e., using dilution probes), which could alter the sample characteristics significantly. Moreover, the presence of chemical compounds, even if they are not directly considered 'odorants' (i.e., carbon dioxide), can chemically alter the sample and its effect on the panellist and, thus, should be investigated.

The aim of the present work if to provide a first evaluation to extend proficiency testing schemes to odour measurements.

In past years, PT exercises on odours were conducted by the Hessian Agency for Nature Conservation, the Environment and Geology (HLUG) in Kassel, Germany [8]. Due to specific design of the experimental device, levels of humidity and composition of the matrix gas (air, in that case) could not be widely modified to reach real concentrations present in stack emissions.

For this scope, in 2012 a full-synthetic bench loop was realised by RSE in Milan; it was based on controlled flow injections of known gases in a closed bench loop, allowing definition of the effluent gases and of the related reference values on a metrological basis [9].

2. Materials and Methods

2.1. General

The data presented in this publication were obtained during preliminary experimental work, using RSE's bench loop (Figure 1), used to generate an odorous sample with a known odour concentration. N-butanol, indicated as reference material in the standard EN 13,725, was used as the first component investigated.

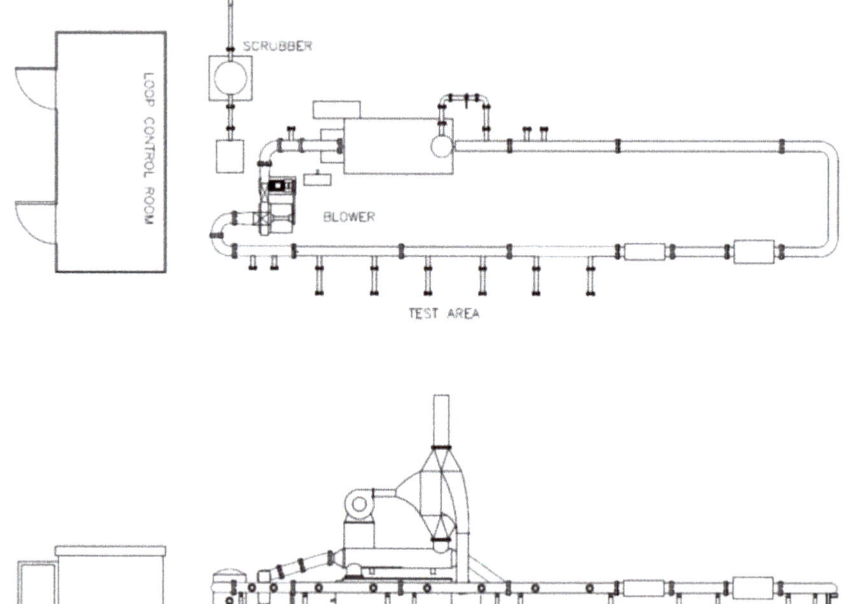

Figure 1. Mechanical dimensions of the LOOP facility.

This first experiment was conducted in order to verify the possibility of implementing odour PTs using RSE's loop test bench.

Using this bench is advantageous—it means the possibility of generating more 're-alistic' mixtures, with high values of humidity (up to 10% v/v), temperatures (between 60 °C and 180 °C), and gas matrix composition (CO_2, O_2, and other compounds) where, even if they are not considered 'odorants', they could alter human perceptions and be considered 'interferents'.

The aim of the first experiment was to verify the possibility of creating a synthetic atmosphere containing n-butanol, N_2, and water vapour. As gas cylinders with the desired n-butanol concentrations were not available, it was decided to inject a liquid solution of water and n-butanol into the bench loop using a vaporisation system designed for such a scope.

Samples taken from the bench loop were analysed by using a GC–FID (DANI Master GC Fast Gas Chromatograph System, DANI Instruments SpA, Milano, Italy) equipped with a capillary column (20 m × 0.18 mm × 1 μm). Helium carrier gas was maintained at a constant flow of 0.8 mL/min.

Moreover, olfactometry determinations were conducted using an olfactometer (T08, ECOMA GmbH, Kiel, Germany) using the same samples.

2.2. The Bench Loop

The basic principle of a PT is to feed the measuring system, under test, with a reference material, i.e., a reference gas mixture, and check the abilities of the participating laboratories to analyse the sample and provide consistent results. The challenge in emission monitoring is obtaining a reference material that follows the common definitions normally provided by ISO Guide 35 [10]—as the real matrix is chemically very complex and physically unstable— while fulfilling the requirements of ISO/IEC 17043 [11]. The experimental facility 'LOOP' is sketched in Figure 1.

Some of its characteristics are:

- Total length of the wind gallery tunnel: about 40 m.
- Internal diameter of the tunnel: 273 mm.
- Internal material: AISI 316.

The LOOP facility is able to generate and maintain, with 'traceable' accuracy and precision reference, atmospheres containing main macro pollutants of interest, with different oxygen and water vapour levels. Full size sampling ports are available to allow participants to use real sampling systems. To obtain the required gaseous mix in the tunnel, a dedicated gas mixing station was realised, making wide use of thermal mass flow controllers using gas cylinders (or from evaporation of liquid reference materials).

Concentrations of various effluents are controlled by means of extractive analysers, which are periodically calibrated; the expanded uncertainty (at a 95% level) on the various measurements of the measured are from 2 to 5%, respectively.

Together with continuous analysis of gaseous compounds, gas velocity inside the LOOP is monitored using a Pitot tube, connected to absolute and differential pressure gauges, and two 4-wire Pt100 sensors.

The facility has DN 100 standard sampling ports that allows up to five measuring teams to work simultaneously, with a total flow extraction up to 50 L/min. Pressure inside the test loop is kept over the ambient value.

2.3. The Vaporisation System

The solution of n-butanol in water is injected into the LOOP system using a vaporisation system, shown in Figure 2.

The liquid solution rate is regulated and continuously maintained using a calibrated peristaltic pump. Subsequently, it is injected into a heat exchanger that is able to reach temperatures from 130 °C to 400 °C.

The vapours generated are fluxed inside the LOOP using a heated transfer line.

Then, the vapours are injected into the facility of the area, following the passage of gases in the heater.

The vaporisation system was designed and realised in 2019. It can reach a maximum vaporisation flow rate of 10 mL/min.

Figure 2. The butanol vaporisation system.

2.4. Reference Value

The reference value for n-butanol was calculated on the base of total input flow rate, using the following equation:

$$C_{buOH}[ppm_{mol}] = \frac{F_{buOH}[mol/min]}{F_{gas_{tot}}[mol/min]}\,10^6 \tag{1}$$

where C_{buOH} is the reference value of n-butanol concentration, expressed in ppm_{mol}; F_{buOH} is the n-butanol flow rate, expressed in mol/min; $F_{gas_{tot}}$ is the total input flow rate, expressed in mol/min.

In order to calculate the odour concentration, the following equation can be used:

$$C_{od}\left[ouE/m^3\right] = \frac{C_{buOH}[ppm_{mol}]}{0.04} \tag{2}$$

where C_{od} is the reference value of the odour concentration, expressed in ouE/m^3. This can be assumed as indicated in the standard EN 13725, which establishes that 0.04 ppm_{mol} of n-butanol is equivalent to 1 ouE/m^3.

2.5. Vaporisation Temperature

First, the right temperature for the vaporisation of the solution was evaluated. Two different temperatures were selected based on the known characteristics of n-butanol and on experiences with other compounds: 270 °C and 140 °C.

In Table 1, the expected values of n-butanol concentration for the samples generated in each case and the measured values are shown. The measured values were obtained by chemical analysis, carried out with GC-FID gas chromatograph.

Table 1. Vaporisation temperature tests results.

Sample	Tvap (°C)	CbuOH Expected (ppm)	CbuOH Measured [1] (ppm)
1	270	23.04	14.30
2	270	38.17	5.69
3	270	76.03	15.22
4	270	76.03	30.33
5	270	78.35	37.01
6	270	154.29	55.61
7	140	24.62	18.02
8	140	74.27	74.74
9	140	77.99	85.52
10	140	150.82	141.74

Note 1: uncertainty due to GC analysis is evaluated lower than 10% or 1 ppm (the higher of the two).

The values obtained by chemical analysis of the samples generated with vaporisation temperature equal to 270 °C are lower than the expected values; whereas the values obtained by chemical analysis of the samples generated with vaporisation temperature equal to 140 °C are consistent to the expected values, as shown in Figure 3.

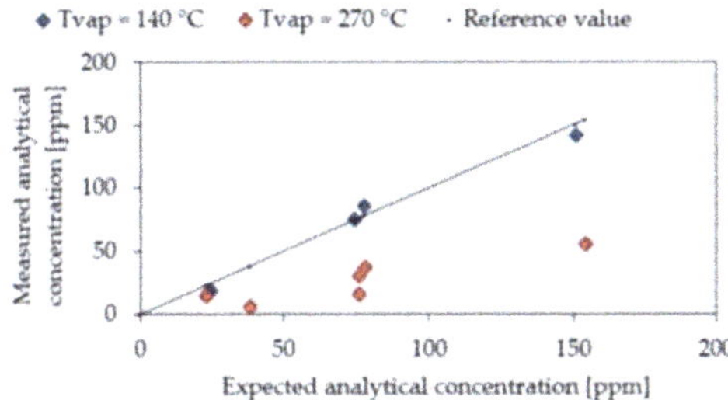

Figure 3. Scatter diagram of between concentration of butanol measured by the chromatograph (C_{yOH} expressed in ppm of butanol) and the concentration expected due the generation process (C_{xOH} expressed in ppm of butanol); data obtained at 140 °C are very near the ideal curve $C_{yOH} = C_{xOH}$.

In fact, the chromatograms obtained in the two cases are very different from each other. Figure 4 presents the characteristic chromatogram of the samples generated with a vaporisation temperature equal to 140 °C. There is only one peak with the characteristic retention time of n-butanol, equal to 1.89 min.

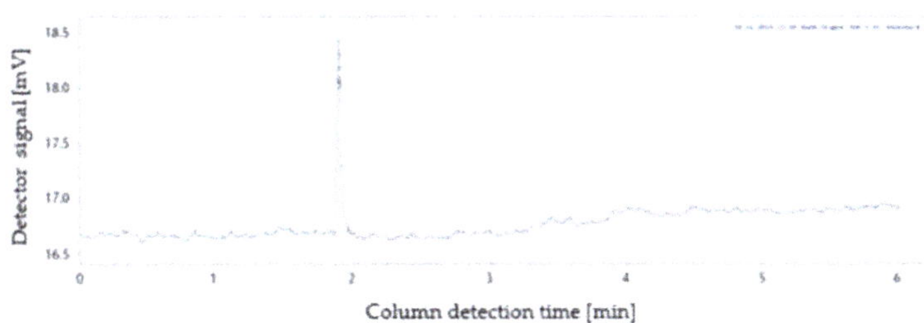

Figure 4. Chromatogram obtained from one of the samples generated with a vaporisation temperature equal to 140 °C; visible pack at 1.89 min is related to butanol and is verified using a traceable gas standard.

Figure 5 presents the characteristic chromatogram of the samples generated with a vaporisation temperature equal to 270 °C, in which it is possible to see:

- A peak with the characteristic retention time of n-butanol, 1.89 min, but with a lower peak area than expected;
- A peak with a lower retention time then that characteristic of n-butanol, equal to 1.46 min.

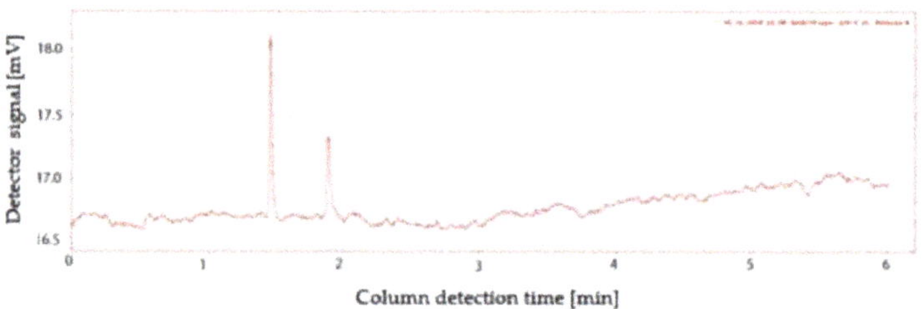

Figure 5. Chromatogram of one of the samples generated with a vaporisation temperature equal to 270 °C; set-up conditions are identical, as in the test showed in Figure 4. In this case, two peaks are present, one at 1.89 min and the second at 1.46 min.

Furthermore, the values obtained from an olfactometry analysis, carried out with an TO8 olfactometer, of the samples generated with a vaporisation temperature equal to 270 °C, were higher than the expected values, as shown in Figure 6.

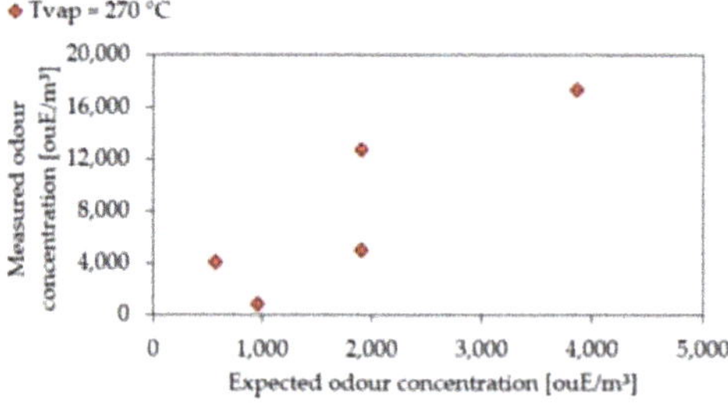

Figure 6. Scatter diagram of olfactometry analysis results for samples with Tvap = 270 °C with respect to the expected values, defined by the analytical values generated.

It is assumed that this phenomenon was due to decomposition of n-butanol caused by high temperatures. The components generated by thermal degradation were not determined, and will be studied in future works.

Based on available data, one of these compounds generated by the decomposition of n-butanol is determined by the FID analysis and is lighter than n-butanol itself.

In subsequent tests, the vaporisation temperature was reduced to 140 °C. This temperature allows obtaining complete vaporisation of the flow rate required for subsequent tests, up to 5 mL/min of liquid solution, and it seems to prevent the decomposition of n-butanol.

The results are not definitive, but they are valid in a case that considered a sample of gases and vapour, containing only nitrogen, N_2, n-butanol, and water vapour.

3. Results

3.1. Execution of the Tests

In October 2019, an experimental campaign was conducted in order to verify the possibility of evaluating sampling uncertainties in odour measurements using a synthetic bench loop.

A sample of nitrogen and water vapour, with a known n-butanol concentration, was generated inside the bench loop. The n-butanol concentration was constantly monitored using a portable FID (Mercury 901, N.I.R.A., Biassono (MB), Italy)

The sampling was carried out from one of the five sampling ports, shown in Figure 7.

Figure 7. Sampling port.

The samples were analysed within 30 h after sampling, conforming to standard EN 13725. Olfactometry and a chemical analysis were carried out on the samples taken in order to obtain, respectively:

- n-Butanol concentration value, expressed in ppm;
- Odour concentration value, expressed in ouE/m^3.

The measured values were compared with the reference values, which were calculated, as described above.

3.2. Measurement Uncertainties

Quality criteria for the overall performance of the sensory measurement method are indicated in European standard EN 13725. Compliance with the quality criteria has to be assessed by performance testing to demonstrate and ensure compliance on a regular basis. According to this standard, the European Reference Odour Mass, EROM, for n-butanol, is used as the conventional quantity value when assessing trueness and precision.

If those criteria are respected in the analysis of reference materials, the standard provides for the use of n-butanol; this quality level can be considered transferable to other odorous substances [12].

The accuracy reflects both the trueness, expressed as bias, and the precision and random error. The test variable for accuracy is A_{od}.

The criterion for accuracy of the odour concentration is:

$$A_{od} = d_w + A_w * r \leq 0.217 \tag{3}$$

where A_W is a statistical factor; d_w is the trueness, expressed as the estimate of within-laboratory bias; r is the repeatability limit.

In addition to the overall accuracy criterion, the precision, expressed as repeatability limit, complies with:

$$r \leq 0.477 \tag{4}$$

This requirement implies that the factor that expresses the difference between two consecutive single measurements, performed on the same testing material in one laboratory under repeatability conditions, will not be larger than a factor 3 in 95% of cases.

The standard EN 13725 indicates that the geometric mean of the individual threshold estimates ITE$_{substance}$—expressed in mass concentration units of the reference gas—has to fall between 0.5 and 2 times the conventional quantity value for that reference material, for n-butanol, from 62 μg/m^3 to 246 μg/m^3 = 0.020 μmol/mol to 0.080 μmol/mol. This means that 95% of the measurements can be found between 50 and 200% of the expected concentration.

These performance criteria cannot be used to express the odour measurement uncertainty for the laboratory because it does not consider the sampling phase.

When a laboratory estimates a general uncertainty for all ordinary measurements carried out by the laboratory, then the measurement uncertainty shall be calculated during an ordinary measurement. This means that the usual analysis condition must be respected, and non-optimised procedures, which tend to reduce uncertainty, cannot be used.

The expanded uncertainty U is expressed as a two-sided confidence interval [13]:

$$\log_{10}(C_{od}) \geq \log_{10}(C_{od}) - \delta_{w,CRM} - U\left(\log_{10}(C_{od})\right) \tag{5}$$

$$\log_{10}(C_{od}) \leq \log_{10}(C_{od}) - \delta_{w,CRM} + U\left(\log_{10}(C_{od})\right) \tag{6}$$

In antilog terms, the coverage interval, confidence interval, is:

$$C_{od} * 10^{-\delta_{w-CRM}-U} \leq C_{od} \leq C_{od} * 10^{-\delta_{w-CRM}+U} \tag{7}$$

The expanded uncertainty of measurement of the logarithm of the odour concentration of environmental samples is:

$$U\left(\log_{10}(C_{od})\right) = k * u_c \tag{8}$$

where k is the coverage factor; an appropriate coverage factor is $k = 2$ to express a 95% coverage probability; u_c is the combined standard uncertainty of measurement of the logarithm of the odour concentration of environmental samples.

In an investigation of a single sampling target, if the sources of variation are independent, the measurement variance σ^2_{meas} is given by:

$$\sigma^2_{meas} = \sigma^2_{sampling} + \sigma^2_{analytical} \tag{9}$$

where $\sigma^2_{sampling}$ is the between-sample variance on one target, and $\sigma^2_{analytical}$ is the between-analysis variance on one sample.

If statistical estimates of variance, s^2, are used to approximate these parameters, the standard uncertainty, u_c, can be estimated [14]:

$$u_c = s_{meas} = \sqrt{s^2_{sampling} + s^2_{analyical}} \tag{10}$$

It is not possible to distinguish these two contributions, but both must be considered in order to have an analysis that is, as much as possible, representative of the real measurement uncertainty.

In dynamic olfactometry, to verify laboratory compliance with performance quality criteria, PTs are carried out using dry samples containing air and buthanol. In these cases, only the contribution to uncertainty related to the analysis of the sample is investigated.

The tests, shown in this publication, were carried out to verify the possibility of using the bench loop to generate the odour sample with a known odour concentration; with this procedure, it would also be possible to investigate the contribution to uncertainty related to the sampling phase.

3.3. Experimental Results

In 2019, tests were conducted to verify the capability of the procedure to generate an odour sample with a known odour concentration, in order to evaluate measurement uncertainty as accurate as possible.

During the same testing day, gas samples containing gradually increasing odour concentrations were generated and sampled. The samples were analysed both from a chemical and olfactometry point of view; the results are shown in Table 2.

Table 2. Chemical and olfactometry results.

Sample	CbuOH Expected (ppm)	CbuOH Measured (ppm)	Cod Expected (ouE/m^3)	Cod Measured (ouE/m^3)	Confidence Interval
1	23.23	18.66	580	542	262–1084
2	40.67	36.81	1016	1085	543–2170
3	56.57	56.94	1414	1218	609–2436
4	74.36	69.52	1859	1934	495–1934
5	106.92	102.24	2673	2436	609–2436

In this test, the vaporisation temperature was maintained at 140 °C, in accordance with the results of previous tests. Moreover, in the bench loop, a sample of nitrogen, n-butanol, and water vapour was injected; so the results shown refer to this particular scenario in which interfering gases, e.g., CO_2, are not injected.

Figure 8 presents the results of the chemical analysis of the samples taken; the values measured are consistent with the expected values.

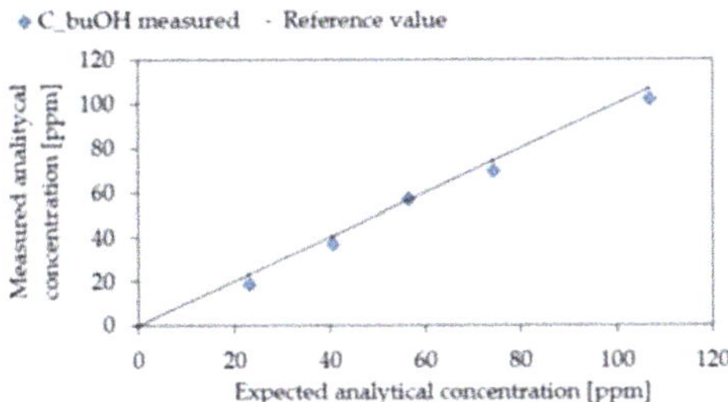

Figure 8. Scatter diagram of between concentration of butanol measured by the chromatograph (C_{yOH} expressed in ppm of butanol) and the concentration expected due the generation process at 140 °C (C_{xOH} expressed in ppm of butanol) obtained during the second field trial.

Figure 9 presents the results of the olfactometry analysis. For each measured value, the confidence interval, calculated in compliance with the standard EN 13725, is indicated.

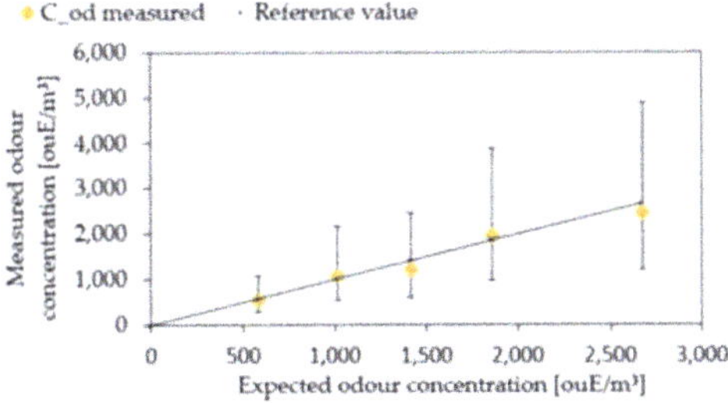

Figure 9. Scatter diagram of between odour concentrations determined using the EN 13752 method and the expected values, due to the generation process at 140 °C (C_{xOH} expressed in ppm of butanol), obtained during the second field trial.

Considering the confidence interval, the measured values are consistent with the expected values.

Furthermore, the measured data were subjected to basic statistical processing. The z-score method was used to assess the results, as indicated in the standard EN 13528 [15], using Equation (11).

$$Z^* = \frac{X_{lab\_norm} - 1}{\sigma} \tag{11}$$

$$X_{lab\_norm} = \frac{X_{lab}}{X_{ref}} \tag{12}$$

The measured values were normalised, as shown in Equation (12), with respect to the reference value. The standard deviation of the proficiency assessment used is equal to:

$$\sigma = 0.10(10\%) \tag{13}$$

The conventional interpretation of z-scores is as follows: a result that gives $|z| \leq 2$ is considered acceptable; a result that gives $2 < |z| \leq 3$ is considered a warning signal; a result that gives $|z| \geq 3$ is considered unacceptable.

The results are shown in Figure 10; all measured values can be considered acceptable.

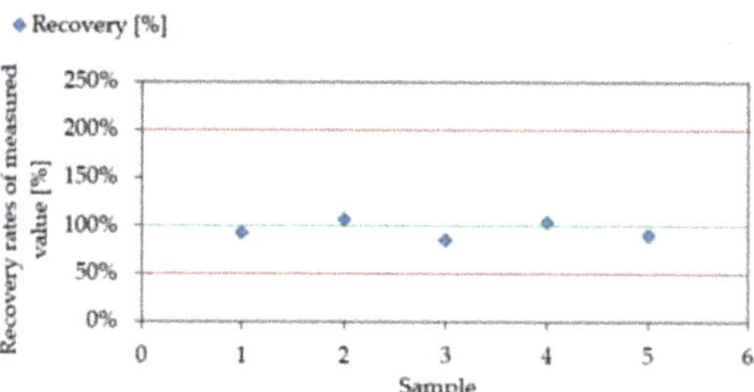

Figure 10. Recovery rates of measured values obtained during the second field trial, considering the relative difference between the measured concentrations in relation with the expected ones, obtained during the second field trial.

Moreover, in order to evaluate the variability of the results, the recovery rate was calculated. Figure 11 shows the recovery rates of the measured values. The variability is in a range between 86% and 107%; these results are located between 50% and 200% of the expected concentration as indicated in the standard EN 13725.

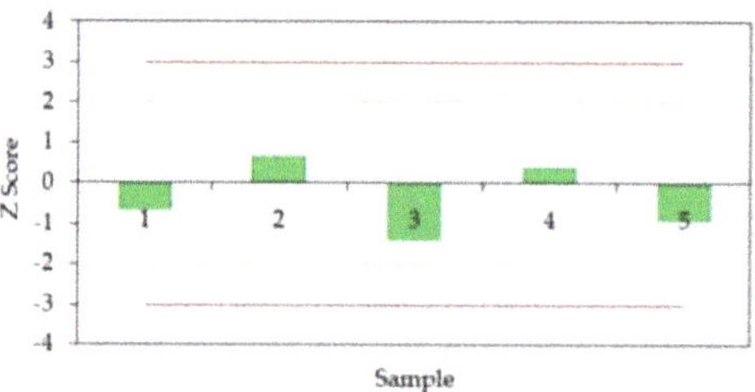

Figure 11. Z Score values obtained for the different concentrations during the second field trial, using the formulation given by EN 13528; in this case, the value of standard deviation selected is 10% of the expected value.

4. Conclusions

The tests were carried out for the essential purpose of designing and upgrading experimental protocols to evaluate sampling uncertainties in odour measurements using synthetic bench loops.

To generate the high levels of butanol needed (up to 500 ppm was necessary prior to dilution into the bench), a vaporisation system was realised and tested, as gas cylinders are not available at such concentrations.

A first test was carried out in order to verify the possibility of generating an odour atmosphere with the target composition.

Results obtained are consistent with the expected values. Moreover, z-scores and recovery rates for each value were calculated. Data obtained showed variability of the values in the range of 86% to 107%, verifying that all results could be considered acceptable, as indicated by the standard EN 13725.

Such an experimental protocol seems to offer the possibility of evaluating sampling uncertainties in odour measurements. Therefore, it allows combining the traditional scheme of inter-laboratory comparison, in which the sampling process is not investigated, and the real measurement, in which the reference value is unknown, using the bench loop. In fact, in the tests carried out with this procedure, it was possible to generate an odorous sample with a known odour concentration.

This procedure would also allow evaluating the transport and storage phases, verifying the sample stability in this period. Moreover, new sampling methods could be analysed using the bench loop.

One interesting future scenario would be injecting interfering gases, e.g., CO_2, in the bench loop in order to evaluate the influence that the presence of those gases could have on the odour concentration measurement.

Another interesting aspect would be evaluating the influence of a change in temperature and humidity on the measured odour concentration in samples with a known n-butanol concentration. In fact, the bench loop allows regulating and maintaining a constant temperature and humidity inside the conduit. It would be significant to investigate the impact these factors have on sampling uncertainty.

Author Contributions: Conceptualization, D.C.; methodology, D.C.; validation, D.C.; data curation, D.C., A.M.C., M.A.; writing—original draft preparation, D.C.; writing—review and editing, D.C., A.M.C. All authors have read and agreed to the published version of the manuscript.

Funding: This research received no external founding.

Institutional Review Board Statement: Not applicable.

Informed Consent Statement: Not applicable.

Data Availability Statement: The datasets used and/or analysed during the current study are available from the corresponding author on reasonable request.

Conflicts of Interest: The authors declare no conflict of interest. The funders had no role in the design of the study; in the collection, analyses, or interpretation of data; in the writing of the manuscript, or in the decision to publish the results.

References

1. CEN. *EN13725, Air Quality-Determination of Odour Concentration by Dynamic Olfactometry and Odour Emission Rate from Stationary by Sources*; CEN: Brussels, Belgium, 2004.
2. Freeman, T.; Needham, C.; Schulz, C. *CH2M BECA Analysis of Options for Odour Evaluation for Industrial or Trade Processes*; Auckland Regional Council CH2M BECA Ltd.: Auckland, New Zealand, 2000.
3. Gwynne, C.; Philips, L.; Turner, N.; Goldstone, M. *Temporal and Spatial Variability of Odours at Wastewater Plants in Western Australia*; Water Corporation: Melbourne, Australia, 2001.
4. Van Harreveld, A.P.; Heeres, P. The validation of the draft European CEN standard for dynamic olfactometry by an interlaboratory comparison on n-butanol. *Gefahrst. Reinhalt. Luft* **1997**, *57*, 393–398.
5. Ramsdale, S.L.; Baillie, C.P. Interlaboratory test program: Odour detection threshold for hydrogen sulphide. *J. Clean Air Soc. Aust. N. Z.* **1996**, *30*, 36–37.
6. Defoer, N.; Van Langenhove, H. Variability and repeatability of olfactometric results of n-butanol, pig odour and synthetic gas mixture. *Water Sci. Technol.* **2004**. [CrossRef]
7. Clanton, C.J.; Schmidt, D.R.; Nicolai, R.E.; Goodrich, P.R.; Jacobson, L.D.; Janni, K.A.; Weisberg, S.; Buckel, J.A. Dynamic olfactometry variability in determining odor dilutions-to-threshold. *Trans. ASAE* **1999**, *42*, 1103–1112. [CrossRef]
8. Stöckel, S.; Cordes, J.; Stoffels, B.; Wildanger, D. Scents in the stack: Olfactometric proficiency testing with an emission simulation apparatus. *Environ. Sci. Pollut. Res.* **2018**, *25*, 24787–24797. [CrossRef] [PubMed]
9. Cipriano, D.; Fialdini, L. Definition of reference values in synthetic emission monitoring bench loop. *Accredit. Qual. Assur.* **2018**. [CrossRef]

10. ISO. *ISO Guide 35, Reference Materials—Guidance for Characterization and Assessment of Homogeneity and Stability*; ISO: Geneva, Switzerland, 2017.
11. ISO. *ISO/IEC 17043, Conformity Assessment-General Requirements for Proficiency Testing*; ISO: Geneva, Switzerland, 2010.
12. Van Harreveld, A.P.; Heeres, P. Quality control and optimization of dynamic olfactometry using n-Butanol as a standard reference odorant. *Staub Reinhalt. Luft* **1995**, *55*, 45–50.
13. JCGM. *GUM Guide to the Expression of Uncertainty in Measurement*; JCGM: Geneva, Switzerland, 2008.
14. EURACHEM/CITAC. *Measurement Uncertainty Arising from Sampling—A Guide to Method and Approaches*, 2nd ed.; EURACHEM: Teddington, Middlesex, UK, 2019.
15. ISO. *ISO 13528 (2015), Statistical Methods for Use in Proficiency Testing by Interlaboratory Comparison*; ISO: Geneva, Switzerland, 2015.

Article

Environmental Odour Nuisance Assessment in Urbanized Area: Analysis and Comparison of Different and Integrated Approaches

Tiziano Zarra *, Vincenzo Belgiorno and Vincenzo Naddeo

Sanitary Environmental Engineering Division (SEED), Department of Civil Engineering, Università degli Studi di Salerno, Via Giovanni Paolo II, 132, 84084 Fisciano, Italy; v.belgiorno@unisa.it (V.B.); vnaddeo@unisa.it (V.N.)
* Correspondence: tzarra@unisa.it; Tel.: +39-089-969335

Abstract: Prolonged exposure to odour emissions causes annoyance which leads to nuisance and consequently to complaints. Different methodologies exist in the literature to evaluate odour impacts, but not all are suitable to assess environmental odour nuisance. Information about their applicability criteria and comparison, is scarce and referred to short time analysis. The research presents and discusses the application of different methods to characterize and assess odour nuisance around an industrial plant localized in a sensitive area. Experimental activities are carried out through a long-time analysis programme. Field inspections and predictive methods are investigated and compared. A modification of the traditional dispersion modelling approach is proposed in order to adapt its application for the prediction of the odour nuisance. The offensiveness and location factors are identified as key parameters in the quantification of the perceived nuisance. The integrated dispersion modelling multi-level approach is highlighted as the most suitable for defining the plant strategies. The paper provides useful information to characterize environmental odour problems and identify appropriate solutions for an effective management of odorous sources, with the aim of reducing complaints, restoring the proper relationship between odorous plants and the surrounding communities and increasing the overall quality of the environment.

Keywords: air quality; field inspection; odour impact; odour modelling; wastewater treatment plant

Citation: Zarra, T.; Belgiorno, V.; Naddeo, V. Environmental Odour Nuisance Assessment in Urbanized Area: Analysis and Comparison of Different and Integrated Approaches. *Atmosphere* **2021**, *12*, 690. https://doi.org/10.3390/atmos12060690

Academic Editors: Günther Schauberger, Martin Piringer, Chuandong Wu and Jacek Koziel

Received: 30 April 2021
Accepted: 26 May 2021
Published: 28 May 2021

Publisher's Note: MDPI stays neutral with regard to jurisdictional claims in published maps and institutional affiliations.

1. Introduction

Odour emissions from industrial plants, especially those involved in environmental management (e.g., wastewater treatment plants, landfills, composting plants) are among the leading cause of odour annoyance within the population living in the surrounding areas. Odour annoyance does not directly affect the health but implies a contingent perception of risk [1–3] and consequently leads to complaints [4,5].

Annoyance is the complex of human reactions that occurs as a result of an exposure to an ambient stressor (odour) that causes negative cognitive appraisal once perceived [6,7]. A repeated annoyance event over an extended period of time that leads to modified or altered behaviour, causes nuisance [4,8]. The main effects on human health caused by continuous exposure to odours are irritant properties, as well as cardiovascular, respiratory and psychobiological effects [9–11]. Therefore, appropriate monitoring along with regulatory tools are necessary to prevent, control and mitigate the emissions of odours from industrial plants and their consequent nuisance in ambient air in an objective and repeatable manner [12–14]. The evaluation of the real nuisance is fundamental to identify the appropriate solutions to adopt to manage the odour problem without underestimating or providing wrong information [15,16].

In the current literature, there are few studies which deal with the identification of the odour nuisance in ambient air. Information about their applicability conditions and comparison between the different approaches, is scarce and referred only to short time-periods. Moreover, no investigation between the perceived odour nuisance and the odour sources and management is reported. Almost all of the studies present in the literature focus on the evaluation of the odour impact [4,5,17–21] and present comparisons between the different methodologies used [12,22–24]. However, the odour impact should not be confused with the odour nuisance and not all the methodologies applied to evaluate the odour impact are suitable to assess odour nuisance [25]. The likelihood of an odour to create nuisance is evaluated by using the FIDOL factors: frequency (F), intensity (I), duration (D), offensiveness (O), and location (L) [4,17,26]. According to this approach, the definition of odour concentration is not complete with a view of defining odour nuisance [27,28].

In order to evaluate the occurrence of odour nuisances in an area where there are odours and/or odour sources, the use of field methods that allow the participation of the communities (also known as Citizen Science methods [29]) would be particularly effective, but they are generally expensive and require a long application time [4]. At the current state, for the assessment of odour nuisances, the methods that use field inspections are carried out through two standardized approaches: using trained assessors (human panel) [30,31] and by means of questionnaires distributed to the local population [32]. These approaches rely on the determination of all the FIDOL factors. On the other hand, the most used approach to predict odour impact from a plant is based on the simulation of their odour emissions using atmospheric dispersion models [4,33,34]. This method is time-saving to deliver the results, but, generally is not useful for evaluating the community odour annoyance and nuisance, as it does not consider properly the factors offensiveness (O) and location (L) [25,35].

The offensiveness (O) factor of an odour can be assumed coincident with its hedonic tone [4]. In Europe, the German guideline VDI 3882/2 (1994) [36] describes one of the most widely used methods to determine the hedonic tone. While in the United Kingdom, different concentration thresholds are set, in relation to the types of industrial plants, so as to take into account the offensiveness parameter in the conventional dispersion model approach [37].

The location (L) factor can be represented by the sensitivity of the area where the odour is dispersed and perceived [4]. Recent studies propose taking into account the different territorial sensitivities to odours through the definition of a "Nuisance Action Plan" (NAP) in urban planning [28]. Indicators and criteria, such as the "land use destination or class of location" (L), "abundance of receptors" (R) and the existing "environmental pressures" (P) are suggested to elaborate the NAP. According to the definition of the location parameter, an area with a major sensitivity is less inclined to support an odour presence.

This paper presents and discusses the application of different methods to assess the odour nuisance perceived by the communities and caused by an industrial plant. The comparison of the methods is highlighted through an application to a real case study over the long time period of one full year. Field inspection methods through trained assessors (grid method) and with questionnaires as well as the predictive dispersion modelling approach are investigated. A multi-percentile and multi-odour concentration threshold criterion approach is proposed and explored in the paper to allow the application of the dispersion model for the prediction of the community odour nuisance. The study aims to provide relevant information and tools to develop effective control or warning strategies for the plant managers, the local control bodies and the annoyed population, in order to manage the odour nuisance problem in an effective manner, reducing the possibility of complaints and increasing the environmental air quality. Moreover, the results can be usefully applied in the design and testing of odor control technologies in odour nuisance problems.

2. Materials and Methods

2.1. Investigated Area

Research activities were carried out by considering the odour nuisance to the community linked to a Wastewater Treatment Plant (WWTP). The surrounding area in which the plant is located is characterized by the presence of a consistent number of sensitive receptors and citizens. A medical diagnostic centre (DC) and a large shopping centre (SC) are situated close to the plant. While slightly further away from the WWTP, there are various shops (SH), bathing beaches (BB) and residential buildings (RB) (Figure 1). A large number of people attend the area for work or leisure on a daily basis, in addition to those who permanently live there.

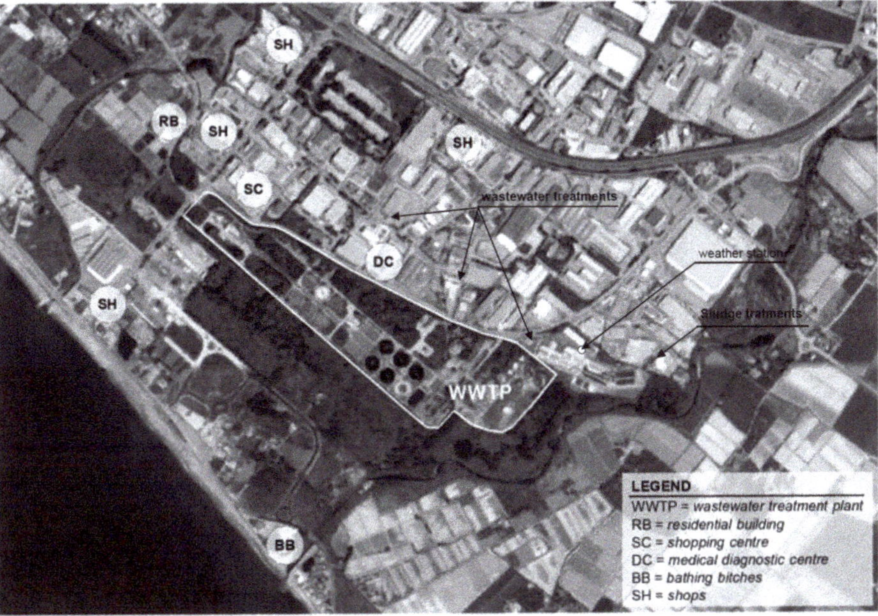

Figure 1. Investigated area and WWTP with principal surrounding sensitive targets (Maps Data: Goggle, ©2019 CNES/Airbus, European Space Imaging, Maxar Technologies).

The investigated WWTP is a conventional activated sludge plant, designed for 700,000 PE (population equivalent). The plant treats civil wastewater from eleven municipalities with an hourly average flow rate of about 8000 m^3/h. The layout of the plant highlights two parallel lines for the wastewater treatments and one single line for the sludge treatment.

2.2. Odour Nuisance Characterization by Trained Assessors

The field inspection technique by trained assessors was carried out according to the EN16841-1:2016 [30]. The grid method was applied covering an area of 2.625 km^2 (1.750 × 1.500 km^2), with the WWTP located in the central position. The selected area included the surroundings of the WWTP, where the odour receptors are prevalent, and it extended in the prevailing wind direction. A total of 42 cells were identified in the investigated area, dividing them into a square mesh of 250 m per side. Ten assessors, divided into teams of four persons, selected according to EN 13725:2003 and familiarized with the on-site specific wastewater odour quality category, identified as only one, were used in the field analysis. According to EN16 841-1:2016, for each corner point, the "odour hours" (n_h) were calculated, along with the "odour load" (OL) for each mesh. For the study, one odour-hour was considered if, by sniffing 1 time every 10 s for an interval of 10 min, the investigated odour quality category was positive perceived for 6 or more times (10% of the defined measurement interval); while the odour load was calculated as the ratio between the sum of odor hours calculated for the 4 corner points, divided by the total number of inspections for the mesh. The weather condition was continuously detected during the surveys, using a weather station located inside the investigated plant (Figure 1) and a portable Kestrel 4000 Pocket Wind Meter (Nielsen-Kellerman, PA, USA) anemometer, in order to take into consideration, the effects of atmospheric dispersion of the substances emitted and perform the field inspection measurement in optimal conditions under wind (wind speed between 1 to 3 m/s, temperature below 35 °C), avoiding possible false positives [38]. A total of 104 single measurements for each assessment point were carried out, over the annual monitoring period, situated generally at a point of intersection of the grid lines. No points were selected within the plant fence line. In detail, two field inspections per week were carried out, according to the EN16841-1:2016 with a survey on the scale of 104 field inspections in a full year. No consecutive days were used for the measurements; throughout the annual period, all days of the week were given roughly equal priority (each day of the week was represented about four times for 104 measurement days); for each assessor during a measurement round, no more than 10 to 12 points were included; the scheduling of the daily start of inspection was elaborated in order to have, after 26 single measurements per measurement point, field inspection data for different times on the day, for all measurement points, thus allowing a statement representative of the course of the day.

The odor frequency in terms of "odour hours" (n_h), assessed in the 42 assessment squares, was represented graphically through isofrequency odour maps, generated by means of the kriging interpolation geostatistical method, performed using the Surfer software [39]. Four isofrequency levels were performed in the study to analyse the odour nuisances: $n_h < 10\%$, no odour nuisance; $10\% \leq n_h < 15\%$, odour nuisance for residential area; $15\% \leq n_h < 25\%$, odour nuisance for industrial area; $n_h \geq 25\%$, strong nuisance [40].

2.3. Odour Nuisance Characterization by Questionnaires Survey

The sociological survey implemented through questionnaires was prepared according to the German guidelines VDI 3883—Part 1 (2015) [32] in order to characterize the odour nuisance in the investigated area caused by the WWTP odour type. The elaborated survey form was a closed type, consisting of a series of multiple questions, structured so as to consider the FIDOL factors (frequency, intensity, duration, offensiveness and location) as well as to characterize the general information and personal data of the subjects interviewed. Even if the questionnaire was inspired by some existing guidelines, it did not correspond to any standard survey form, but was specifically tailored to refer to only the wastewater treatment odour quality. Each question included five possible responses.

Twenty different investigation points were selected in an area of about 3 km² around the WWTP in which the questionnaires were submitted verbally to the respondents by an expert assessor.

The questionnaires were addressed specifically to residents and employees who have been working for many years in the survey area and who are able to describe what a WWTP does, in order to increase the reliability of the answers. The sample was randomly selected in the investigated area, trying to respect the demographic profile of the investigated zone [41].

A total of 1000 observations were recorded during the monitoring period of one year, 50 for each point. The choice of the distribution points was made in the prevailing wind direction and included the zone where the odour receptors were prevalent, around the WWTP. Different distances from the source were selected according to the Miedema study [27], for the individuation of the investigation points, ranging from as close to the plant as possible to a distance where the odour was expected to be hardly detectable [42,43].

Twelve investigated points were included at a distance of 500 m from the nearest border of the WWTP area.

The collected data were encoded in Excel and statistically processed. The odour annoyance index (OAI) was calculated for each investigated point, using the data referring to the questions related to the FIDOL factors [35,44] by applying the following equation:

$$OAI_{Py} = \frac{1}{N_t} \sum w_i N_i,$$

(1)

where:

- P_y is the investigated point (y = 1–20);
- N_t is the total number of recorded observations for the investigated point;
- w_i are the weighting factors of the answer i (i = 1–5) of the question n (n = 1–5);
- N_i are the number of observations recorded at the answer i of the question n.

Five annoyance levels were defined in the study to analyse the odour nuisances: OAI = O, no odour nuisance; 0 < OAI $\leq$ 1.25, slight odour nuisance; 1.25 < OAI $\leq$ 2.50, distinct nuisance; 2.50 < OAI $\leq$ 3.75, strong nuisance; 3.75 < OAI $\leq$ 5.00, extreme nuisance.

2.4. Odour Nuisance Characterization by Atmospheric Dispersion Modelling

The Calpuff Modelling System was used to simulate the dispersion of the odours emitted by the WWTP in the atmosphere, in order to assess the odour impact due to the presence of the WWTP. The Calpuff Modelling System consists of three basic components: a meteorological preprocessor (Calmet) [45], the dispersion model (Calpuff) [46] and a graphical post-processor (Calpost). For the application, three different types of input data are provided: topographical, meteorological and emission.

A modelling domain of 16 km^2 (4.0 km for each side) was selected, with a square grid of receptors every 100 m. Hourly average data of the meteorological parameters (e.g., wind direction, wind speed, atmospheric stability, temperature, precipitation, etc.) covering one year were given as input. These data were recorded by a meteorological station located in the WWTP area (Figure 1) and elaborated by the preprocessor Calmet.

Seven different odour sources (P_i, i = 1–7) (Figure 2) were investigated in terms of emission data, calculating their Odour Emission Rate (OER). Four of the investigated sources belong to the wastewater treatment lines, while three sampling points are related to the sludge treatment line. Eight samplings for each source were carried out, over a monitoring period of two months, with a weekly frequency. In total, 56 samples were collected. The OER values used in the simulations were calculated using the highest values of the Specific Odour Emission Rate (SOER) detected over the entire analysis period to simulate the worst case scenario. The SOER values were determined in relation to the source typology as a function of the measured odour concentration [4]. A vacuum sampler (Ecoma, D) equipped with a 10 L Nalophan bag and a SF450 flux chamber (Scentroid, IDES Canada Inc., Whitchurch-Stouffville, ON, Canda) was used for the sampling. The operating parameters of the flux chamber were: diameter 450 mm, an enclosed surface area of 0.155 m^2 and neutral air flux equal to 3.9 lpm.

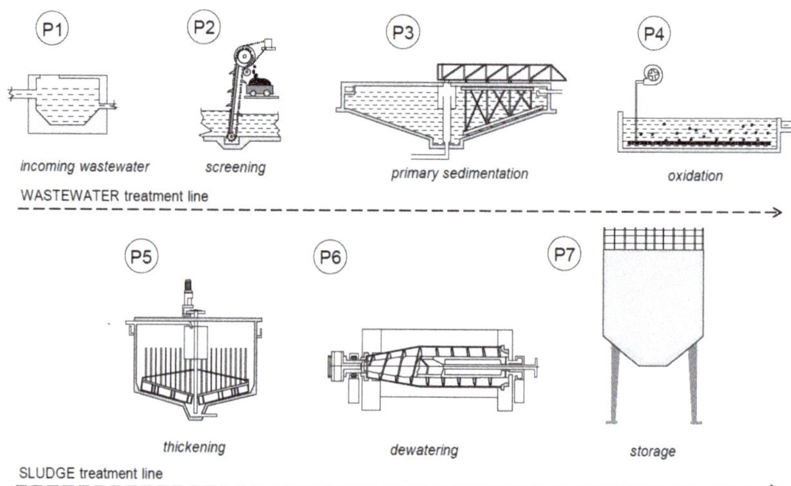

Figure 2. Investigated odour sources (P_i, i = 1–7).

Odour concentrations of the collected samples were carried out by dynamic olfactometry analyses, conducted at the Olfactometric Laboratory of the Sanitary Environmental Engineering Division (SEED) of the University of Salerno, according to the EN 13725:2003. A TO8 olfactometer (ECOMA, D), based on the "yes/no" method, was used relying on a panel composed of four trained experts. All the measurements were analysed within 14 h after sampling, according to Zarra et al. [47].

A peak-to-mean ratio (P/M) equal to 2.3 was adopted, according to Italian guidelines [5,46,48], in order to obtain the hourly peak concentration as the output of the Calpuff dispersion model.

The simulation period was set equal to one year. The meteorological data used refer to the same year in which the field inspections were carried out.

In order to adapt the method to assess the odour nuisance, the offensiveness (O) and location (L) factors were implemented in the conventional dispersion modelling application approach. The offensiveness of the odour was considered by assuming that its influence is directly related to the use of a different odour threshold concentrations values adopted to calculate the results, according to the approach carried out by the UK Environmental Agency in the H4 guidance. The location factor was introduced by investigating its variability on the percentiles of calculation (equal to the hours of odour exposure), taking into consideration that an area with major sensitivity is less inclined to support an odour presence [4].

Hourly average odour concentrations thresholds (C_T) values of 1.0 OU_E/m^3 and 1.5 OU_E/m^3 and the 85th and the 98th percentile (p_T) were investigated; the values were selected according to those indicated by the German guidelines GOAA [40] and the UK Environmental Agency H4 guidance [37] for the odour impact evaluation. A total of four long-term simulations of one year duration were carried out.

3. Results and Discussions

3.1. Odour Nuisance Characterization by Trained Assessors

Figure 3 reports the three isofrequency levels of odour hours calculated over the overall monitoring period for the investigated area.

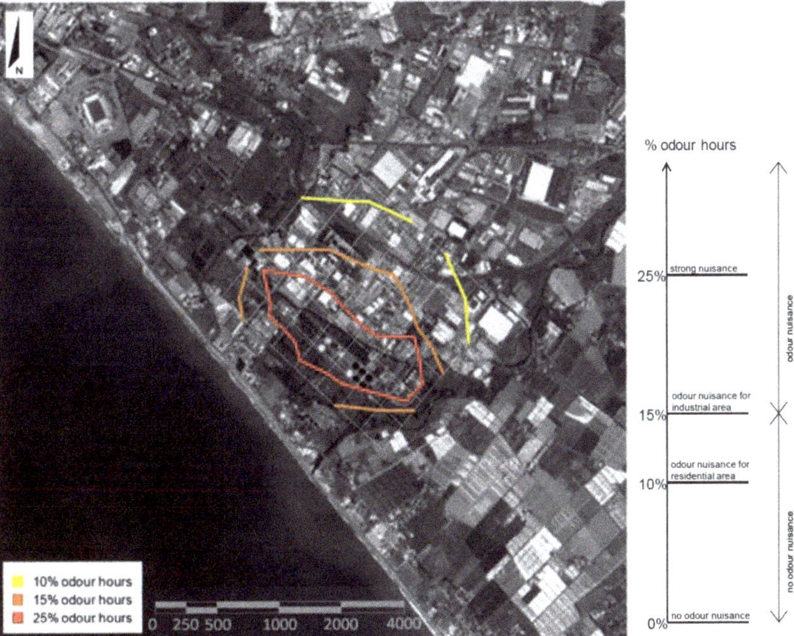

Figure 3. Isofrequency levels of odour hours (Maps Data: Goggle, ©2019 CNES/Airbus, European Space Imaging, Maxar Technologies).

The results show that the area affected by a strong nuisance of the considered odour quality type (wastewater odour) is basically coincident with the area of the WWTP location (Figure 1). The form of the isofrequency level representing the strong nuisance odour provides an indication of the areas of the plant that contribute most to the odour nuisance and on which it is therefore necessary to act in order to reduce the olfactory nuisances for the neighbouring communities. For the investigated case study, the sludge treatments (Figure 2) (located on the almost extreme right of the area enclosed by the red line in the Figure 3) are highlighted as the sources that require greater attention in the management of odour emissions.

Considering the whole covered area by the grid method (equal to a surface of 1.5 km × 1.75 km = 2.625 km^2), the study highlights furthermore how a surface of 1424 km^2, corresponding to 54%, is annoyed if the area is considered as industrial, while a percentage of almost 93% is affected by odour nuisance if the area would be assumed as residential. In any case, only a very little area of the investigated surface was characterized by a percentage of odour hours less than the 10%. This result seems to indicate that the monitored area for the study is too small and that it would be advisable to extend this to have a greater overview. However, an extension of the area clearly involves more time and costs for the assessment.

Another result provided by this type of approach could be the identification of the minimum separation distance from the odour sources to be respected in order to not perceive the nuisance. For the analysed plant, a distance between 0.5 km and 1.10 km from the border of the location of the WWTP was identified.

3.2. Odour Nuisance Characterization by Questionnaires Survey

The personal data of the subjects interviewed, collected through the survey form, show that the gender distribution is almost equal (52% male—48% female), their age ranges mostly from 30 to 65 (59%) and from 18 to 29 (29%). Only 12% are in the range 66–72.

Of the 1000 total recorded individual observations, 226 questionnaires (22.6% of the total observations) report not having perceived odours of the wastewater treatment quality or not having perceived odours in the area. Figure 4 summarizes the results of the collected data of the persons that perceived WWTP odours, related to the part of the questionnaire which dealt with the frequency (F), duration (D) and offensiveness (O) of odours.

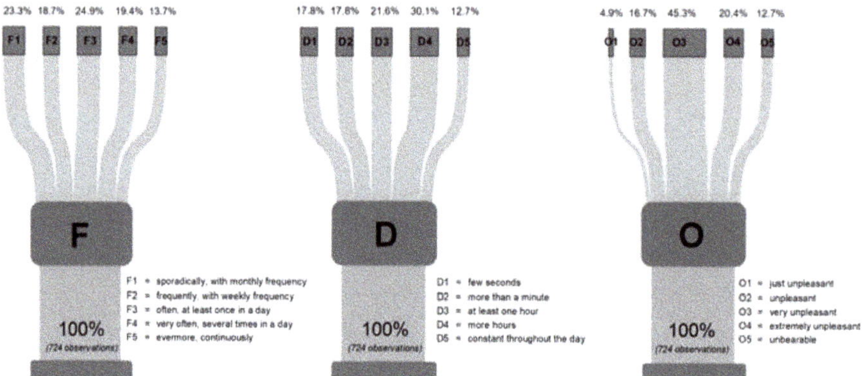

Figure 4. Frequency (**F**), duration (**D**) and offensiveness (**O**) factors results of the questionnaires survey.

As reported in Figure 4, there is no significantly predominant frequency where the odour is perceived. However, the responses highlight a preponderance of lasting odour (more hours) and their "very unpleasant" hedonic tone (equal to 45.3% of the respondents).

Figure 5 shows the histogram of the distances, measured on the Google Earth map between the selected investigation points and the closest border point of the WWTP area, with reference to the number of observations. The distances are in the range of 40 m (point ID number XVII)–1550 m (point ID number I) (Figure 6).

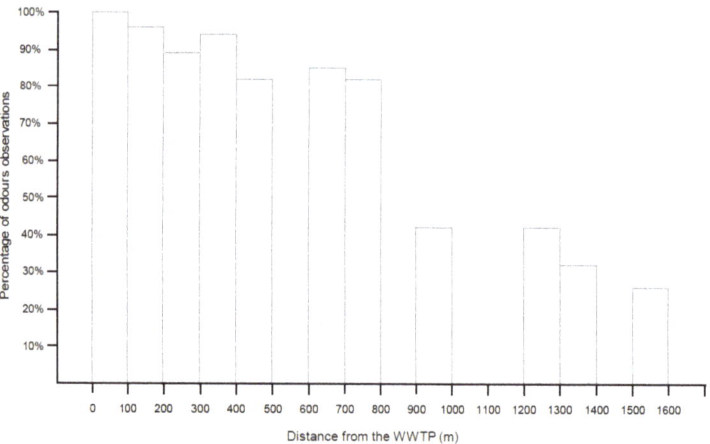

Figure 5. Relation between the "odour" observations and the distances from the WWTP.

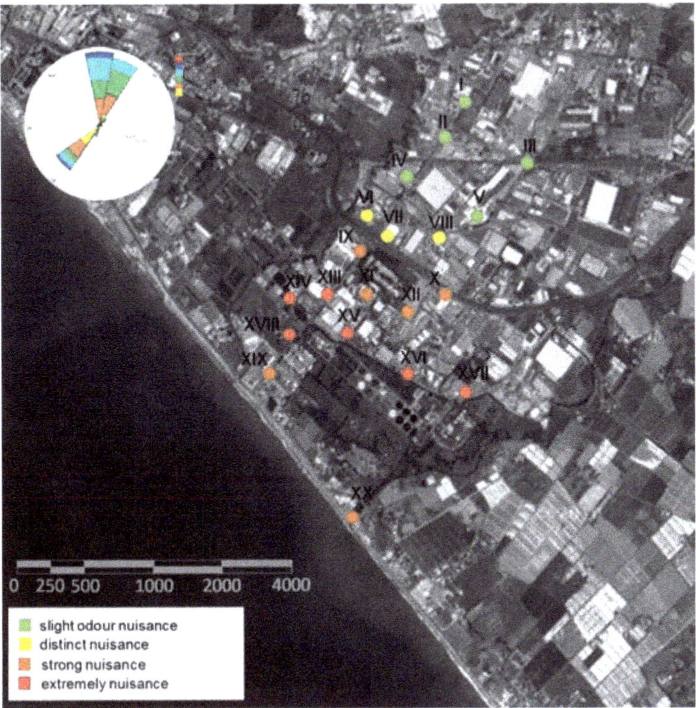

Figure 6. Odour nuisance level for each investigated point complete with wind rose (Maps Data: Goggle, ©2019 CNES/Airbus, European Space Imaging, Maxar Technologies).

The results show that the odours emitted from the WWTP are perceived in their immediate surroundings. However, a decrease of the perception of the odour is detected upon increasing the distance from the plant. At distances greater than 800 m, there is a drastic drop in the number of interviewees who declare to have perceived WWTP odour.

Figure 6 shows the odour nuisance level assumed for each investigated point. The level was defined by calculating the Odour Annoyance Index (OAI) using all the collected data over the monitoring period.

The highest values of odour nuisance were recorded by individuals located closer the WWTP. The maximum OIA, equal to 4.51, was calculated for the medical diagnostic centre, identified with the investigation point number XVI in the field assessment survey. A strong nuisance was recorded up to maximum distances of 500 m from the WWTP fence line, where the bathing beaches (BB) and some residential buildings (RB) (investigation point identified with the number XX in the assessment scheme) are located.

The quantification of the OIA index allows to identify the real community odour nuisance perception at the investigated points and, consequently, provides helpful indications of the areas where the odour sources are located that require the definition of effective odour control strategies in order to decrease the nuisance. The main weaknesses of this type of approach are that it cannot be applied where people are not permanently present and that the representation of the results, in fact linked to the distribution of people, is fragmented. Another limitation of this technique concerns the application in areas with sufficient population density to achieve statistically significant results. However, its major strength should be represented by the direct involvement of the community.

3.3. Odour Nuisance Characterization by Atmospheric Dispersion Modelling

Figure 7 highlights, for each investigated source, the odour concentrations and the odour emissions values detected over the monitored period and used as input in the dispersion model.

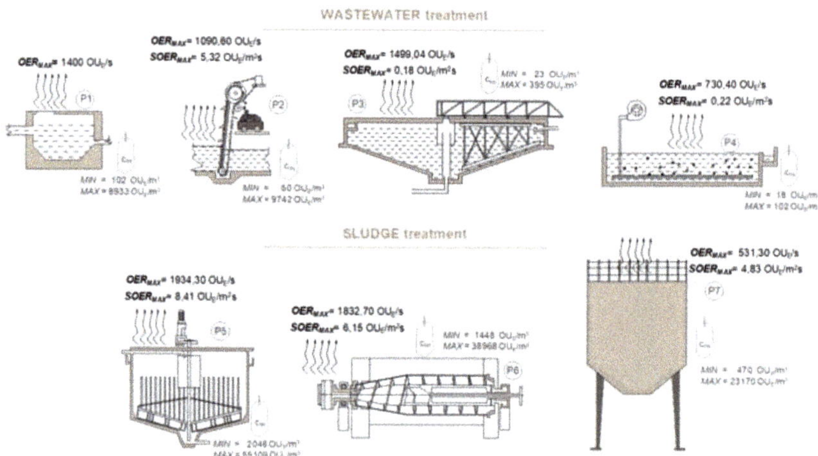

Figure 7. Odour characterization data of the investigated sources of the WWTP.

In terms of odour concentration, the highest values were detected at the sludge thickening ($55'109\ \mathrm{OU_E/m^3}$) while the lowest in the oxidation unit ($18\ \mathrm{OU_E/m^3}$). Similarly, by analysing the odour emission rate data (OER), on the basis of the values of maximum specific odour emission rate (SOERmax (i = 1–8)) calculated over the whole analysis period, the treatment of thickening of the sludge highlights the maximum value ($1'934.3\ \mathrm{OU_E/s}$).

Considering equal to 100% the value of the sum of the calculated odour emissions of the investigated sources, in terms of OER, the analysis shows how the treatment of thickening contributes with 21.45% to the total amount of emissions. Likewise, considering only the emissions from the investigated treatment processes of the WWTP, it is worth noting how the sludge treatments (P5, P6) represent 53.15% of the total odour emissions, while the wastewater treatments (P2, P3, P4) contribute with 46.85%.

The detected data also allow to highlight how the odour emissions related to a sludge treatment unit (on average equal to 26.58% of the total contribution, for treatment) is about 70% higher than those produced by a wastewater treatment unit (on average equal to 15.62% of the total odour emissions).

Figure 8 shows the results of the atmospheric dispersion model application. The isopleths relevant to the 98th and 85th percentile of the hourly peak concentrations of an odour threshold of $1.0\ \mathrm{OU_E/m^3}$ and $1.5\ \mathrm{OU_E/m^3}$ are represented in the maps. The combination of generally suggested values for example on the Guideline on Odour in Ambient Air (GOAA) in Germany ($1\ \mathrm{OU_E/m^3}$ and 85th) and on the EA in the UK ($1.5\ \mathrm{OU_E/m^3}$ and 98th) [5] is presented and elaborated in order to analyse their influence on the variation of the isopleths in terms of shape and impacted area.

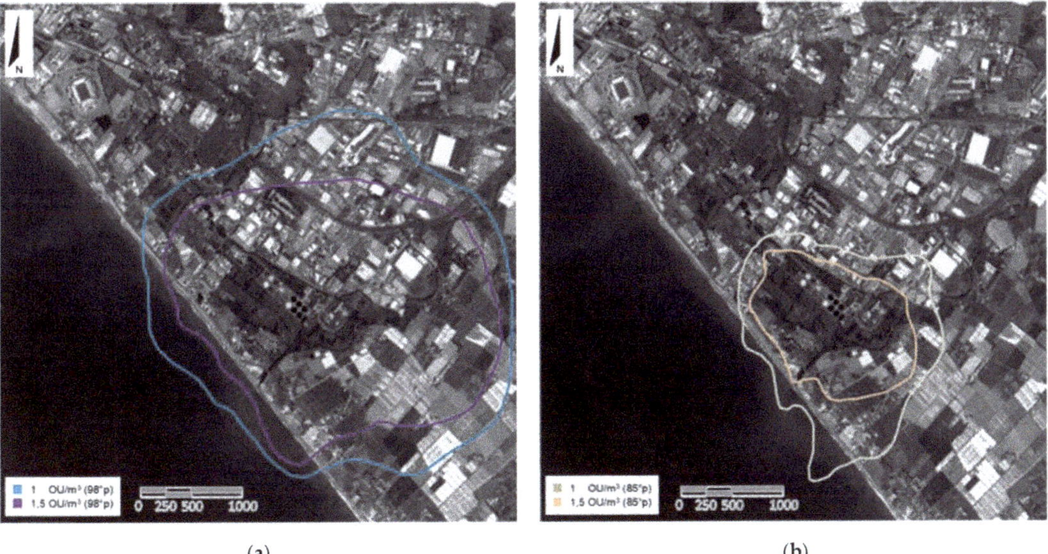

Figure 8. Isopleths relevant to the 98th (**a**) and 85th (**b**) percentile of the hourly peak concentrations of an odour threshold of 1.0 OU$_E$/m^3 and 1.5 OU$_E$/m^3 (Maps Data: Goggle, ©2019 CNES/Airbus, European Space Imaging, Maxar Technologies).

The results show relevant differences of the calculated isopleths, especially in terms of the covered surface. The greatest odour annoyed area is obtained using an odour concentration threshold equal to 1.0 OU$_E$/m^3 and the 98th percentile. In this case, the isopleths cover an area of 8747 km^2. While the smallest odour annoyed area (equal to 1533 km^2) is identified by applying an hourly average odour concentration threshold of 1.5 OU$_E$/m^3 at the 85th percentile (i.e., that exceeds 15% of the hours in a year).

The isopleths relevant to a higher percentile (98th percentile instead the 85th percentile), as well as the use of a lower odour concentration threshold (1.0 OU$_E$/m^3 instead of 1.5 OU$_E$/m^3) confirm to be a more conservative approach, in accordance with other studies [5,49] in terms of odour nuisance identification, covering generally a larger area. In the identification of the odour annoyed area by the atmospheric dispersion model application, it is possible to note that it has a greater influence on the variation of the considered percentile compared to the variation of the hourly peak concentrations of the odour threshold. At the same time, it can be seen how the prevailing meteoclimatic conditions monitored in the area (see wind rose in Figure 6) produce a greater impact on the shape of the isopleths, adopting a lower percentile for their calculation.

3.4. Odour Nuisance Levels Comparison

Figure 9 reports the overlay mapping of the odour nuisance levels elaborated through the conventional field inspections methods as well as the results obtained by the integrated odour dispersion modelling approach, in order to consider the offensiveness (O) and location (L) factors, over the annual monitoring period. The calculated isopleths relevant to the 98th and 85th percentile of the hourly peak odour concentration thresholds of 1.0 OU$_E$/m^3 and 1.5 OU$_E$/m^3 are compared, respectively, with the odour nuisance levels defined by the application of the investigated field inspection methods.

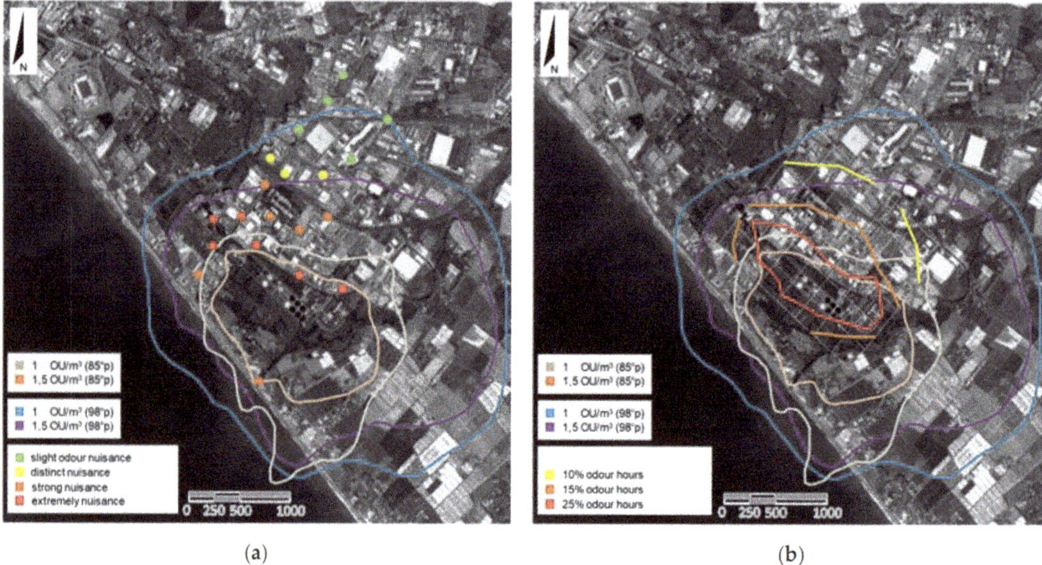

Figure 9. Comparison of the odour nuisance levels calculated by adopting different approaches: (**a**) atmospheric dispersion modelling vs. field inspection method using the questionnaire-based survey; (**b**) atmospheric dispersion modelling vs. field inspection method using the survey based on trained assessors (Maps Data: Goggle, ©2019 CNES/Airbus, European Space Imaging, Maxar Technologies).

The different approaches define the odour nuisances in the investigated area; however, they offer different accessory information. The forecast modelling approach provides information of large surfaces and identifies the odour nuisances with a continuous "contour line". The field approaches give information about the presence or absence of people in the area and the location of the sensitive receptors.

Although the dispersion modelling approach is less time-consuming, its results are strictly linked to the choice of the odour concentration threshold (C_T) and its exceedance probability (p_T). For the analysed case, large discrepancies are highlighted between the modelling and the field inspection approaches. The use of the 98th percentile and an odour concentration threshold of 1.5 OU_E/m^3 for the computation of the isopleths matches best with the field inspection method using the questionnaire-based survey. While the criterion with a lower exceedance probability (p_T = 85%) underestimates clearly the odour hours in the north-west direction of the WWTP and, on the contrary, overestimates the frequency of odour hours in the south-east direction. Analogously, the use of the impact criterion with an odour concentration threshold of 1.0 OU_E/m^3 and the use of the 98th percentile overestimates the annoyed area by the WWTP quality odour, detected by the field inspection technique using the questionnaire-based survey, covering a wider surface everywhere around the plant.

The results of applying the 98th percentile and an odour concentration threshold of 1.5 OU_E/m^3 also allows to highlight the following interesting considerations. For the investigated WWTP, the nuisance in the surrounding area is caused by an overrun of a moderate odour concentration threshold (1.5 OU_E/m^3) for a few hours in the year (98th equals to about 175 h of one year with 365 days). The results, therefore, demonstrate the influence of the "location" factor in the identification of the nuisance area. The presence of a different type and number of receptors (such as the medical diagnostic centre and the large shopping centre) close to the WWTP area, its proximity to the residential centre and the large number of people who frequent the area for work or leisure on a daily basis, in addition to those who live permanently, are all factors that make the area particularly sensitive, for which a few episodes of overcoming are enough to create nuisance.

Furthermore, the outcome confirms the criteria reported in the "IPPC-H4" Guidance for Odour elaborated by the UK-Environmental Agency (2011) and proves that the odour emitted from a WWTP is very unpleasant (offensiveness characteristics), since a low odour concentration threshold is enough to cause annoyance. However, the results show also how the offensiveness of the WWTP odour, in the investigated case, does not assume the highest value (corresponding to an odour concentration threshold value of 1 OU_E/m^3, which defines in general the maximum protection level and/or the minimum separation distances for the receptors from the odour source). This may be related to the different odour sources present in the WWTP, attributing the odour nuisance above all to the odour sources of the sludge treatments, characterized by higher odour emissions in terms of OU_E/m^3.

The implementation of the offensiveness (O) and location (L) factors in the dispersion modelling approach is therefore a key action in order to generate results comparable with those of the field inspections and, consequently, for the evaluation of the nuisance.

The need to move from the approach which uses a default single-percentile and single-odour concentration threshold to a multi-criteria evaluation procedure, which takes into account the offensiveness and location factors to define the odour nuisance area, is therefore highlighted.

The choice of a multi-percentile approach leads also to solving other limitations related by the application of the traditional methodology, such as the restriction of the meteorological conditions that can be modelled [25]. Moreover, the appropriateness of a multi-criteria investigation proves that odour nuisance is not related to a univocal relationship between the intensity and the duration of the event. In fact, as observed by Cavalini [50], the nuisance for the inhabitants exposed to permanent weak odours are the same as the nuisance perceived to the exposure of discontinuous episodes of strong odours.

4. Conclusions

The application and comparison of different odour assessment approaches for the odour nuisance characterization in an urbanized area highlight the offensiveness and location factors as key factors in the analysis, especially in the dispersion modelling use.

For the atmospheric dispersion modelling approach, the application of a multi-percentile (p_T) scale, correlated to the sensitivity of the area in which the odour is dispersed, proves necessary to consider the influence of the location (L) factor in the nuisance evaluation. The adoption of different odour concentration thresholds (C_T), chosen on the basis of the hedonic tone of the investigated odour, turns out to be effective in implementing the factor offensiveness (O). A multi-criteria procedure, instead of one with predefined and pre-regulated criteria and relative values (percentile and odour concentration threshold) in any scenarios, that may overestimate or underestimate odour nuisance, is therefore to use to assess the odour nuisance. Furthermore, the adoption of a multi-criteria procedure will allow the introduction of an odour risk-assessment approach, supporting the definition of required protection level and/or minimum separation distances between the odour source and the nearest receptors [51–54].

On the other hand, for the field assessment techniques, the research shows a good correspondence between the assessed odour nuisance levels for the investigated methods. The field inspection with questionnaires provides results of an area in relation to the people availability and the collected data. The method can be affected by the emotional involvement in the problem of the interviewed persons. An extensive social participation and strong community involvement is therefore proven necessary in order to have objective evaluations.

The field inspection approach through trained assessors appears to be the most realistic approach to assess the odour nuisance. However, the method is expensive in terms of costs and human resources and requires long implementation times. An optimization of these factors can be implemented by analysing trends, according to the studies of Zarra et al. [38].

An ideal arrangement to assess the odour nuisances could be the use of the air dispersion modelling approach, determining the offensiveness (O) and location (L) factors to be implemented through questionnaires and validating the effectiveness of the results with the use of a restricted (in time and on measurement points) field inspection assessment through trained assessors.

The research provides useful information for the selection of the most appropriate odour nuisance assessment approach in urbanized areas, as well as to promote the development of integrated methods, in order to emphasize the strengths of each methodology, as well as optimize the overall reliability.

Author Contributions: Conceptualization, methodology, validation, formal analysis, investigation, T.Z.; writing, T.Z. and V.N.; resources, T.Z. and V.B.; supervision, T.Z., V.N. and V.B. All authors have read and agreed to the published version of the manuscript.

Funding: This research received no external funding.

Institutional Review Board Statement: Not applicable.

Informed Consent Statement: Not applicable.

Acknowledgments: The University of Salerno, the SEED Laboratory and SPONGE srl are acknowledged for the administrative and technical support.

Conflicts of Interest: The authors declare no conflict of interest.

References

1. Aatamila, M.; Verkasalo, P.K.; Korhonen, M.J.; Suominen, A.L.; Hirvonen, M.; Viluksela, M.K.; Nevalainen, A. Odour annoyance and physical symptoms among residents living near waste treatment centres. *Environ. Res.* **2011**, *111*, 164–170. [CrossRef] [PubMed]
2. Zarra, T.; Reiser, M.; Naddeo, V.; Belgiorno, V.; Kranert, M. Odour emissions characterization from wastewater treatment plants by different measurement methods. *Chem. Eng. Trans.* **2014**, *40*, 37–42.
3. Oliva, G.; Zarra, T.; Pittoni, G.; Senatore, V.; Galang, M.G.; Castellani, M.; Belgiorno, V.; Naddeo, V. Next-generation of instrumental odour monitoring system (IOMS) for the gaseous emissions control in complex industrial plants. *Chemosphere* **2021**, *271*, 129768. [CrossRef]
4. Belgiorno, V.; Naddeo, V.; Zarra, T. *Odour Impact Assessment Handbook*; John Wiley & Sons, Inc.: New York, NY, USA, 2012; pp. 1–312, ISBN 9781119969280.
5. Brancher, M.; Griffiths, K.D.; Franco, D.; Lisboa, H.M. A review of odour impact criteria in selected countries around the world. *Chemosphere* **2017**, *168*, 1531–1570. [CrossRef]
6. Naddeo, V.; Belgiorno, V.; Zarra, T. *Odour Characterization and Exposure Effects*; John Wiley & Sons, Inc.: New York, NY, USA, 2012. [CrossRef]
7. Zarra, T.; Giuliani, S.; Naddeo, V.; Belgiorno, V. Control of odour emission in wastewater treatment plants by direct and undirected measurement of odour emission capacity. *Water Sci. Technol.* **2012**, *66*, 1627–1633. [CrossRef]
8. Van Harreveld, A.P. From odorant formation to odour nuisance: New definition for discussing a complex process. *Water Sci. Technol.* **2001**, *44*, 9–15. [CrossRef] [PubMed]
9. Claeson, A.S.; Lidén, E.; Nordin, M.; Nordin, S. The role of perceived pollution and health risk perception in annoyance and health symptoms: A population-based study of odorous air pollution. *Int. Arch. Occup. Environ. Health* **2013**, *86*, 367–374. [CrossRef] [PubMed]
10. Jacquemin, B.; Sunyer, J.; Forsberg, B.; Götschi, T.; Bayer-Oglesby, L.; Ackermann-Liebrich, U.; De Marco, R.; Heinrich, J.; Jarvis, D.; Torén, K.; et al. Annoyance due to air pollution in Europe. *Int. J. Epidemiol.* **2007**, *36*, 809–820. [CrossRef] [PubMed]
11. Schiffman, S.S.; Graham, B.G.; Williams, C.M. Journal of the Air & Waste Management Association Dispersion Modeling to Compare Alternative Technologies for Odor Remediation at Swine Facilities Modeling to Compare Alternative Technologies for Odor Remediation at Swine Facilities Dispersion Modeling to Compare Alternative Technologies for Odor Remediation at Swine Facilities. *J. Air Waste Manage. Assoc.* **2012**, *58*, 1166–1176.
12. Ranzato, L.; Barausse, A.; Mantovani, A.; Pittarello, A.; Benzo, M.; Palmeri, L. A comparison of methods for the assessment of odor impacts on air quality: Field inspection (VDI 3940) and the air dispersion model CALPUFF. *Atmos. Environ.* **2012**, *61*, 570–579. [CrossRef]
13. Wojnarowska, M.; Ilba, M.; Szakiel, J.; Turek, P.; Soltysik, M. Identifying the location of odour nuisance emitters using spatial GIS analyses. *Chemosphere* **2021**, *263*, 128252. [CrossRef]

14. Zarra, T.; Galang, M.G.; Ballesteros, F.; Naddeo, V.; Belgiorno, V. Environmental odour management by artificial neural network—A review. *Environ. Int.* **2019**, *133*, 105189. [CrossRef] [PubMed]

15. Muñoz, R.; Sivret, E.C.; Parcsi, G.; Lebrero, R.; Wang, X.; Suffet, I.H.; Stuetz, R.M. Monitoring techniques for odour abatement assessment. *Water Res.* **2010**, *44*, 5129–5149. [CrossRef] [PubMed]

16. Giuliani, S.; Zarra, T.; Nicolas, J.; Naddeo, V.; Belgiorno, V.; Romain, A.C. An alternative approach of the E-Nose Training Phase in odour Impact Assessment. *Chem. Eng. Trans.* **2012**, *30*, 139–144.

17. Nicell, J.A. Assessment and regulation of odour impacts. *Atmos. Environ.* **2009**, *43*, 196–206. [CrossRef]

18. Sommer-Quabach, E.; Piringer, M.; Petz, E.; Schauberger, G. Comparability of separation distances between odour sources and residential areas determined by various national odour impact criteria. *Atmos. Environ.* **2014**, *95*, 20–28. [CrossRef]

19. Badach, J.; Kolasińsk, P.; Paciorek, M.; Wojnowski, W.; Dymerski, T.; Gębicki, J.; Dymnicka, M.; Namieśnik, J. A case study of odour nuisance evaluation in the context of integrated urban planning. *J. Environ. Manag.* **2018**, *213*, 417–424. [CrossRef]

20. Ravina, M.; Bruzzese, S.; Panepinto, D.; Zanetti, M. Analysis of Separation Distances under Varying Odour Emission Rates and Meteorology: A WWTP Case Study. *Atmosphere* **2020**, *11*, 962. [CrossRef]

21. Zhang, Y.; Ning, X.; Li, Y.; Wang, J.; Cui, H.; Meng, J.; Teng, C.; Wanga, G.; Shang, X. Impact assessment of odor nuisance, health risk and variation originating from the landfill surface. *Waste Manag.* **2021**, *126*, 771–780. [CrossRef]

22. Sironi, S.; Capelli, L.; Céntola, P.; Del Rosso, R.; Pierucci, S. Odour impact assessment by means of dynamic olfactometry, dispersion modelling and social participation. *Atmos. Environ.* **2010**, *44*, 354–360. [CrossRef]

23. Capelli, L.; Sironi, S.; Del Rosso, R.; Guillot, J.M. Measuring odours in the environment vs. dispersion modelling: A review. *Atmos. Environ.* **2013**, *79*, 731–743. [CrossRef]

24. Oettl, D.; Kropsch, M.; Mandl, M. Odour assessment in the vicinity of a pig-fatting farm using field inspections (EN 16841-1) and dispersion modelling. *Atmos. Environ.* **2018**, *181*, 54–60. [CrossRef]

25. Griffiths, K.D. Disentangling the frequency and intensity dimensions of nuisance odour, and implications for jurisdictional odour impact criteria. *Atmos. Environ.* **2014**, *90*, 125–132. [CrossRef]

26. Freeman, T.; Cudmore, R. Review of Odour Management in New Zealand. *Air Qual. Tech. Rep.* **2002**, *24*, 1–163.

27. Miedema, H.M.E.; Walpot, J.I.; Vos, H.; Steunenberg, C.F. Exposure-annoyance relationships for odour from industrial sources. *Atmos. Environ.* **2000**, *34*, 2927–2936. [CrossRef]

28. Fasolino, I.; Grimaldi, M.; Zarra, T.; Naddeo, V. Odour control strategies for a sustainable nuisances action plan. *Glob. Nest J.* **2016**, *18*, 734–741.

29. Arias, R.; Capelli, L.; Diaz, C. A New Methodology Based on Citizen Science to Improve Environmental Odour Management. *Chem. Eng. Trans.* **2018**, *68*, 7–12.

30. EN 16481-1. *Ambient Air-Determination of Odour in Ambient Air by Using Field Inspection-Part 1: Grid Method*; CEN: Brussels, Belgium, 2016.

31. EN 16481-2. *Ambient Air-Determination of Odour in Ambient Air by Using Field Inspection-Part 2: Plume Method*; CEN: Brussels, Belgium, 2016.

32. VDI 3883-1. *Effects and Assessment of Odours—Assessment of Odour Annoyance—Questionnaires*; Beuth Verlag GmbH: Berlin, Germany, 2015.

33. Chemel, C.; Riesenmey, C.; Batton-Hubert, M.; Vaillant, H. Odour-impact assessment around a landfill site from weather-type classification, complaint inventory and numerical simulation. *J. Environ. Manag.* **2012**, *93*, 85–94. [CrossRef] [PubMed]

34. Hayes, J.E.; Stevenson, R.J.; Stuetz, R.M. The impact of malodour on communities: A review of assessment techniques. *Sci. Total Environ.* **2014**, *500–501*, 395–407. [CrossRef] [PubMed]

35. Nicolas, J.; Cors, M.; Romain, A.C.; Delva, J. Identification of odour sources in an industrial park from resident diaries statistics. *Atmos. Environ.* **2010**, *44*, 1623–1631. [CrossRef]

36. VDI 3882-2. *Olfactometry—Determination of Hedonic Odour Tone*; Beuth Verlag GmbH: Berlin, Germany, 1994.

37. Environment Agency. *H4 Odour Management—How to Comply with Your Environmental Permit*; European Environment Agency (EEA): Copenhagen, Denmark, 2011; p. 46. Available online: https://assets.publishing.service.gov.uk/government/uploads/system/uploads/attachment_data/file/296737/geho0411btqm-e-e.pdf (accessed on 27 May 2021).

38. Zarra, T.; Naddeo, V.; Giuliani, S.; Belgiorno, V. Optimization of field inspection method for odour impact assessment. *Chem. Eng. Trans.* **2010**, *23*, 93–98.

39. *Golden Software Surfer 12. User 's Guide*; Golden Software, Inc.: Golden, CO, USA, 2014; pp. 1–124. Available online: http://downloads.goldensoftware.com/guides/Surfer12_Users_Guide_Preview.pdf (accessed on 27 May 2021).

40. Administrative regulation. Detection and Assessment of Odour in Ambient Air—Guideline on Odour in Ambient Air GOAA (1994/1999/2004/2008). *Länderausschuss Immiss. GOAA* 2008. Available online: https://www.lanuv.nrw.de/fileadmin/lanuv/luft/gerueche/pdf/GOAA10Sept08.pdf (accessed on 27 May 2021).

41. Sucker, K.; Both, R.; Bischov, M.; Guski, R.; Winneke, G. Odor frequency and odor annoyance. Part I: Assessment of frequency, intensity and hedonic tone of environmental odors in the Weld. *Int. Arch. Occup. Env. Health* **2008**, *81*, 671–682. [CrossRef]

42. Piringer, M.; Schauberger, G. Dipsersion Modelling for Odour Exposure Assessment. In *Odour Impact Assessment Handbook*; Sons, J.W., Ed.; Wiley: Hoboken, NJ, USA, 2012; pp. 125–174, ISBN 9780813801261.

43. Nicolas, J.; Delva, J.; Cobut, P.; Romain, A.C. Development and validating procedure of a formula to calculate a minimum separation distance from piggeries and poultry facilities to sensitive receptors. *Atmos. Environ.* **2008**, *42*, 7087–7095. [CrossRef]

44. VDI 3883-2. *Effects and Assessment of Odours—Determination of Annoyance Parameters by Questioning—Repeated Brief Questioning of Neighbour Panellists*; Beuth Verlag GmbH: Berlin, Germany, 1993.
45. Scire, J.S.; Robe, F.R.; Fernau, M.E.; Yamartino, R.J. *A User's Guide for the CALMET Meteorological Model*; Earth Tech. Inc.: Concord, MA, USA, 2000; p. 37.
46. Scire, J.S.; Strimaitis, D.G.; Yamartino, R.J. *A User's Guide for the CALPUFF Dispersion Model*; Earth Tech. Inc.: Concord, MA, USA, 2000; p. 521.
47. Zarra, T.; Reiser, M.; Naddeo, V.; Belgiorno, V.; Kranert, M. A comparative and critical evaluation of different sampling materials in the measurement of odour concentration by dynamic olfactometry. *Chem. Eng. Trans.* **2012**, *30*, 307–312.
48. Schauberger, G.; Piringer, M.; Petz, E. Diurnal and annual variation of the sensation distance of odour emitted by livestock buildings calculated by the Austrian odour dispersion model (AODM). *Atmos. Environ.* **2000**, *34*, 4839–4851. [CrossRef]
49. Brancher, M.; Piringer, M.; Grauer, A.F.; Schauberger, G. Do odour impact criteria of different jurisdictions ensure analogous separation distances for an equivalent level of protection? *J. Environ. Manag.* **2019**, *240*, 394–403. [CrossRef] [PubMed]
50. Cavalini, P.M.; Koeter-Kemmerling, L.G.; Pulles, M.P.J. Coping with odour annoyance and odour concentrations: Three field studies. *J. Environ. Psychol.* **1991**, *11*, 123–142. [CrossRef]
51. Schauberger, G.; Piringer, M.; Jovanovic, O.; Petz, E. A new empirical model to calculate separation distances between livestock buildings and residential areas applied to the Austrian guideline to avoid odour nuisance. *Atmos. Environ.* **2012**, *47*, 341–347. [CrossRef]
52. Piringer, M.; Knauder, W.; Petz, E.; Schauberger, G. A comparison of separation distances against odour annoyance calculated with two models. *Atmos. Environ.* **2015**, *116*, 22–35. [CrossRef]
53. Piringer, M.; Knauder, W.; Petz, E.; Schauberger, G. Factors influencing separation distances against odour annoyance calculated by Gaussian and Lagrangian dispersion models. *Atmos. Environ.* **2016**, *140*, 69–83. [CrossRef]
54. Brancher, M.; Piringer, M.; Knauder, W.; Wu, C.; Griffiths, K.D.; Schauberger, G. Are Empirical Equations an Appropriate Tool to Assess Separation Distances to Avoid Odour Annoyance? *Atmosphere* **2020**, *11*, 678. [CrossRef]

Article

The Use of the Odor Profile Method with an "Odor Patrol" Panel to Evaluate an Odor Impacted Site near a Landfill

Yuge Bian *, Haoning Gong and I. H. (Mel) Suffet

Department of Environmental Health Science, UCLA, School of Public Health, Los Angeles, CA 90095, USA; feron.gong@outlook.com (H.G.); msuffet@ucla.edu (I.H.S.)
* Correspondence: gloriabian@g.ucla.edu; Tel.: +213-716-0960

Abstract: A third-party-trained "Odor Patrol" program was conducted at a school that is about a one-mile distance from a landfill to clarify the odor nuisance problems from the landfill. Every 20 min from 6 to 9 a.m. on school days, the "Odor Profile Method" (OPM) was used with the landfill odor wheel to identify the odor type and intensity of each odor type. This study showed that an Odor Patrol using the OPM can accurately define odor nuisance changes over time and can be used as a method to confirm changes of odor nuisances in a field study. The Odor Patrol only found 13 data inputs of the 1000 data inputs (1.3%) for the 100-day odor monitoring with a landfill odor or trash odor that could cause odor complaints. The Odor Patrol data and the Odor Complaint data compared well. The OPM by an "Odor Patrol" could determine the contribution of the nuisance odors from 6 to 9 a.m. at the school site, about one mile away from the landfill. The study demonstrated a novel approach for odor monitoring by using the Odor Profile Method with an Odor Patrol. The OPM not only confirmed the mitigation of a landfill odor problem, but it also determined odor character, odor intensity, odor frequency and odor duration during this study period. "Landfill gas" was determined to be primarily a rotten vegetable odor with a secondary sewery/fecal odor of lower intensity, and "trash odors" were primarily a rancid and sweet odor with a secondary sewery/fecal and/or rotten vegetable odor of lower intensities generated from trash reaching the landfill. The order of intensity observed from high to low was: Trash odor (Rancid–Sweet) > Rotten vegetable > Sewery/Fecal > Rancid. Thus, trash odor is the major problematic odor from the landfill site. Quality assurance methods were used to remove local odors from the evaluation.

Keywords: Odors; Odor Patrol; Odor Profile Method; monitoring Odors

Citation: Bian, Y.; Gong, H.; Suffet, I.H. The Use of the Odor Profile Method with an "Odor Patrol" Panel to Evaluate an Odor Impacted Site near a Landfill. *Atmosphere* **2021**, *12*, 472. https://doi.org/10.3390/atmos12040472

Academic Editors: Günther Schauberger, Martin Piringer, Chuandong Wu, Jacek Koziel and Roberto Bellasio

Received: 27 January 2021
Accepted: 5 April 2021
Published: 9 April 2021

Publisher's Note: MDPI stays neutral with regard to jurisdictional claims in published maps and institutional affiliations.

1. Introduction

Since 2009, California's South Coast Air Quality Management District (SCAQMD) has seen a dramatic increase in the number of complaints related to nuisance odors from a landfill. This landfill processes one-third of the daily waste from Los Angeles County per day. This is about 8300 tons of waste daily or more than 2.3 million tons annually. The landfill primarily accepts trash that has been separated from other waste streams, such as, green wastes, metals and plastics.

The SCAQMD method for odor detection relies on counting consumer complaints and sending an odor inspector to investigate the odor problem at the complaint location. When the consumer complaints reach an excessive number, and these are verified by the odor inspectors, SCAQMD considers it an odor nuisance problem.

Figure 1 shows the Odor Complaint history between 6 and 9 a.m. in an area about a mile away from the landfill between the years 2008 and 2018 [1]. Figure 1 shows that thousands of complaints were most frequently received between 6 and 9 a.m. on weekdays.

Atmosphere **2021**, *12*, 472. https://doi.org/10.3390/atmos12040472   https://www.mdpi.com/journal/atmosphere

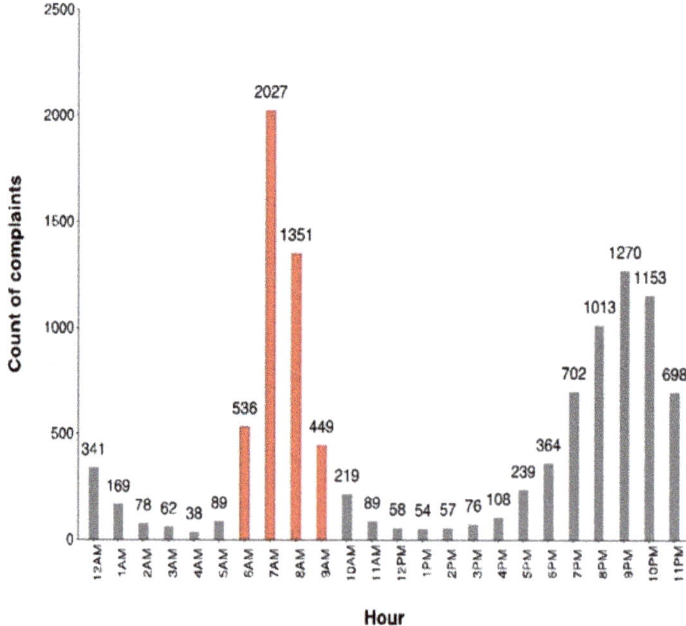

Figure 1. Odor complaints from January 2008 to April 2018 near the school site [1].

There were some concerns related to the complaint data. The general population has not been trained with professional knowledge about identifying odors. Thus, they might be mistaking other environmental odors from the surrounding area with the odors from the landfill. The population may make false odor reports because of continuous odor concerns. The inspectors developed a general knowledge of the source of odors characteristics as landfill gas and trash odor. However, they were not professionally trained to characterize multiple odor characteristics such as rancid and sweet at a site, unlike the food and drinking water professional panelists, who use a standard method: Flavor Profile Analysis (FPA) [2,3]. The inspectors, apparently, did not have an odor intensity scale to rate the intensity of each odor characteristic. Additionally, another major concern was the duration of an odor event. The residents might not be observing an odor at the same time as when an inspector arrives at the complaint site.

The landfill was required by SCAQMD not to accept waste between 6 and 9 a.m. on weekdays due to a large number of complaints. SCAQMD decided that a Third-Party Odor-Monitoring Program should be conducted at the school site from 30 October 2017 to 2 May 2018 (100-week days) to evaluate the effects of closing the landfill between 6 and 9 a.m. weekdays. The Odor Monitoring Program selected was a human panel called an "Odor Patrol" that used the unique "Odor Profile Method" (OPM) [4–7] for air analysis based upon the well-documented Flavor Profile Analysis [2,3] that uses trained odor professionals.

The objectives of the study were:

1. Determine if the "Odor Profile Method" by an "Odor Patrol" can evaluate the odor nuisance caused by a particular odor source: i.e., determine the odor character, odor intensity of each odorant and determine the frequency and duration of the odor problem.
2. Can the "Odor Profile Method" by an "Odor Patrol" be used to determine if closure of the landfill between 6 and 9 a.m. can minimize odor complaints at an impacted site about one mile away?

The novelty of this study is the use of OPM to describe the odors and the intensity of each odor that can be defined as being from a landfill.

2. Materials and Methods

2.1. Experimental Method–Odor Profile Method

The Odor Profile Method was based upon the FPA method for the food and drinking water industries, specifically Method 2170, the "Flavor Profile Analysis" of "Standard Methods of Water and Wastewater" [3]. The FPA has a 7-level sugar scale for the intensity of an odorant. The FPA was developed for the headspace of drinking water. It has been used for over 30 years and is a standard method to evaluate odor problems and for quality control of specific odors in drinking water. The method was developed for air analysis studies by Burlingame [4], Burlingame et al. [5], Burlingame [6] and Curren et al. [7]. The landfill odor wheel (Figure 2) was used to identify each odor type and intensity [8]. Training and quality assurance measures were used [9].

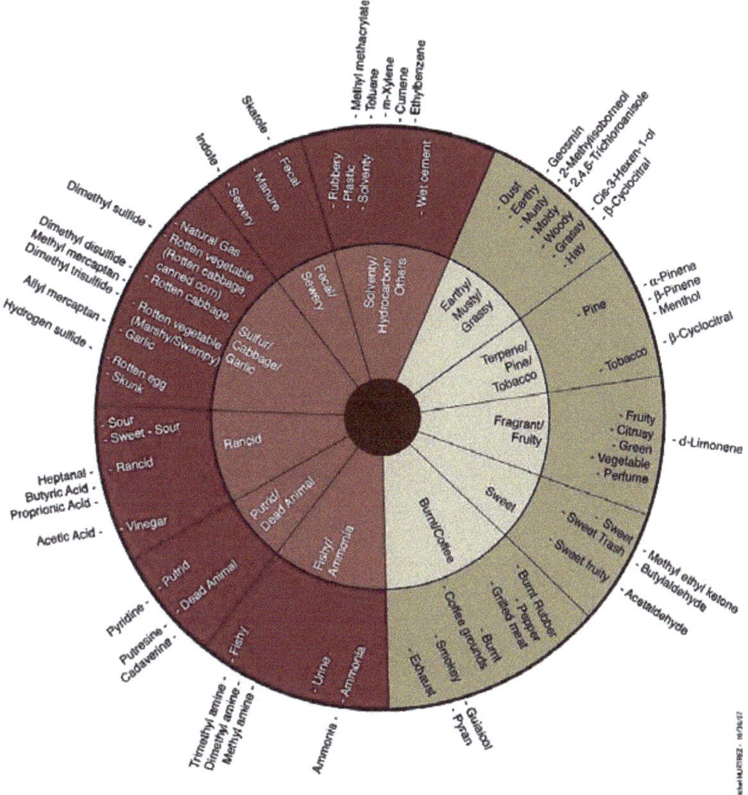

Figure 2. Landfill Odor Wheel [8].

The OPM includes (1) identifying one or more odor notes in the air sampled and (2) determining the odor intensity for each odor note [4–7]. The OPM method includes:

- Screening panelists for anosmia (lack of the sense of smell) using a "scratch-n-sniff" test [10].
- Using a minimum of 4 trained panelists of the 10 trained panelists for each OPM sample evaluation.

- Training odor panelists with the primary odorants of an odor wheel and mixtures of 2, 3 and 4 of these standard odorants over a 6-week period. Additionally, including training panelists on the background odors that could be found around the school as grassy.
- Teaching a standardized odor note vocabulary to panelists using the "landfill "odor wheels" (Figure 2) that consist of three rings: an inner ring of general odor categories, a middle ring of specific odor notes within each segment and an outer ring of known or potential odorants associated with each odor note.
- Panelists are calibrated to the odor intensity scale (Table 1)—threshold (1), slight (2), weak (4), medium (6), medium–strong (8), strong (10) and very strong (12)—using sugar-in-water solutions tasted by mouth that represent weak (5% sugar), medium–strong (10% sugar) and very strong (15% sugar) as defined by the FPA mehod (3).
- Group discussions are permitted after the individual odor evaluations to help panelists define their responses; however, panelists are ultimately instructed to work independently.
- Overall panel results for an "odor note" (an odor character with an associated odor intensity) require at least 50% agreement among panelists. The odor notes are calculated as the panel average mean with a standard deviation reported. If a panelist does not report the odor note, a zero is included in the calculation of the mean.
- If less than 50% of the panel agrees on an odor note, an "other odor note" is stated without an odor intensity.

Table 1. Flavor Profile Analysis: Odor Intensity Strength Scale [5].

Intensity Rating	Flavor Standard (% Sugar in Water)	Intensity Description
0	0	No odor
1	Threshold	Can detect odor but cannot describe the odor character
2	Very Weak	Odor barely perceptible
3	Recognized	Action Level
4	5	Odor clearly exists but takes time to describe
6	Weak–Moderate	Odor readily perceived and identified
8	10	Odor is uncomfortable to smell for extended periods of time
10	Moderate–Strong	Odor is uncomfortable to smell for extended periods of time
12	15	Odor is unbearable to smell for even short periods of time

The OPM intensity scale is based upon the Weber–Fechner Law [11], in which a single odorant's intensity is proportional to the Log of the odorant's concentration.

$$\text{Odor Intensity} = k \, \text{Log (Concentration)} + b \tag{1}$$

Whereas the concentrations are units such as ppb or $\mu g/m^3$, and k is a constant (called the Weber–Fechner coefficient) that is unique to each odorant. Figure 3 shows the relationship between Odor Intensity and the Log (Concentration). The odor level of detection (1) and recognition (4) of the odor name are shown. An action level of 3 below recognition is suggested to minimize an odor problem.

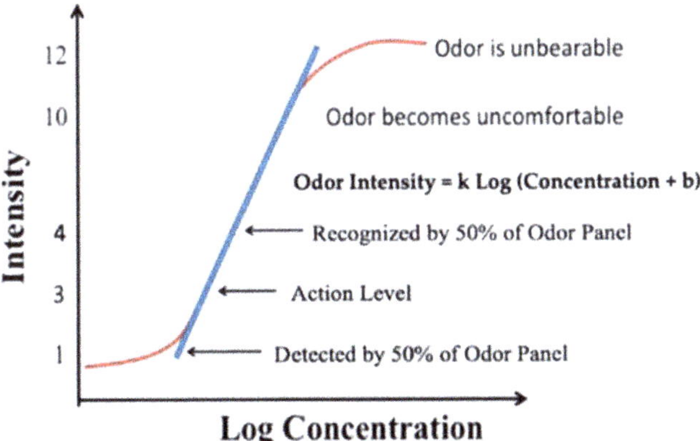

Figure 3. Weber–Fechner curve showing intensity vs. concentration. Note: The straight line represents the Weber–Fechner Law.

2.2. Background Information on Odors Related to a Landfill

Reference [1] indicated that during the Odor Patrol time period in the previous year to this study (from 2016 to 2017) "trash odor" was the primary odor. Additionally, a lower intensity, "rotten vegetable" and/or a "sewage/fecal" odor were also identified either from the trash gas or from landfill gas.

- Landfill Gas (LG) is described by the landfill community and inspectors for gases that are produced within the landfill from anaerobic reactions. Landfill gas, according to the Landfill Wheel (Figure 2), was within the general category (Sulfur/Cabbage/Garlic). Primary landfill gas odors in this category are rotten vegetable, rotten cabbage, and garlic. These odors are caused by the anaerobic production of sulfur compounds by microorganisms using sulfate instead of oxygen as the electron acceptors within the landfill [12] (see Figure 4). A secondary odor of lower intensity in the Landfill Wheel (Figure 2) was within the general category (Sewage/Fecal). These secondary odors in this category are also named sewage/fecal. This odor is generated by microorganisms under the same low-oxygen conditions by degrading nitrogen compounds (e.g., proteins) to yield compounds such as indole and skatole, which have a sewery/fecal odor character described in Figure 2. Figure 5 shows the microbiological origin of these odors [13]. In this paper this will be referred to as a sewery odor.
- Trash Odors (TRs) are described by the landfill community generically occur when trucks filled with trash are waiting and while dumping trash at the landfill. Although soil is used to cover the trash after it is dumped, the "trash odors" can still escape. Trash Odors, according to the Landfill Wheel (Figure 2), are within the general categories (Rancid and Sweet). The "trash odor" is primarily "rancid" from the air oxidation of fats to fatty acids. The fatty acids can be further oxidized in air to aldehydes and ketones. The aldehydes and ketones add a "sweet" note to the "rancid" odor as presented in the "Landfill Odor Wheel" (Rancid and Sweet) general categories (Figure 2). Secondary odors of lower intensity, "rotten vegetable" and/or "sewage/fecal" odors are produced from reduced sulfur and nitrogen compounds generated in low-oxygen (anaerobic) pockets by microbes within the load of trash that is transported to the landfill as described in the "landfill gas" odor section.

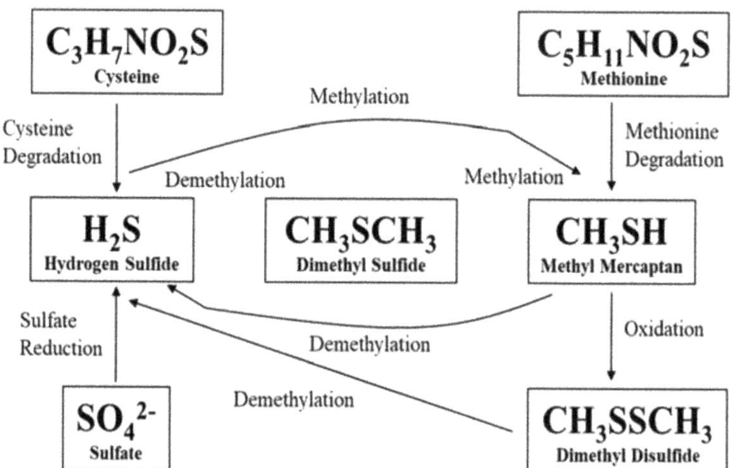

Figure 4. Production of odorous volatile sulfur compounds in a landfill. Based upon studies of anaerobically digested biosolids during cake storage [12].

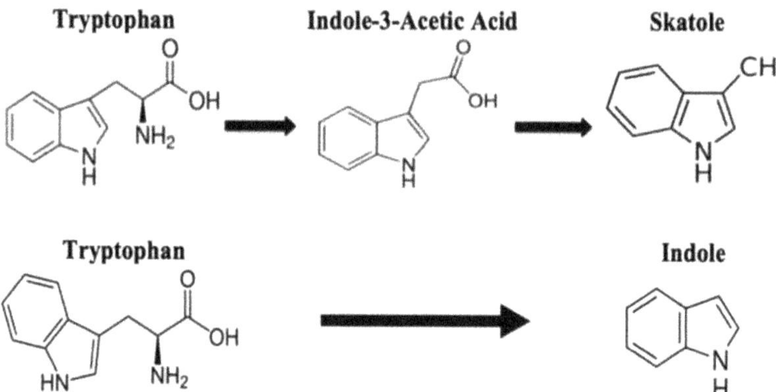

Figure 5. Microbial products of landfill gas sewery/fecal odors from amino acids. Based upon studies of anaerobically digested biosolids [13].

2.3. Experimental Procedure

The study was conducted from 31 October 2017 to 2 May 2018 for 100 school days (except weekends, holidays and school closing days). OPM analysis was completed in an open area school playground every 20 min from 6 to 9 a.m., e.g., 6:00, 6:20 and 6:40, respectively, to 9:00. Each of the four panelists had their own data sheets. At each sampling time, the panelist logged in their own data without talking to each other. After an average of a 3-min evaluation, the panel went to an air-conditioned, odor-free room to refresh their sense of smell before the next sampling period. If needed, the panelists discussed the sample results.

A walk around the campus before, during and after the 6–9 a.m. period was performed for the detection of environmental odors. At 9 a.m., as a quality assurance measurement, all panelists were required to open the dumpster outside of the school site, smell it and write down the odor data on their sample sheet. All data sheets were collected each day by the project manager and logged into the computer.

2.4. Quality Control

Each panelist was screened for anosmia (lack of the sense of smell) using a "scratch-n-sniff" test [10]. All of the panelists passed the test with over 70% correct answers, which means they all had a normal sense of smell. Prior to the start date of the study, all panelists went through complete odor training on the Odor Profiling Method. The training included (1) distinguishing different odor categories by a single standard gas sample; (2) distinguishing the different intensities on the odor intensity scale by sugar water standard; (3) distinguishing odor categories from mixed standard gas samples containing 2–3 chemical mixtures. The panelists were required to repeat the training to assure they were confident to identify the odors of importance. The duration of the study was about 5 months. Thus, the panelist completed a refresher training course halfway through the study.

In the air-conditioned classroom where the panelist stayed, no one was permitted to bring food, coffee or anything that smelled. The panelists did not wear perfume or use strongly scented shampoo or body wash. Therefore, panelists returning to the classroom could relax their noses and get prepared for the next sampling time period. Panelists were required to report any nose cold and withdrew from the panel for that day. A stuffed nose dramatically affects the sense of smell.

The odor panel performed" quality control" by monitoring trash odor produced at a covered waste trash container (dumpster) for food waste near the school site. Quality control data were collected after the Odor Patrol data were completed for that day. The odors from the dumpster represented some odors expected from the trash sent to the Sunshine Landfill. These odors are the rancid–sweet and rotten vegetable odors. Thus, this refreshed the Odor Patrol's ability to recognize the odors from the landfill. This also confirmed the claim that if the school is experiencing these odors, they are coming from the trash in the landfill.

The school site is about a 1-mile distance from the landfill site. The odors that can be perceived at this school site are not limited to only the landfill. Many types of environmental odors can affect the school area. Thus, it is critical to distinguish the environmental odors from the targeted odors from the landfill. Therefore, it is noted that the panelists were trained on environmental odors, such as grassy and gasoline. The panelists were required to walk around the perimeter of the school to determine if any odor were from the nearby environment before leaving the school site. Local environmental odors, such as "smoke" during a wildfire in the area, "gasoline" from vehicle and traffic, "grassy" from mowing the lawn in the neighborhood nearby, etc., were not included in the data set.

3. Results

3.1. OPM Data by the Odor Patrol

Table 2 shows the total odor panel results for Week 2 on a weekly data sheet including the background odors. As Table 2 shows, each day of sampling has 10 sample time slots. The yellow highlights are for samples where landfill odors were observed. The legend of Table 2 indicates which odors are the background odor at the site.

Table 2. Exemplary Weekly Data Sheet for Week 2.

WEEK 2. Odor Characteristics and Intensities of Each Odor Characteristic Observed by the Odor Profile Method Panel					
Sample Location: School Yard—Facing the Landfill Site including background odors that were observed					
		11/6/17	**11/7/17**	**11/8/17**	**11/9/17**
Sample #	**Time: AM**	**ODOR**	**ODOR**	**ODOR**	**ODOR**
1	6:00	Other odor notes: musty, fecal, rotten vegetable	Other odor notes: sewery, sweet, lemon, rancid, pine	**Rotten vegetable 3.6 ± 2.2, rancid 2.4 ± 1.7. Other odor notes: sweet, detergent**	Other odor notes: musty
2	6:20	Other odor notes: musty	**Sewery 2.8 ± 3.0, rancid 1.0 ± 0.9; Other odor notes: sweet, pine, rotten vegetable**	Other odor notes: rotten vegetable, rancid, sweet, sewery	Other odor notes: musty, rotten vegetable, pine, rancid
3	6:40	**Rotten vegetable 2.8 ± 1.1; Other odor notes: pine, grassy, rancid**	Other odor notes: sweet, pine, vegetable	Not detected	Other odor notes: rancid, sweet
4	7:00	Other odor notes: pine, grassy	Other odor notes: sweet, pine, gasoline	Not detected	Not detected
5	7:20	Not detected	Grassy 2.0 ± 2.0; Other odor notes: pine	Other odor notes: rotten vegetable	Other odor notes: musty
6	7:40	Not detected	Other odor notes: pine	Not detected	Other odor notes: sweet
7	8:00	Not detected	Other odor notes: sweet, lemon, rancid	Not detected	Not detected
8	8:20	Other odor notes: burnt	Other odor notes: gasoline, detergent, burnt	Not detected	Not detected
9	8:40	Other odor notes: lemon	Other odor notes: sewery, gasoline, burnt	Not detected	Not detected
10	9:00	Other odor notes: grassy, detergent	Other odor notes: pine, musty	Other odor notes: sweet grass	Other odor notes: rancid, sweet, sweet grass, detergent
Weather Condition		**Partly cloudy, 72°/53° F**	**Partly cloudy, 76°/54° F**	**Cloudy, 77°/51° F**	**Cloudy, 72°/54° F**
	Table 2 Legend: Odor Characteristics—Odor Wheel—Landfill Odor Wheel Used.				
	Intensity Scale: 0—odor free; 1—threshold, 2—very weak; 4—weak—recognition; 6—moderate, perceived; 8—moderate.—strong; 10—strong; 12- very strong				
	Odor Profile Method				
	Odor Character—Panel Average ± Standard Deviation				
	Other Odor Note = < 50% of an odor was reported by the odor panel. This is an odor that should be below a complain level.				
	Comments—Odor Notes—Primarily background odors that are not from the landfill				
1	Garlic -REAL if after checking suspected odor from the kitchen area, when actually walking toward kitchen				
2	Grassy—suspected from a lawn mower operation close to school; could hear the lawn mower machine.				
3	Detergent- suspected from cleaning at the school				
4	Gasoline- suspected from a truck or car outside of school				
5	Musty- usually during cloudy weather, wet ground odor or after a rain event				
6	Burnt -during the wildfire period				
7	Lemon—from plants possibly				
8	Solvent -unknown source at the school				
9	Perfume—suspected from teachers, parents or flowers				
10	Flowery -from flower blooming near the school				
11	Pine—probably from trees nearby school				
12	Ammonia—suspected animal urine				

Table 3 shows only the odor panel sample data when landfill odors were observed during the study from 30 October 2017 to 2 May 2018 (100-week days over 22 weeks). The environmental background odors were not included in these sample results. Only 13 times out of 1000 odor samples evaluated in the 100-week study showed significant landfill-related odors. Each data point shown in Table 3 presents information on the week, date and time of analysis. The values shown under "Odor Character" represent the average odor intensity and the standard deviation for each odor note that was detected by over 50% of the odor panel. For odor characters that were detected by less than 50% of the odor panel an "other odor note" was recorded without an odor intensity.

Table 3. The 13 Odor Panel Sample Data When Landfill Odors Were Observed during the "Odor Patrol Study "of 100 midweek days. Note: environmental background odors were not included in this data set.

13 data points of Significant Landfill Odor Observed			
Week	Date	Time AM	Odor Character
2	6 Nov 2017	6:00	Rotten vegetable 2.8 $\pm$ 1.1; Other odor note: rancid
2	7 Nov 2017	6:20	Sewery 2.8 $\pm$ 3.0, rancid 1.0 $\pm$ 0.9; Other odor note: sweet, rotten veg
2	8 Nov 2017	6:40	Rotten vegetable 3.6 $\pm$ 2.2, rancid 2.4 $\pm$ 1.7; Other odor note: sweet
6	11 Dec 2017	6:20	Sewery 1.0 $\pm$ 1.2
8	17 Jan 2018	7:40	Rancid 1.5 $\pm$ 1.9, Other odor note: sewery, rotten veg
9	22 Jan 2018	6:00	Rancid 2.0 $\pm$ 2.8; Other odor note: sewery
12	12 Feb 2018	7:00	Sewery 1.5 $\pm$ 1.9
12	12 Feb 2018	7:20	Sewery 2.0 $\pm$ 2.3;
12	13 Feb 2018	7:40	Sewery 2.0 $\pm$ 2.3; Other odor note: rancid
13	21 Feb 2018	7:20	Rotten vegetable 1.0 $\pm$ 1.2
17	20 Mar 2018	8:40	Sweet trash 4.0 $\pm$ 1.6
17	20 Mar 2018	9:00	Sweet trash 2.5 $\pm$ 3.0
19	12 Apr 2018	6:40	Sewery 1.0 $\pm$ 1.2

Figure 6 shows the time and frequency of the 13 detected odors in Table 3. For example, the first column shows that between the 6 and 7 a.m. time slot, rancid odor was detected 3 times, rotten vegetable odor was detected 2 times and sewery odor was detected 3 times. The primary times of observation were 6 and 7 a.m.

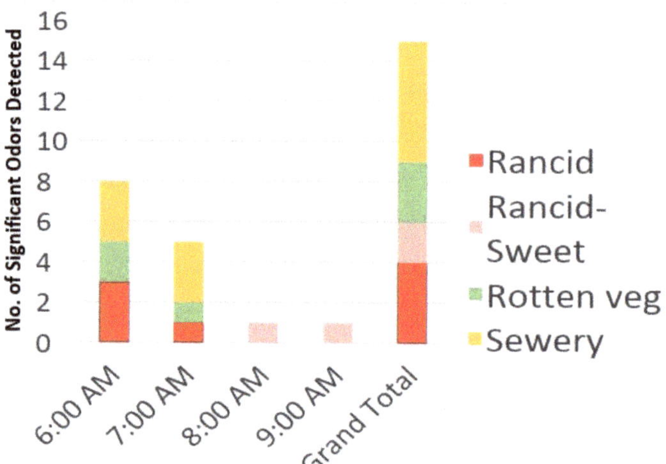

Figure 6. Hours of significant odors detected by the Odor Patrol. Note: if two different odors were detected at same time, it is considered two odor events.

Figure 7 shows the intensities of landfill odors detected by the Odor Patrol. The intensities of the Trash Odor (rancid), and Landfill Odor (rotten vegetables) and sewery odors predominate. Most of the odors were recorded with intensity between one and four. The order of intensity across types from high to low is: Trash odor (Rancid–Sweet) > Rotten vegetable > Sewery > Rancid alone.

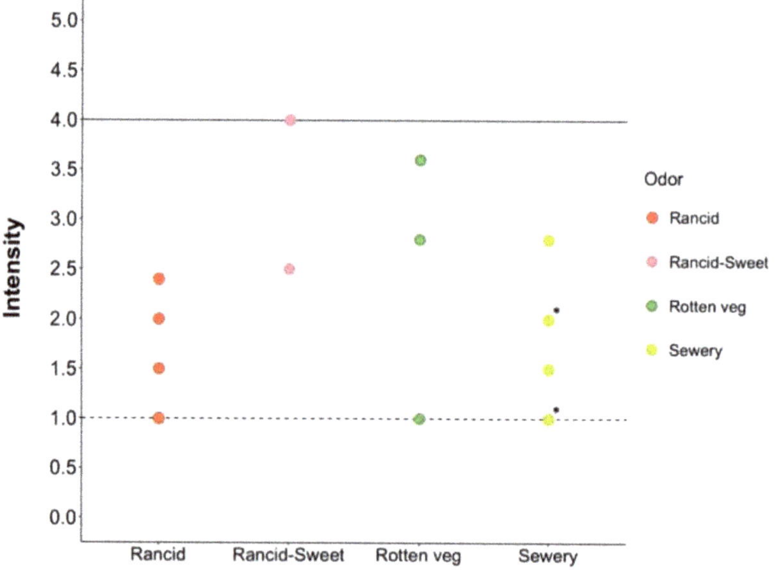

Figure 7. Average intensity and types of significant odors detected by the Odor Patrol. * Represents two results at the same intensity.

3.2. Complaint Data from SCAQMD

Figures 8 and 9 show the primary locations of complaint data from SCAQMD during the Odor Patrol study and at the same time period the year before, respectively. The dots in Figures 8 and 9 show the primary locations of complaints within one mile of the school during equivalent Odor Patrol periods in 2017–2018 vs. 2016–2017, respectively. The dots do not represent each complaint, only the general location of complaints.

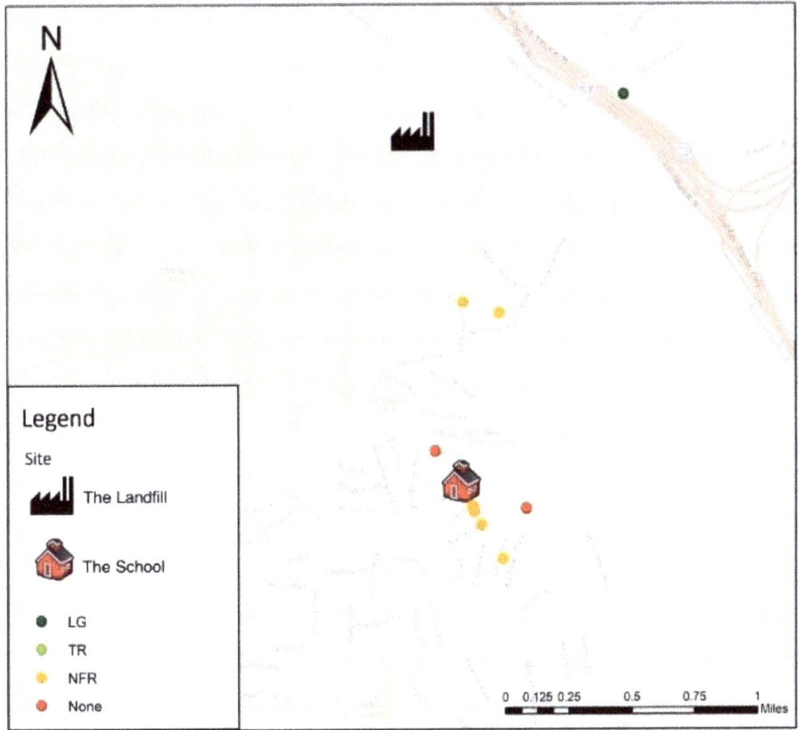

Figure 8. Odor complaints near the school during the Odor Patrol time period: 6–9 a.m., 2017–2018. The dots do not represent each complaint, only the general location of complaints. LG–Landfill Gas, TR–Trash Odor, NFR–No Field Response. The South Coast Air Quality Management District (SCAQMD) inspector did not go to the location. None: No odor was detected by SCAQMD inspectors at the complaint site.

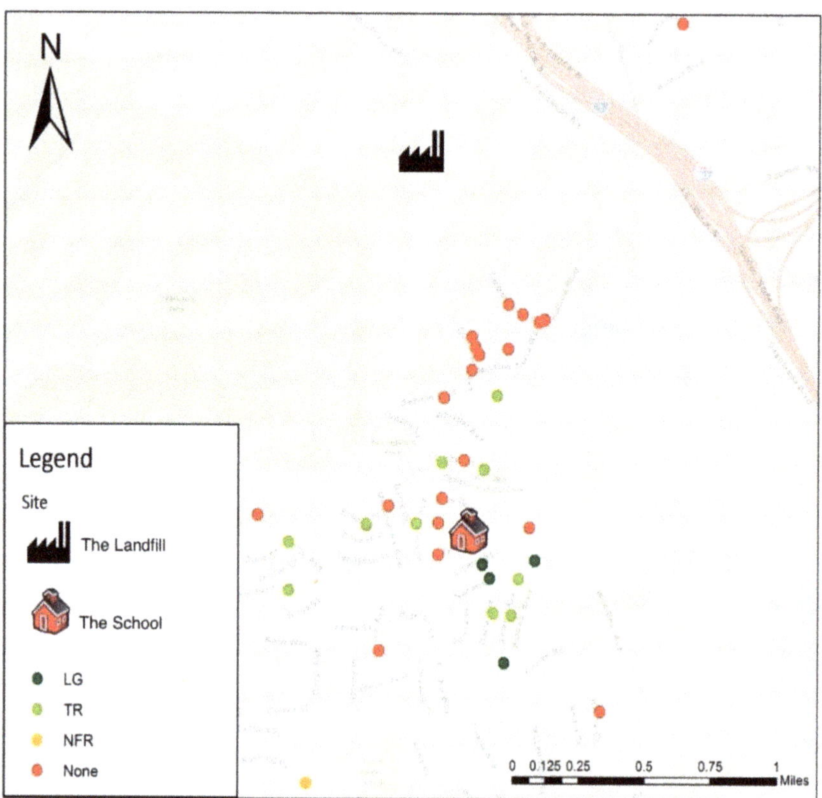

Figure 9. Odor Complaints near the school during the Odor Patrol time period: 6–9 a.m., 2016–2017. The dots do not represent each complaint, only the general location of complaints. LG–Landfill Gas, TR–Trash Odor, NFR–No Field Response:. The SCAQMD inspector did not go to the location, None: no odor was detected by the SCAQMD inspectors at the complaint site. Complaints near the school during the Odor Patrol time period: 6–9 a.m.

The total complaints reported to SCAQMD during the Odor Patrol time period in 2017–2018 were 21 (Figure 8). There were 264 complaints reported to SCAQMD in 2016–2017 (Figure 9). In the year 2016 to 2017, there were 8 land field gas (LG) odors and 170 trash odor (TR) confirmed, while only 2 LG and no TR were reported in 2017–2018 during the Odor Patrol period.

4. Discussion

4.1. Odor Profiling Method vs. European Standard 16841–1:2016

The European Standard EN 16481-1:2016 [14] describes a grid field inspection method that uses a direct assessment of ambient air by trained panel members to characterize odor exposure in a defined area. The panelists determine whether they recognize an odor note selected from a list. The panelists write down their observations every 10 s for 10 min (60 observations). If six of those observations are a recognized odor note, then the label "odor hour" is applied (although not a full hour of odor exposure occurred) [14]. The frequency of "odor hours" is usually completed for a square of four measurement points that map an area. The grid method has been used to study industrial areas in German cities and provided a representative map of the exposure of the population to recognizable odors [15]. These studies took six months to one year, each with 21 panelists. Bax et al. [16] reviewed the grid field inspection method and stated that the method "cannot be used for the assessment of the odor concentration." Thus, the grid field assessment method can recognize odor notes from a selected list but cannot determine the intensity of each odor note.

Different from the EN 16481-1:2016, the Odor Profile Method trains panelists to not only recognize odor notes from a selected list, i.e., the odor wheel, but also to detect all odor notes and define an intensity for each odor note. The panel specifies each odor character with its odor intensity. A minimum of four trained panelists are used. If over 50% of the panel agrees on an odor note, the average intensity +/−, a standard deviation, as shown in Tables 2 and 3, was calculated. The intensity is based upon the Odor Intensity Strength Scale (Table 1). Thus, the Odor Profile Method can be considered an expansion of the European Standard EN 16481-1:2016 [14]. This paper used the OPM at one location. However, the OPM method can also be used over a grid as well.

4.2. Odor Patrol Data

The first objectives of the study were to determine if the "Odor Profile Method" by an "Odor Patrol" can evaluate the odor nuisance caused by a particular odor source; i.e., determine the odor character, odor intensity of each odorant and frequency and duration of the odor problem. The first part of the objective to determine the odor character and odor intensity was accomplished. Table 3 shows only the Odor Patrol sample data when landfill odors were observed during the study from 30 October 2017 to 2 May 2018 (100-week days over 22 weeks). Only 13 out of 1000 samples collected in the 100-week study showed significant landfill odors. Thus, only 1.3% of the samples were potentially attributable to the landfill during the Odor Patrol study.

Tables 2 and 3 show the second part of the first objective was capable of providing information on odor frequency and odor duration. For example, the 3 positive OPM data in Table 2 and the 13 positive samples in Table 3 showed the frequency of OPM data. Additionally, these tables clearly show the duration of an odor event was always within a 20 min time period. These results appear to explain why SCAQMD odor inspectors may have missed some odor events after receiving a complaint call and the time it took to visit a complaint site.

The study demonstrated a novel approach for odor monitoring by using the Odor Profile Method with an "Odor Patrol." The OPM not only confirmed the mitigation of a landfill odor problem, but it also determined odor character, odor intensity, odor frequency and odor duration during this study period. "Landfill gas" was determined to be primarily a rotten vegetable odor with a secondary sewery/fecal odor of lower intensity, and "trash odors" were primarily a rancid and sweet odor with a secondary sewery/fecal and/or rotten vegetable odor of lower intensities generated from trash at the landfill. Quality assurance methods were used to remove local odors from the evaluation.

The OPM detected the occurrences of these specific odors. For example, Figure 6 shows that from 6:00 to 6:40 a.m., a rancid odor was detected three times, rotten vegetable was detected two times and sewery/fecal was detected three times. The occurrence of most of the landfill-related odors was detected from 6:00 a.m. to 7:40 a.m. The order of intensity across types from high to low in Figure 7 is: trash odor (rancid–sweet) > rotten vegetable > sewery/fecal > rancid. Thus, trash odor is the major problematic odor from the landfill site.

The results from the OPM data by the Odor Patrol show the method is capable of determining the odors from a specific odor source without mistaking the odors from the surrounding background. OPM not only can provide odor type but can also provide odor intensity. This can help develop guidelines to understand the sources of a major odor nuisance. OPM can also provide information on the time and duration of a specific odor nuisance.

4.3. Complaint Data vs. Odor Patrol Data

The second objective of the study was to determine if the OPM by an "Odor Patrol" could be used to determine if the closure of a landfill between 6 and 9 a.m. could minimize odor complaints at an impacted site one mile away? The results indicate that the landfill did not contribute much to nuisance odors in 2017–2018 during the hours of 6 to 9 a.m. Figures 8 and 9 show only the complaint data from SCAQMD. To make a reasonable comparison, Figure 9 shows the complaint data from 30 October 2016 to 2 May 2017, which is the exact date and month one year prior to our Odor Patrol period. There was a dramatic decrease from the previous year: 264 complaints in 2016–2017 versus only 21 complaints in 2017–2018 from 6 to 9 a.m. [1]. Thus, the complaint data show that the odor nuisance from the landfill source has been mitigated.

The Odor Patrol data results agree with the complaint data as only 1.3% of the data points showed landfill-related odor detection, and the intensity was at low levels between one and four (see Table 1 and Figure 3). Therefore, the scientifically based OPM used by the Odor Patrol can be used as a confirmation of the absence of odor complaints in 2017–2018.

4.4. Limitations and Future Improvement

Without the knowledge of the Odor Patrol, the owners of the landfill before and during the study period were completing operational changes to mitigate the odors from the landfill (1). Figure 10 shows the timeline of operational changes at the landfill gas collection system, utilization of compacted soil for intermediate trash cover, the application of "PosiShell," the utilization of "Closure Turf," Vegetative Covers, etc. Thus, odor mitigation approaches were being evaluated together, i.e., closure of the landfill from 6 to 9 a.m. and operational changes at the landfill.

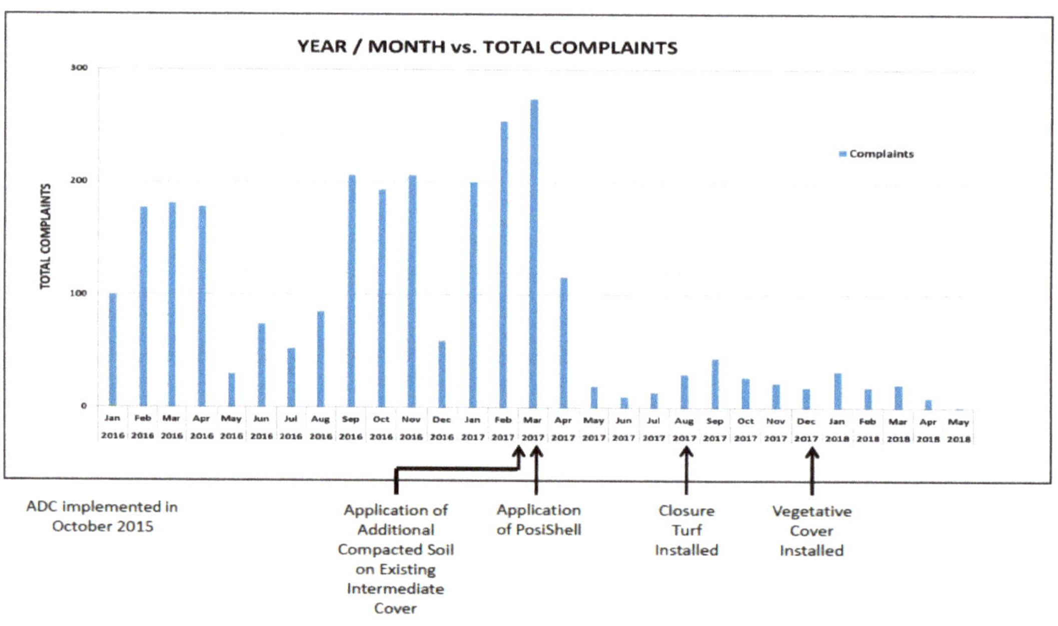

Figure 10. The Intermediate Cover Enhancement (ICE) timeline at the landfill [1]. The "Odor Patrol" study was from 30 October 2017 to 2 May 2018 (100 weekdays) when the Closure Turf and Vegetative Covers were installed and the total complaints dropped significantly.

Figure 10 shows a timeline of improvements at the landfill [1]. The "Odor Patrol" study was from 30 October 2017 to 2 May 2018 (100-week days) when the Closure Turf and Vegetative Covers were installed and the total complaints dropped significantly. Thus, either (A) optimizing the landfill operating approach, (B) closing the landfill from receiving trash at 6–9 a.m. or (C) both together lowering the odors from the landfill and decreasing odor complaints about a mile away near the school site.

The angles between wind direction and direction from complaint locations to the landfill were calculated from an air monitoring station at the landfill facing the direction of the complaint location. During the Odor Patrol time period in 2016–2017, over 65% of the winds coming from landfill to complaint locations were within ±15°, and about 90% of the winds coming from source to complaint locations were within ±30°. During the Odor Patrol time period in 2017–2018, 50% of the winds coming from landfill to complaint locations were within ±15°, and 76% of the winds coming from landfill to complaint locations were within ±30°. Due to the lack of weather station data at the landfill and the school location, a thorough meteorological study was not performed. Thus, Odor Patrol data from 30 October 2017 to 2 May 2018 were compared to the complaint data in the same period of time (30 October 2016 to 2 May 2017) in the previous year, and minimum meteorological differences were observed.

At present, it cannot be stated definitively that the closure of the landfill between 6 and 9 a.m. had no effect on the Odor Patrol data at the school or Odor Complaint data to SCAQMD. An Odor Patrol study as completed in this study could be completed with trucks delivering and dumping to the landfill at 6–9 a.m. to develop evidence that the landfill could receive trash between 6 and 9 a.m. This also should be compared to complaint data from SCAQMD. A complete meteorological study could be done during the 100 days of the Odor Patrol event to provide further information on weather influence on nuisance odor complaints.

5. Conclusions

This study shows that the "Odor Profile Method" by an "Odor Patrol" can evaluate the odor nuisance caused by a particular odor source; i.e., determine the specific character and odor intensity of each odorant and determine the frequency and duration of the odor problem. Additionally, the "Odor Profile Method" by an "Odor Patrol" could determine the contribution of the nuisance odors from 6 to 9 a.m. at the school site, about one mile away from the landfill. Quality assurance was maintained for the odor panel. The data show that the odor control panelists' sense of smell was consistent throughout the 100 days of odor monitoring.

The Odor Patrol only found 13 data of the 1000 data inputs (1.3%) for the 100-day odor monitoring with landfill odor or trash odors that could cause odor complaints. This indicates the landfill did not generate much odor nuisance to the school site from 6 to 9 a.m. when the landfill was not receiving trash. The complaint data also showed a dramatic decrease in odor complaints from 2016–2017 to 2017–2018. Therefore, the Odor Patrol data and the Odor Complaint data correlate well, and the Odor Profile Method can be used to confirm the mitigation of a landfill odor problem.

The study demonstrated a novel approach for odor monitoring by using the Odor Profile Method with an Odor Patrol. The OPM defines the odors to study that come from the landfill. The OPM determined "landfill gas" was primarily a rotten vegetable odor with a secondary sewery/fecal odor of lower intensity and the "Trash odors" were primarily a rancid and sweet odor with a secondary sewery/fecal and/or rotten vegetable odor of lower intensities generated from trash reaching the landfill. The order of intensity observed from high to low was: trash odor (rancid–sweet) > rotten vegetable > sewery/fecal > rancid alone. Thus, trash odor was the major problematic odor from the landfill site. Quality assurance methods were used to remove local odors from the evaluation.

In conclusion, despite other factors such as weather and temperature differences, the closing operation of the landfill from 6 to 9 a.m. and also the implementation of odor control measures at the landfill did help mitigate the problems of unpleasant odors during the Odor Patrol time period of 6–9 a.m. in 2017–2018.

Author Contributions: Conceptualization I.H.S.; methodology, Y.B.; data curation, H.G.; writing—original draft preparation, Y.B.; writing—review and editing, I.H.S.; visualization, Y.B.; supervision, I.H.S. All authors have read and agreed to the published version of the manuscript.

Funding: This research was funded by South Bay Air Quality Management District via Republic Services, Browning-Ferris Industries of California, Inc.

Institutional Review Board Statement: UCLA Institutional Review Board Approval No. 11-002514.

Informed Consent Statement: Informed consent was obtained from all odor panel subjects involved in the study. We thank the students who participated as panelists in the Odor Patrol.

Data Availability Statement: The data presented in this study are available on request from the corresponding author. The raw data is in the authors files. The data of importance to the results of the study are presented in the article.

Conflicts of Interest: The authors declare no conflict of interest.

Presented: 29 September–5 October 2019 the 17th Sardinia, International Waste Management and Landfill Symposium Forte Village, Santa Margherita de Pula (CA), Italy.

References

1. Tseng, Eugene, Presentation at South Coast Hearing Board Meeting. Available online: https://m.youtube.com/channel/UCKLwLnQt6svrAeL9DgAAFqg (accessed on 26 June 2018).
2. Lawless, H.T.; Heymann, H. *Sensory Evaluation of Food; Principals and Practice*, 2nd ed.; Springer Pubishing: New York, NY, USA, 2010.
3. Rice, W.; Baird, R.B.; Eaton, A.D. (Eds.) Method 2170, Flavor Profile Analysis (FPA). In *Standard Methods for the Evaluation of Water and Wastewater*, 23rd ed.; American Public Health Association (APHA), American Water Works Association (AWWA) and Water Environment Federation (WEF): Washington, DC, USA, 2017.
4. Burlingame, G.A. Odor Profiling of Environmental Odors. *Water. Sci. Technol.* **1999**, *40*, 31–38. [CrossRef]
5. Burlingame, G.A.; Suffet, I.H.; Khiari, D.; Bruchet, A.L. Development of an Odor Wheel Classification Scheme for Wastewater. *Water Sci. Technol.* **2003**, *49*, 201–209. [CrossRef]
6. Burlingame, G.A. A Practical Framework Using Odor Survey Data to Prioritize Nuisance Odors. *Water Sci. Technol.* **2009**, *59*, 595–602. [CrossRef] [PubMed]
7. Curren, J.; Snyder, C.; Abraham, S.; Suffet, I.H. Comparison of Two Standard Odor Intensity Evaluation Methods for Odor Problems in Air or Water. *Water Sci. Tech.* **2014**, *69*, 142–146. [CrossRef] [PubMed]
8. Decottignies, V.; Bruchet, A.; Suffet, I.H. New Landfill Odour Wheel: A New Approach to Characterize Odour Emissions at Landfill Sites. In Proceedings of the Sardinia, Twelfth International Waste Management and Landfill Symposium, CISA, Environmental Sanitary Engineering Centre, Sardinia, Italy, 5–9 October 2009.
9. Wittes, J.; Turk, A. *The Selection of Judges for Odor Discrimination Panels, Correlation of Subjective-Objective Methods in the Study of Odors and Taste, STP 440-EB*; Stahl, W., Ed.; ASTM International: West Conshohocken, PA, USA, 1968; pp. 49–70. [CrossRef]
10. Doty, R.L.; Shaman, P.; Kimmelman, C.P.; Dann, M.S. University of Pennsylvania Smell Identification Test: A Rapid Quantitative Olfactory Function Test for the Clinic. *Laryngoscope* **1984**, *94*, 176–178. [CrossRef] [PubMed]
11. Fechner, G. *Elemente de Psychophysik*; Leipzig University: Leipzig, Germany, 1859.
12. Higgins, M.J.; Yarosz, D.P.; Chen, Y.C.; Murthy, S.N.; Maas, N.; Cooney, J.; Glindemann, D.; Novak, J.T. Cycling of Volatile Organic Sulfur Compounds in Anaerobically Digested Biosolids and Its Implications for Odors. *Water Environ. Res.* **2006**, *78*, 243–252. [CrossRef] [PubMed]
13. Chen, Y.; Higgins, M.J.; Murthy, S.N.; Maas, N.A.; Covert, K.J.; Toffey, W.E. Production of Odorous Indole, Skatole, p-Cresol, Toluene, Styrene, and Ethyl Benzene in Biosolids. *J. Residuals Sci. Tech.* **2006**, *3*, 193–202.
14. CEN. *EN 16841-1:2016 Ambient Air—Determination of Odour in Ambient Air by Using Field Inspection—Part 1: Grid Method*; CEN: Brussels, Belgium, 2016.
15. Mannebeck, B.; Mannebeck, C.; Mannebeck, D.; Hauschildt, H.; Van Den Burg, A. Field inspections according to EN16841-1:2015 in a naturally evolved neighborhood of industry and living areas. State-of-the-art-technology of a comprehensive data collection, interaction of different sources and effects on the perceiving citizens. *Chem. Eng. Trans.* **2016**, *54*, 181–186.
16. Carmen, B.; Selena, S.; Laura, C. How Can Odors Be Measured? An Overview of Methods and Their Applications. *Atmosphere* **2020**, *11*, 92. [CrossRef]

Article

Determination of Dose–Response Relationship to Derive Odor Impact Criteria for a Wastewater Treatment Plant

Yan Zhang [1,2], Weihua Yang [1,2,*], Günther Schauberger [3], Jianzhuang Wang [1], Jing Geng [2], Gen Wang [2] and Jie Meng [1,2]

[1] Tianjin Sinodour Environmental Technology Co., Ltd., Tianjin 300191, China; zhangyan_510@126.com (Y.Z.); iven519@163.com (J.W.); sabrina_meng@126.com (J.M.)
[2] State Environmental Protection Key Laboratory of Odor Pollution Control, Tianjin Academy of Eco-Environmental Sciences, Tianjin 300191, China; jane_gj@126.com (J.G.); wanggen1978@126.com (G.W.)
[3] WG Environmental Health, Unit for Physiology and Biophysics, University of Veterinary Medicine, A-1210 Vienna, Austria; Gunther.Schauberger@vetmeduni.ac.at
* Correspondence: yangwh65@126.com; Tel.: +86-022-87671627

Abstract: Municipal wastewater treatment plants (WWTPs) inside cities have been the major complained sources of odor pollution in China, whereas there is little knowledge about the dose–response relationship to describe the resident complaints caused by odor exposure. This study explored a dose–response relationship between the modelled exposure and the annoyance surveyed by questionnaires. Firstly, the time series of odor concentrations were preliminarily simulated by a dispersion model. Secondly, the perception-related odor exposures were further calculated by combining with the peak to mean factors (constant value 4 (Germany) and 2.3 (Italy)), different time periods of "a whole year", "summer", and "nighttime of summer", and two approaches of odor impact criterion (OIC) ("odor-hour" and "odor concentration"). Thirdly, binomial logistic regression models were used to compare kinds of perception-related odor exposures and odor annoyance by odds ratio, goodness of fit and predictive ability. All perception-related odor exposures were positively associated with odor annoyance. The best goodness of fit was found when using "nighttime of summer" in predicting odor-annoyance responses, which highlights the importance of the time of the day and the time of the year weighting. The best predictive performance for odor perception was determined when the OIC was 4 ou/m^3 at the 99th percentile for the odor exposure over time periods of nighttime of summer. The study of dose–response relationship could be useful for the odor management and control of WWTP to maximize the satisfaction of air quality for the residents inside city.

Keywords: air dispersion model; dose–response relationship; odor impact criterion (OIC); perception-related odor exposure; wastewater treatment plant (WWTP)

Citation: Zhang, Y.; Yang, W.; Schauberger, G.; Wang, J.; Geng, J.; Wang, G.; Meng, J. Determination of Dose–Response Relationship to Derive Odor Impact Criteria for a Wastewater Treatment Plant. *Atmosphere* **2021**, *12*, 371. https://doi.org/10.3390/atmos12030371

Academic Editor: Ashok Luhar

Received: 13 January 2021
Accepted: 4 March 2021
Published: 12 March 2021

Publisher's Note: MDPI stays neutral with regard to jurisdictional claims in published maps and institutional affiliations.

1. Introduction

Municipal wastewater treatment plants (WWTPs) as important facilities for urban management have been constructed fast in recent years in China [1]. Unpleasant smells emitted from WWTP may cause both ecological and social problems. Odors are released from pretreatment operations, such as screen and sand filter, primary settler, aeration facility, and sludge de-watering unit in the WWTP. Odor complaints for WWTPs continue to increase by surrounding residents in cities [2]. Although odors are commonly treated as a kind of nuisance, rather than being considered a direct risk for human health, they affect the quality of life and can even cause physical effects on human health [3].

There are many tools to assess odor pollution, some of which include observing the current odor impacts or effects by measurement and monitoring, such as sensorial, analytical, and combined-sensorial technique [4,5] or underlying empirical long-term experiences by community assessment techniques [6,7]. In contrast, other tools make use of a "model" to predict what the impact might be [8–10]. Nowadays, most odor assessment regulations all over the world were determined based on the application of dispersion model [11]. In general, different types of models can be used to simulate the dispersion of odorants into the atmosphere, such as empirical model, Gaussian model, Lagrangian particle model and so on. One of the main advantages of Lagrangian particle model is the ability to treat wind calms, it simulates the dispersion of the emitted odorants with computational particles moving in the wind field and three-dimensional turbulence field. [12,13].

Meanwhile, odor impact criterion (OIC) is a jurisdictional standard according to the desired protection level of general population, which aims to compare with simulation exposure in ambient air by air dispersion model [11]. Whereas the limit values of OIC are highly variable in different countries, which related to the national habits, such as 0.25 ou/m^3 at the 90th percentile for residential and mixed areas in Germany and 2 ou/m^3 for residential areas in Manitoba Canada. This means that the transfer of OIC from other jurisdictions is meaningfulness.

The definition of the OIC depends on several factors, collectively known as the FIDOL factors (frequency, intensity, duration, offensiveness, and location) [14]. In many countries, OICs were defined as the combination of odor concentration threshold (in ou/m^3) and exceedance probability (in %), also called percentile [15]. To mimic the odor sensation of the human nose, short-time peak concentrations, which are derived from one hour mean values simulated by dispersion models, can also be included in the criteria [16]. Besides, many factors have been considered to more precisely represent odor annoyance, such as hedonic tones, human psychosocial health, living quality, other environmental stress, age, work, and so on [17]. Furthermore, odor might be perceived more often at specific times, which represent the time of the day and the year can be weighted in odor episodes regarding their annoyance potential [18].

Exposure-annoyance relationship is an important method to study OIC, which has been analysed in industrial sources [19], livestock, agriculture or farming sources [20–23] and other sources [24,25]. However, there is little knowledge of dose–response relationship describing the resident complaints caused by odor exposure in China. In this work, six odor emitting units of the WWTP were under measurement to identify the time series of odor concentrations for the surrounding residents by an appropriate air dispersion model. Peak to mean factors, temperature and daytime weightings were primarily taken as confounders for coupling with odor concentrations, then two approaches ("odor-hour" and "odor concentration") of OIC were calculated to transform the time series of odor concentration to the perception-related odor exposures. Meanwhile, community question-naires were investigated from twelve urban regions surrounding the WWTP to obtain odor annoyance. Dose–response relationship between the perception-related odor exposure and the investigated annoyance was studied by binomial univariate logistic models. It is noticed that the study of dose–response relationship should be useful to determine the OIC for WWTP and other industries in the future.

2. Materials and Methods

2.1. Site Description and Its Surroundings

The study was conducted at a WWTP located in Northern China, and more precisely in the region of Tianjin. The WWTP was identified as a possible source of nuisance odors, influencing normal life of people living in this area. The distribution of odor complaint incidents during 2017 was shown in Figure 1, derived from environmental protection hotline "12369" of Tianjin, China. On this background, we decided to carry out a specific study by determining a dose–response relationship to derive odor impact criteria of the WWTP.

Figure 1. The location of the wastewater treatment plant (WWTP) and twelve surrounding residential areas (**A–L**). The yellow stars which located in seven residential areas (**A**), (**C–G**), and (**J**) represent the major off-site locations of odor complaints. The distance between the boundary of plant and the southwest corner of residential area K is about 1.2 km. WWTP: Municipal wastewater treatment plant.

The WWTP covered about 295,000 m^2 and the designed treatment capacity was 400,000 m^3/d. It collected wastewater from four administrative regions of Tianjin, served a population of about 1.11 million and 730 enterprises. Six odor emitted units inside this WWTP were selected for analysis. Twelve off-site locations of surrounding residential areas were selected to evaluate odor impact.

2.2. Questionnaire Data Collection

Cross-sectional questionnaire data were obtained from twelve urban regions (Figure 1). A total number of 126 persons were randomly selected and contacted by face-to-face in June 2018. Adult residents (>18 years old) living more than 1 year as being representatives were requested to anonymously participate in the study.

The questionnaire was developed based upon a number of prior investigations [26–28] and consisted of two main sections. The first part included general socio-demographic data (i.e., age, gender, address, and years living in the region), while the second part referred to environmental stressors, including satisfaction of living environment and origin of pollution (i.e., noise, traffic, catering, waste, sewage, or others). Regarding the unpleasant smells of sewage, the questions included: degree of perceived odor intensity (estimated using the 6-point scale, i.e., "0 = no odor","1 = very faint strength" "2 = faint strength", "3 = moderate strength", "4 = strong strength", and "5 = very strong strength"), degree of perceived odor annoyance (estimated using the 5-point scale, i.e., "0 = not annoyed", "1 = slightly annoyed", "2 = moderately annoyed", "3 = very annoyed", and "4 = extremely annoyed"), occurrence time (separated the time of the day into 5 periods, a multiple choice question, i.e., in the morning, at noon, in the afternoon, in the evening, and in the middle of the night), and occurrence season (a multiple choice question, i.e., spring, summer, autumn, and winter).

2.3. Odor Expoure

2.3.1. Sampling Campaign

The air samples were collected according to all odor emitting units of this WWTP, including six treatment processes of screen, sand filter, primary settler, aeration tank, secondary sedimentation, and sludge dewatering unit. Screen and sludge dewatering unit are regarded as point sources, due to these two workshops are completely closed with sealing measures, and the exhaust gas is discharged by chimneys. Sand filter, primary settler, aeration tank, and secondary sedimentation units are regarded as area sources due to the unsealed surfaces (Figure 1). The sampling campaigns were conducted for two days of May and June respectively in 2018, during 8:00 to 18:00, and the total numbers of samples were 24 during the investigation.

The odor samples from point sources were collected by the SOC-01 sampler with "lung" principle (Tianjin Sinodour Environmental Technology Co., Ltd., China) and deposited to a 10 L bio-oriented polyester sample bag equipped with a Teflon TM inlet tube [29]. The odor samplings on area sources were carried out by a wind tunnel system, which consists of a PET hood positioned over the emitting surface. The wind tunnel has a rectangular section inlet and outlet duct (0.042 m × 0.024 m). The central body of the wind tunnel is a 0.5 m wide, 1.0 m long, and 0.13m high rectangular section chamber. The sample stream was filtered through activated carbon at a specific sweep air velocity by a fan, and air sample was collected at the outlet duct with a vacuum pump in 10 L sampling bag [30]. The sweep air velocity inside the wind tunnel remained fixed at 0.064 m s^{-1}.

The gas temperature and exit velocity were measured by thermal anemometer with flow probe (Testo, Germany). The bags were cleaned twice using sample gas before sampling for avoiding the interference of background odor concentration. All samples were sent to the laboratory to analysis within 24 h. Temperature (26–36 °C), humidity (60–90%), and pressure (990–1000 hPa) were measured during the sampling periods by hand-held wind speed and direction indicator (Kestrel, Palo Alto, CA, USA).

2.3.2. Determination of Odor Concentration and Odor Emission Rate

Odor emission rate (OER) is the essential input data for air dispersion modeling and its value directly determines the impact degree on the environmental odor [31]. To evaluate OER, first the calculation of the odor concentration is required.

Odor concentration was measured by the triangle odor bag method in accord with Chinese regulation: Air quality—Determination of odor -Triangle Odor Bag Method [32]. A sniff team of six trained panelists, distinguished the odor from three bags with one odor sample and two odor free air samples. When a given panel member provided an incorrect answer and a correct answer in adjacent dilution ratio, the test for this panel was considered finished, then the personal olfactory threshold was calculated. Finally, the odor concentration was calculated according to the personal olfactory threshold of the six sniff members.

The calculation methods of OER for point source and area source were shown in Equations (1) and (2).

$$OER_1 = C \cdot V \tag{1}$$

$$OER_2 = (C \cdot L / S_1) \cdot S \tag{2}$$

where OER_1, OER_2 is the odor emission rate for point source and area source respectively, OU/s; C is the odor concentration, ou/m^3; V is the flow rate measured by the gas flow meter, m^3/s; L is the flow rate in the outlet duct of wind tunnel system, m^3/s; S_1 is the base area of wind tunnel, m^2; S is the total area of emitting surface, m^2.

2.3.3. Odor Dispersion Model

The odor release was calculated by a Lagrange puff model, the CALPUFF model. There were several scientific studies proved the possibility of applying CALPUFF for modelling the dispersion of odorants [33,34]. Puff models represent a continuous plume as a number of discrete packets of pollutant and evaluate the contribution of a puff to the concentration at a receptor by a "snapshot" approach. The basic equation for the contribution of a puff to the concentration at a receptor is expressed in Equations (3) and (4):

$$C = \frac{Q}{2\pi\sigma_y\sigma_z} g\ exp\left(-\frac{d_a^2}{2\sigma_x^2}\right)exp\left(-\frac{d_c^2}{2\sigma_y^2}\right) \tag{3}$$

$$g = \frac{2}{(2\pi)^{1/2}\sigma_z}\sum_{n=-\infty}^{\infty} exp\left[-(H_c+2nh)^2/\left(2\sigma_z^2\right)\right] \tag{4}$$

where C is the ground-level odor or pollutant concentration, ou/m³ or mg/m³, Q is the odor or pollutant mass in the puff, OU or mg, σ_x is the standard deviation of the Gaussian distribution in the along-wind direction, m, σ_y is the standard deviation of the Gaussian distribution in the cross-wind direction, m, σ_z is the standard deviation of the Gaussian distribution in the vertical direction, m, d_a is the distance from the puff center to the receptor in the along-wind direction, m, d_c is the distance from the puff center to the receptor in the cross-wind direction, m, g is the vertical term of the Gaussian equation, s/m, H is the effective height above the ground of the puff center, m, and h is the mixing height, m.

The meteorological data used for the study area consisted of two parts, surface meteorological data and upper-air meteorological data, both were from 31 December 2016 to 31 December 2017. Surface meteorological data such as wind direction, wind speed, air pressure, temperature, relative humidity, and cloud cover were obtained from the weather station, which was the nearest one from the WWTP. Upper-air meteorological data were generated by WRF (Weather Research and Forecasting) model with 1 km resolution relevant to the studied area. The wind rose diagram at the WWTP was shown in Figure 2.

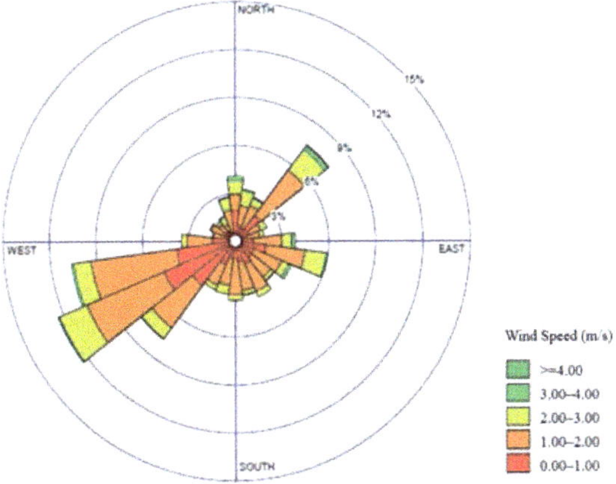

Figure 2. Wind rose diagram in the atmospheric vertical altitude of 10 m at the WWTP during 2017. Legend denotes wind speed categories and their associated colors. WWTP: Municipal wastewater treatment plant.

Geophysical data (flat terrain, urban area) and grids (nested grids, assessment squares defined by 32 × 32 km² with the smallest 50 m spacing) were taken as input data in the model. 126 sensitive points were set in the model according to urban resident sites investigated by questionnaires. Then a time serious of one hour mean values of the odor concentrations were calculated at each household, where the corresponding questionnaire was available.

2.4. Perception-Related Odor Exposure Analysis

2.4.1. Preliminary Perception-Related Odor Exposure Variables

Preliminary perception-related exposure variables were determined for weighting the odor concentrations by the following confounders:

(1) Peak-to-mean factor (F): In regard to the duration of one single human breath, the short-term concentration fluctuations were transformed from one hour mean values of the odor concentrations (e.g., constant value 4 (Germany), 2.3 (Italy) or 1 (UK)) [15];

(2) Temperature and daytime: The annoyed time period of the year and time period of the day were obtained by the community questionnaires to emphasize those hourly values, when residents are more sensitive to odor.

By these confounders, the odor concentrations were transformed to the perception-related odor exposure preliminarily at each site.

2.4.2. Perception-Related Odor Exposures by OICs

In order to determine which OIC shows a better performance, the "odor-hour" metric approach, expressed as the threshold of a certain percentile (like in Ireland [15]) and the "odor concentration "approach, expressed as the threshold of a certain concentration (like in Germany [15]) were used to calculate the further perception-related odor exposures, which are shown as follows.

(1) The threshold of a certain percentile at a certain site: The odor concentrations at 98, 95, 90, 85, 80, and 70 percentiles were selected, based on the time series of the preliminary perception-related odor concentrations, expressed as C98, C95, C90, C85, C80, and C70, respectively;

(2) The threshold of a certain concentration at a certain site: The probabilities exceeding odor concentration thresholds of 1, 2, 3, 4, and 5 ou/m^3 were selected, based on the time series of the preliminary perception-related odor concentrations, expressed as P1, P2, P3, P4, and P5, respectively.

Then the further perception-related odor exposures by different OICs were calculated at 126 sensitive points (households).

2.5. Dose–Response Relationship Analysis

Binomial logistic regression models basing on log-logit sigmoid equations were used to estimate the association between the perception-related odor exposures (i.e., C98, C95, C90, C85, C80, C70, P1, P2, P3, P4, and P5, respectively) and the odor annoyances which were derived from questionnaires. For binomial models, the outcome variables of odor annoyance degrees were dichotomized into two scores (score = 0, derived from "not annoyed", "slightly annoyed"; and score = 1, derived from "moderately annoyed", "very annoyed", and "extremely annoyed"). As well, the independent variables of perception-related odor exposures were transformed into log_e values, which is due to a log fit between odor exposure and odor annoyance previously was found to be closer than a linear fit [21].

In this analysis, the associations for the dose–response relationships were estimated by OR (odds ratio), 95% CI (confidence interval), and P (significance level). The goodness of fit (Akaike information criterion, AIC; McFadden R^2; Hosmer-Lemeshow test, HL test) were obtained. The predictive abilities were also investigated by using AUC (area under the ROC (receiver operating characteristic) curve) and accuracy parameter. The statistical analyses were performed in both SPSS and MATLAB software.

3. Results and Discussion

3.1. Socio-Demographic Characteristics of Participants

The long-term experiences of the communities were investigated and a total of 126 valid questionnaires were obtained in the study. The number of questionnaires in each investigated residential area was shown in Table 1. In general, 60 respondents were females and 66 respondents were males; About 64% of respondents were over the age of 45, about 36% of respondents were at the age of ranging 18 to 45; 68 respondents were living in the household lower than 5 years, 34 respondents were living between 5 to 10 years, and 24 respondents were living more than 10 years; Besides sewage smells, 7 respondents were influenced by noise impact and 13 respondents experienced environmental stressor of waste smells.

Table 1. The questionnaire number, averaged odor intensity and odor annoyance by respondents in investigated residential areas A–L.

Title Questionnaire Result	Investigated Residential Area											
	A	B	C	D	E	F	G	H	I	G	K	L
Questionnaire number	10	10	12	13	11	11	12	10	11	12	10	15
Averaged odor intensity	2.8	2.5	2.8	3.0	3.0	2.4	2.1	0.8	0.8	2.5	0.7	1.4
Odor annoyance (%)	75	33	64	45	63	55	44	11	10	50	0	29

In the 126 questionnaires, about 40% of the residents were annoyed by sewage odor at their households, which consist of "moderately annoyed", "very annoyed", and "extremely annoyed", and about 23% of the residents were annoyed as "very annoyed" and "extremely annoyed". Besides, summer was the most serious annoyed season when sewage smell occurred (20%, 50%, 18%, and 12% was the proportion of occurrence time in spring, summer, autumn, and winter month, respectively). Nighttime was the most serious annoyed time in the day when sewage smell occurred (13%, 10%, 21%, 46%, and 10% was the proportion of occurrence time in the morning, noon, afternoon, night, and midnight of the day, respectively). Odor complaints occur predominantly in the afternoon and evening hours of the warm season when residents are outside [18].

Respondents who lived in residential area A annoyed the most by sewage smell, followed by residential areas C, E, F, and G; Averaged odor intensities in residential areas A-G were all higher to 2, indicated that many people in these residential areas can perceive sewage smell; A small number of people may perceive sewage smell in residential areas H, I, K, which were located on the southwest and 650 m~1100 m away from the WWTP boundary (Table 1).

3.2. Odor Exposure and Perception-Related Odor Exposure

The time series of the odor concentrations were calculated over 8760 h. The average odor concentrations of the 8760 h for 126 sensitive points ranged from 0.3 ou/m^3 to 4.8 ou/m^3. The order of mean odor concentration from highest to lowest was residential area A (3.7), B (3.6), C (2.7), D (1.3), L (1.2), E (1.1), G (0.8), J (0.8), I (0.5), F (0.5), K (0.5), and H (0.4).

The time series of the odor concentrations were firstly multiplied by peak to mean factors 1, 2.3, and 4, respectively, and divided into three time periods of "a whole year", "summer", and "nighttime of summer", respectively, due to the serious annoyed season of the year and time of the day obtained from the community questionnaires. Then the odor exposures were calculated by "odor concentration" metric approach, expressed as C98, C95, C90, C85, C80, and C70, respectively, and "odor-hour" metric approach, expressed as P1, P2, P3, P4, and P5, respectively. Based on these calculations, the groups of perception-related odor exposures were obtained.

3.3. Dose–Response Relationship by Binomial Univariate Logistic Models

The perception-related odor exposures and investigated odor annoyances were performed to establish dose–response associations by binomial univariate logistic models. Results revealed the associations between odor annoyance and (1) odor concentrations (C98, C95, C90, C85, C80, and C70); (2) odor percentiles (P1, P2, P3, P4, and P5) (Table 2).

Table 2. OR values for odor annoyance by binomial logistic regression models [a,b].

Odor Concentration	Peak to Mean Factor	Variable of Odor Exposure: The Threshold of Concentration					
		Modeled by a Year		Modeled by Summer		Modeled by Nighttime of Summer	
C70	4/2.3/1	2.063	1.433–2.971	1.967	1.440–2.688	1.757	1.347–2.293
C80	4/2.3/1	2.438	1.577–3.770	2.481	1.630–3.714	2.254	1.553–3.273
C85	4/2.3/1	2.279	1.499–3.466	2.308	1.557–3.416	2.402	1.589–3.633
C90	4/2.3/1	2.193	1.398–3.439	2.365	1.534–3.647	2.475	1.597–3.835
C95	4/2.3/1	2.278	1.408–3.687	2.473	1.543–3.964	3.153	1.870–5.316
C98	4/2.3/1	2.652	1.493–4.712	3.448	1.855–6.409	4.085	2.128–7.843

Odor Percentile	Peak to Mean Factor	Variable of Odor Exposure: The Threshold of Percentile					
		Modeled by a Year		Modeled by Summer		Modeled by Nighttime of Summer	
P1	1	7.403	2.674–20.499	6.287	2.659–14.867	8.362	3.135–22.307
	2.3	11.791	3.363–41.343	8.277	3.065–22.356	13.821	3.987–47.902
	4	18.103	4.204–77.954	10.942	3.591–33.338	20.836	5.077–85.516
P2	1	3.814	1.842–7.893	4.257	2.119–8.551	3.840	2.021–7.295
	2.3	7.874	2.767–22.412	6.371	2.677–15.162	8.719	3.163–24.036
	4	10.345	3.134–34.148	7.627	2.936–19.812	12.677	3.833–41.931
P3	1	2.594	1.515–4.440	3.066	1.769–5.313	2.824	1.743–4.576
	2.3	6.902	2.585–18.428	6.014	2.604–13.892	7.110	2.876–17.575
	4	8.536	2.882–25.283	6.681	2.759–16.177	9.735	3.366–28.156
P4	1	2.177	1.383–3.428	2.555	1.601–4.077	2.411	1.606–3.619
	2.3	4.851	2.118–11.107	4.986	2.351–10.574	4.759	2.287–9.901
	4	7.403	2.674–20.499	6.032	2.603–13.981	8.362	3.135–22.307
P5	1	1.950	1.309–2.906	2.231	1.486–3.349	2.143	1.496–3.070
	2.3	3.448	1.747–6.806	3.776	1.975–7.220	3.527	1.934–6.430
	4	6.934	2.589–18.575	5.834	2.557–13.307	7.298	2.916–18.262

[a] Odor exposures were log$_e$ transformed; [b] p Values were all lower than 0.001.

The values of OR were invariant for a certain concentration combined with different constant values of peak to mean factor due to the multiple relations. In regard to the results basing on "a whole year" confounder, all odor exposure variables were positively associated with odor annoyance. C98 as exposure assessment variable seemed to be a slightly better association than other odor concentrations (OR = 2.652; 95% CI =1.493–4.712), and P1 combined with F = 4 showed the greatest correlation (OR = 18.103; 95% CI = 4.204–77.954). Furthermore, the associations were substantially larger when using "odor-hour" metric approach than "odor concentration" approach.

Then the analysis was performed basing on "summer" and "nighttime of summer" confounders. All odor exposure variables were also positively associated with odor annoyance. The strongest association was found when using the combination of P1, F =4, and "nighttime of summer" as exposure assessment variable (e.g., OR = 20.836, 95% CI: 5.077–85.516). Besides, the highest association between odor concentration and odor annoyance was found when using the combination of C98 and "nighttime of summer" (OR = 4.085, 95% CI: 2.128–7.843).

3.4. Goodness of Fit and Predictive Ability of Binomial Logistic Models

The goodness of fit obtained from "nighttime of summer" showed to be preferable in the combination of P2 and F = 1 (AIC = 152.9, McFadden R^2 = 0.131 and HL test = 0.215 and the combination of P4 and F = 1 (AIC = 153.0, McFadden R^2 = 0.131 and HL test = 0.063) (Table 3). The predictive ability of accuracy and AUC of logistic models seemed not accordance with each other (Table 4). The best accuracy was obtained in the combination of C98 and "summer" and the combination of P4, F = 1 and "nighttime of summer" (accuracy = 66.7). However, the best consequence of AUC was obtained in the combination of C95 and "nighttime of summer" (AUC = 0.743). On the whole, goodness of fit (AIC, McFadden R^2) and predictive ability (AUC) showed that the values obtained by "summer" and "nighttime of summer" had better predictive performance than "a year", especially by "nighttime of summer". The results illuminate that the odor episode should be weighted by the time of the day and the time of the year when studying odor annoyance, OICs, and so on.

Table 3. Goodness of fit (AIC, McFadden R^2, HL test) for odor annoyance by binomial logistic regression models [a].

| Odor Concentration | Peak to Mean Factor | Variable of Odor Exposure: The Threshold of Concentration | | | | | | | | |
| | | Modeled by a Year | | | Modeled by Summer | | | Modeled by Nighttime of Summer | | |
		AIC	McFadden R^2	HL Test	AIC	McFadden R^2	HL Test	AIC	McFadden R^2	HL Test
C70	4/2.3/1	157.6	0.105	0.335	154.3	0.124	0.083	154.8	0.121	0.104
C80	4/2.3/1	156.5	0.110	0.045	153.1	0.130	0.028	153.5	0.128	0.107
C85	4/2.3/1	158.6	0.099	0.098	155.1	0.119	0.016	155.5	0.117	0.073
C90	4/2.3/1	162.7	0.075	0.102	162.7	0.075	0.032	156.7	0.109	0.050
C95	4/2.3/1	163.2	0.072	0.144	159.6	0.093	0.220	153.3	0.129	0.274
C98	4/2.3/1	163.2	0.072	0.036	157.6	0.104	0.098	153.6	0.127	0.109

| Odor Percentile | Peak to Mean Factor | Variable of Odor Exposure: The Threshold of Percentile | | | | | | | | |
| | | Modeled by a Year | | | Modeled by Summer | | | Modeled by Nighttime of Summer | | |
		AIC	McFadden R^2	HL Test	AIC	McFadden R^2	HL Test	AIC	McFadden R^2	HL Test
P1	1	158.6	0.098	0.115	155.1	0.119	0.075	154.5	0.123	0.141
	2.3	158.2	0.101	0.112	155.1	0.119	0.016	155.2	0.118	0.560
	4	157.6	0.104	0.248	154.5	0.122	0.289	154.3	0.124	0.699
P2	1	160.0	0.090	0.465	155.2	0.118	0.054	154.0	0.125	0.350
	2.3	158.4	0.100	0.230	155.1	0.119	0.163	155.0	0.119	0.250
	4	158.4	0.099	0.151	155.1	0.119	0.037	155.0	0.120	0.509
P3	1	161.3	0.083	0.168	156.0	0.114	0.026	152.9	0.131	0.215
	2.3	158.5	0.099	0.124	154.8	0.120	0.077	154.1	0.125	0.147
	4	158.3	0.100	0.111	154.8	0.121	0.040	154.7	0.121	0.017
P4	1	162.5	0.075	0.079	156.6	0.110	0.436	153.0	0.131	0.063
	2.3	158.9	0.097	0.750	153.9	0.126	0.076	153.5	0.128	0.398
	4	158.6	0.098	0.115	155.1	0.119	0.385	154.5	0.123	0.141
P5	1	163.3	0.071	0.031	157.1	0.107	0.170	154.5	0.123	0.259
	2.3	160.4	0.088	0.361	155.8	0.115	0.039	154.0	0.125	0.173
	4	158.5	0.099	0.052	155.0	0.120	0.067	154.2	0.124	0.142

[a] Odor exposures were log$_e$ transformed.

Table 4. Predictive ability (accuracy, AUC) for odor annoyance by binomial logistic regression models [a].

| Odor Concentration | Peak to Mean Factor | Variable of Odor Exposure: The Threshold of Percentile Concentration | | | | | |
| | | Modeled by a Year | | Modeled by Summer | | Modeled by Nighttime of Summer | |
		Accuracy (%)	AUC	Accuracy (%)	AUC	Accuracy (%)	AUC
C70	4/2.3/1	64.3	0.711	64.3	0.742	65.9	0.730
C80	4/2.3/1	63.5	0.711	64.3	0.736	62.7	0.738
C85	4/2.3/1	62.7	0.713	63.5	0.738	62.7	0.740
C90	4/2.3/1	63.5	0.684	61.9	0.727	65.1	0.734
C95	4/2.3/1	64.3	0.679	62.7	0.710	63.5	0.743
C98	4/2.3/1	65.1	0.672	66.7	0.717	63.5	0.736

Table 4. *Cont.*

Odor Percentile	Peak to Mean Factor	Variable of Odor Exposure: The Threshold of Percentile					
		Modeled by a Year		Modeled by Summer		Modeled by Nighttime of Summer	
		Accuracy (%)	AUC	Accuracy (%)	AUC	Accuracy (%)	AUC
P1	1	62.7	0.713	64.3	0.740	65.1	0.736
	2.3	64.3	0.708	65.1	0.736	65.1	0.728
	4	64.3	0.711	64.3	0.735	65.1	0.735
P2	1	63.5	0.696	64.3	0.720	65.9	0.727
	2.3	65.1	0.712	64.3	0.734	65.1	0.737
	4	64.3	0.709	63.5	0.739	65.1	0.729
P3	1	64.3	0.688	63.5	0.721	65.1	0.733
	2.3	62.7	0.708	63.5	0.739	65.1	0.732
	4	65.1	0.712	64.3	0.736	65.1	0.736
P4	1	65.1	0.683	62.7	0.722	66.7	0.736
	2.3	62.7	0.701	64.3	0.740	65.1	0.730
	4	62.7	0.713	64.3	0.727	65.1	0.736
P5	1	64.3	0.678	63.5	0.716	64.3	0.731
	2.3	64.3	0.695	64.3	0.736	64.3	0.728
	4	62.7	0.708	65.1	0.718	65.1	0.732

[a] Odor exposures were loge transformed.

3.5. Odor Impact Criteria of the WWTP

The best predictor of odor exposure was selected as the combination of P4, F = 1, and "nighttime of summer", integrating goodness of fit and predictive ability. The univariate binomial logistic function was shown in Equation (5). Furthermore, in order to visualize the results of the logistic function, the dose–response curve was shown in Figure 3.

$$p = 1 + \exp{(-1.595 - 0.880 \ln P4)} \tag{5}$$

where p is the probability of odor annoyance, 0–1; P4 is the probability exceeded odor concentration thresholds of 4 ou/m^3, calculated by air dispersion model over the time period of nighttime of summer, %.

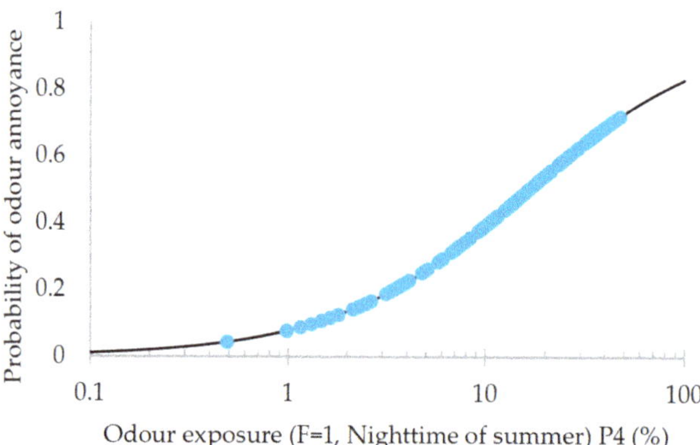

Figure 3. Exposure–response univariate binomial logistic model between odor exposure and probability of odor annoyance; The odor exposure of P4 is the probability exceeded odor concentration thresholds of 4 ou/m^3, calculated by air dispersion model over the time period of nighttime of summer, %; F: Peak-to-mean factor.

Aiming to limit the percentage of people experiencing some form of odor-induced annoyance to 10% or less [35,36], the target value of OIC was calculated as follows: 4 ou/m^3 at the 99th percentile for the odor exposure calculated by air dispersion model over the time period of nighttime of summer.

3.6. Lagrange Dispersion Model and Separation Distances

A generalized non-steady-state air quality modeling system for regulatory use, Sigma Research Corporation developed the CALPUFF dispersion model and programs. The model contains algorithms for near-source effects such as building downwash, transitional plume rise, partial plume penetration, subgrid scale terrain interactions, as well as longer range effects such as pollutant removal (wet scavenging and dry deposition), chemical transformation, vertical wind shear, overwater transport, and coastal interaction effects. Most of the algorithms contain options to treat the physical processes at different levels of detail depending on the model application [37]. CALPUFF model is driven by temporally and spatially varying meteorological data on hourly basis, which can handle continuous puffs of pollutants being emitted from a source into the ambient wind flow.

Capelli et al. [13] discussed the preference of the Lagrangian CALPUFF model, due to the limitations of Gaussian models (inability to handle calm conditions, lack of three-dimensional meteorology, and steady-state assumption). In a calibration study of Rood [38], the CALPUFF model showed the smallest variance, highest correlation, and highest number of predictions within a factor of two compared to the AERMOD model. Even for odorous substances, CALPUFF was compared with other dispersion models with good results [39].

Direction-dependent separation distances are commonly used procedure to avoid odor annoyance between emission sources and residential areas, calculated by air dispersion model [40]. The separation distances were simulated by CALPUFF model, basing on the OIC (4 ou/m^3 at the 99th percentile for the odor exposure over the time period of nighttime of summer). As shown in Figure 4, residential areas A, B, C, D, E, G, and J were completely annoyed by sewage odors, which was basically accorded with the results of odor complaints in 2017. Besides, a number of people in residential areas F, H, I, K, and L were annoyed, and the separation distance in west-south direction was about 900 m from the WWTP boundary.

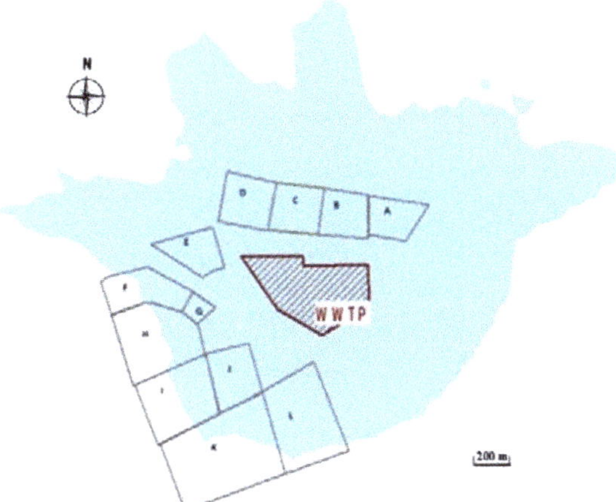

Figure 4. Direction-dependent separation distances based on the odor impact criterion (OIC) by air dispersion model. OIC: 4 ou/m^3 at the 99th percentile for the odor exposure calculated over the time period of nighttime of summer.

4. Conclusions

In conclusion, this was supposed be the first study aimed at determining a certain OIC using different perception-related odor exposure approaches to study dose–response relationship by binomial logistic regression models. The odor exposures calculated over time period of "nighttime of summer" showed better predictive performance than "a whole year" and "summer" in predicting odor-annoyance responses. OIC was taken as 4 ou/m^3 at the 99th percentile for the odor exposure calculated by air dispersion model over the time period of nighttime of summer. Furthermore, the separation distances of the WWTP were calculated by CALPUFF model basing on the OIC, which was about 900 m from the WWTP boundary in the west-south direction.

Author Contributions: Conceptualization, G.S.; methodology, W.Y., G.W. and J.M.; formal analysis, Y.Z., W.Y., J.G. and J.W.; data curation, Y.Z., W.Y. and J.W.; writing—original draft preparation, Y.Z.; writing—review and editing, Y.Z. and G.S.; visualization, Y.Z.; supervision, W.Y.; funding acquisition, W.Y. All authors have read and agreed to the published version of the manuscript.

Funding: This research was funded by Natural Science Foundation of Tianjin City, grant number 18JCQNJC08800.

Informed Consent Statement: Informed consent was obtained from all subjects involved in the study.

Conflicts of Interest: The authors declare no conflict of interest.

References

1. Wang, X.D.; Wang, B.G.; Zhao, D.J.; Liu, S.L. Sources and components of MVOC from a municipal sewage treatment plant in Guangzhou. *China Environ. Sci.* **2011**, *31*, 576–583.
2. Stuetz, R.M.; Fenner, R.A.; Engin, G. Assessment of odours from sewage treatment works by an electronic nose, H$_2$S analysis and olfactometry. *Water Res.* **1999**, *33*, 453–461. [CrossRef]
3. Han, Z.; Qi, F.; Wang, H.; Li, R.; Sun, D. Odor assessment of NH$_3$ and volatile sulfide compounds in a full-scale municipal sludge aerobic composting plant. *Bioresour. Technol.* **2019**, *282*, 447–455. [CrossRef] [PubMed]
4. Barczak, R.J.; Kulig, A. Comparison of different measurement methods of odour and odorants used in the odour impact assessment of wastewater treatment plants in Poland. *Water Sci. Technol.* **2017**, *75*, 944–951. [CrossRef]
5. Lim, J.H.; Cha, J.S.; Kong, B.J.; Baek, S.H. Characterization of odorous gases at landfill site and in surrounding areas. *J. Environ. Manag.* **2018**, *206*, 291–303. [CrossRef]
6. Mohamed, E.; Shelly, M. Industrial odor source identification based on wind direction and social participation. *Int. J. Environ. Res. Public Health* **2019**, *16*, 1242.
7. Hayes, J.E.; Fisher, R.M.; Stevenson, R.J.; Mannebeck, C.; Stuetz, R.M. Unrepresented community odour impact: Improving engagement strategies. *Sci. Total Environ.* **2017**, *609*, 1650–1658. [CrossRef]
8. Brancher, M.; Piringer, M.; Franco, D.; Filho, P.B.; Lisboa, H.D.M.; Schauberger, G. Assessing the inter-annual variability of separation distances around odour sources to protect the residents from odour annoyance. *J. Environ. Sci.* **2019**, *79*, 11–24. [CrossRef]
9. Douglas, P.; Hayes, E.T.; Williams, W.B.; Tyrrel, S.F.; Kinnersley, R.P.; Walsh, K.; Driscolle, M.O.; Longhurst, P.J.; Pollard, S.J.T.; Drew, G.H. Use of dispersion modelling for environmental impact assessment of biological air pollution from composting: Progress, problems and prospects. *Waste Manag.* **2017**, *70*, 22–29. [CrossRef] [PubMed]
10. Pandey, G.; Sharan, M. Performance evaluation of dispersion parameterization schemes in the plume simulation of FFT-07 diffusion experiment. *Atmos. Environ.* **2018**, *172*, 32–46. [CrossRef]
11. Piringer, M.; Knauder, W.; Petz, E.; Schauberger, G. A comparison of separation distances against odour annoyance calculated with two models. *Atmos. Environ.* **2015**, *116*, 22–35. [CrossRef]
12. Sironi, S.; Capelli, L.; Centola, P.; Rosso, R.D. Odour impact assessment by means of dynamic olfactometry, dispersion modelling and social participation. *Atmos. Environ.* **2010**, *44*, 354–360. [CrossRef]
13. Capelli, L.; Sironi, S.; Del Rosso, R.; Guillot, J.M. Measuring odours in the environment vs. dispersion modelling: A review. *Atmos. Environ.* **2013**, *79*, 731–743. [CrossRef]
14. Sommer-Quabach, E.; Piringer, M.; Petz, E.; Schauberger, G. Comparability of separation distances between odour sources and residential areas determined by various national odour impact criteria. *Atmos. Environ.* **2014**, *95*, 20–28. [CrossRef]
15. Brancher, M.; Griffiths, K.D.; Franco, D.; De Melo Lisboa, H. A review of odour impact criteria in selected countries around the world. *Chemosphere* **2017**, *168*, 1531–1570. [CrossRef] [PubMed]
16. Piringer, M.; Schauberger, G.; Mikovits, C.; Zollitsch, W.; Hörtenhuber, S.J.; Baumgartner, J.; Schönhart, M. Climate change impact on the dispersion of airborne emissions and the resulting separation distances to avoid odour annoyance. *Atmos. Environ. X* **2019**, *2*, 100021. [CrossRef]

17. Schauberger, G.; Piringer, M.; Schmitzer, R.; Kamp, M.; Sowa, A.; Koch, R.; Eckhof, W.; Grimm, E.; Kypke, J.; Hartung, E. Concept to assess the human perception of odour by estimating short-time peak concentrations from one-hour mean values. Reply to a comment by Janicke et al. *Atmos. Environ.* **2012**, *54*, 624–628. [CrossRef]
18. Schauberger, G.; Piringer, M.; Petz, E. Odour episodes in the vicinity of livestock buildings: A qualitative comparison of odour complaint statistics with model calculations. *Agric. Ecosyst. Environ.* **2006**, *114*, 185–194. [CrossRef]
19. Li, J.; Zou, K.; Li, W.; Wang, G.; Yang, W. Olfactory Characterization of Typical Odorous Pollutants Part I: Relationship between the Hedonic Tone and Odor Concentration. *Atmosphere* **2019**, *10*, 524. [CrossRef]
20. Miedema, H.M.E.; Walpot, J.I.; Vos, H. Exposure-annoyance relationships for odour from industrial sources. *Atmos. Environ.* **2000**, *34*, 2927–2936. [CrossRef]
21. Blanes-Vidal, V.; Baelum, J.; Nadimi, E.S.; Løfstrøm, P.; Christensen, L.P. Chronic exposure to odorous chemicals in residential areas and effects on human psychosocial health: Dose–response relationships. *Sci. Total Environ.* **2014**, *490*, 545–554. [CrossRef]
22. Cantuaria, M.L.; Lofstrom, P.; Blanes-Vidal, V. Comparative analysis of spatio-temporal exposure assessment methods for estimating odor-related responses in non-urban populations. *Sci. Total Environ.* **2017**, *605–606*, 702–712. [CrossRef]
23. Moshammer, H.; Oettl, D.; Mandl, M.; Kropsch, M.; Weitensfelder, L. Comparing annoyance potency assessments for odors from different livestock animals. *Atmosphere* **2019**, *10*, 659. [CrossRef]
24. Sucker, K.; Both, R.; Bischoff, M.; Guski, R.; Krämer, U.; Winneke, G. Odor frequency and odor annoyance Part II: Dose–response associations and their modification by hedonic tone. *Int. Arch. Occup. Environ. Heath* **2008**, *81*, 683–694. [CrossRef] [PubMed]
25. Weitensfelder, L.; Moshammer, H.; Dietmar, O.; Payer, I. Exposure-complaint relationships of various environmental odor sources in Styria, Austria. *Environ. Sci. Pollut. Res.* **2019**, *26*, 9806–9815. [CrossRef] [PubMed]
26. Verein Deutscher Ingenieure. *Determination of Annoyance Parameters by Questioning Repeated Brief Questioning of Neighbour Panelists (VDI3883 Part 2)*; Beuth Verlag GmbH: Berlin, Germany, 1993.
27. Hayes, J.E.; Stevenson, R.J.; Stuetz, R.M. Survey of the effect of odour impact on communities. *J. Environ. Manag.* **2017**, *204*, 349–354. [CrossRef] [PubMed]
28. Miedema, H.M.E.; Ham, J.M. Odour annoyance in residential areas. *Atmos. Environ.* **1988**, *22*, 2501–2507. [CrossRef]
29. Mustafa, M.F.; Liu, Y.; Duan, Z.; Guo, H.; Xu, S.; Wang, H.; Lu, W. Volatile compounds emission and health risk assessment during composting of organic fraction of municipal solid waste. *J. Hazard Mater.* **2016**, *327*, 35–43. [CrossRef]
30. Jiang, K.; Bliss, P.J.; Schulz, T.J. The development of a sampling system for determining odor emission rates from areal surfaces: Part I. aerodynamic performance. *J. Air Waste Manag. Assoc.* **1995**, *45*, 917–922. [CrossRef]
31. Lucernoni, F.; Tapparo, F.; Capelli, L.; Sironi, S. Evaluation of an odour emission factor (OEF) to estimate odour emissions from landfill surfaces. *Atmos. Environ.* **2016**, *144*, 87–99. [CrossRef]
32. Ministry of Ecology and Environment of China (MEEC). *Air Quality-Determination of Odor-Triangle Odor Bag Method (GB 1467593)*; Ministry of Ecology and Environment of China: Beijing, China, 1993.
33. Capelli, L.; Selena, S. Combination of field inspection and dispersion modelling to estimate odour emissions from an Italian landfill. *Atmos. Environ.* **2018**, *191*, 273–290. [CrossRef]
34. Businia, V.; Capelli, L.; Sironi, S. Comparison of CALPUFF and AERMOD models for odour dispersion simulation. *Chem. Eng. Trans.* **2012**, *30*, 205–210.
35. Invernizzi, M.; Brancher, M.; Sironi, S.; Capelli, L.; Piringer, M.; Schauberger, G. Odour impact assessment by considering short-term ambient concentrations: A multi-model and two-site comparison. *Environ. Int.* **2020**, *144*, 105990. [CrossRef]
36. Bull, M.; McIntyre, A.; Hall, D. *Guidance on the Assessment of Odour for Planning v1.1*; IAQM: London, UK, 2018.
37. Scire, J.S.; Strimaitis, D.G.; Yamartino, R.J. A user's guide for the CALPUFF dispersion model. *Earth Tech. Inc.* **2000**, *521*, 1–521.
38. Rood, A. Performance evaluation of AERMOD, CALPUFF, and legacy air dispersion models using the Winter Validation Tracer Study dataset. *Atmos. Environ.* **2014**, *89*, 707–720. [CrossRef]
39. Yu, Z.; Guo, H.; Laguë, C. Development of a livestock odor dispersion model: Part II. Evaluation and validation. *J. Air Waste Manag. Assoc.* **2011**, *61*, 277–284. [CrossRef] [PubMed]
40. Wu, C.; Brancher, M.; Yang, F.; Liu, J.; Qu, C.; Schauberger, G.; Piringer, M. A comparative analysis of methods for determining odour-related separation distances around a dairy farm in Beijing, China. *Atmosphere* **2019**, *10*, 231. [CrossRef]

atmosphere

Article

Limitations of GC-QTOF-MS Technique in Identification of Odorous Compounds from Wastewater: The Application of GC-IMS as Supplement for Odor Profiling

Wei Gao [1,2], Xiaofang Yang [1,*], Xinmeng Zhu [1,2], Ruyuan Jiao [1], Shan Zhao [3], Jianwei Yu [1] and Dongsheng Wang [1,2,4,*]

[1] State Key Laboratory of Environmental Aquatic Chemistry, Research Center for Eco-Environmental Sciences, Chinese Academy of Sciences, Beijing 100085, China; weigao_st@rcees.ac.cn (W.G.); xmzhu_st@rcees.ac.cn (X.Z.); ryjiao@rcees.ac.cn (R.J.); jwyu@rcees.ac.cn (J.Y.)

[2] University of Chinese Academy of Sciences, Beijing 100049, China

[3] Research and Development Center, Beijing Drainage Group Co., Ltd., Beijing 100124, China; zhaoshan@bdc.cn

[4] School of Environmental Studies, China University of Geosciences, Wuhan 430074, China

[*] Correspondence: xfyang@rcees.ac.cn (X.Y.); wgds@rcees.ac.cn (D.W.)

Abstract: Odorous emissions from wastewater treatment plants (WWTPs) cause negative impacts on the surrounding areas and possible health risks on nearby residents. However, the efficient and reliable identification of WWTPs' odorants is still challenging. In this study, odorous volatile organic compounds (VOCs) from domestic wastewater at different processing units were profiled and identified using gas chromatography-ion mobility spectrometry (GC-IMS) and gas chromatography quadrupole-time-of-flight mass spectrometry (GC-QTOF-MS). The GC-QTOF-MS results confirmed the odor contribution of sulfur organic compounds in wastewater before primary sedimentation and ruled out the significance of most of the hydrocarbons in wastewater odor. The problems in odorous compounds analysis using GC-QTOF-MS were discussed. GC-IMS was developed for visualized analysis on composition characteristics of odorants. Varied volatile compounds were detected by GC-IMS, mainly oxygen-containing VOCs including alcohols, fatty acids, aldehydes and ketones with low odor threshold values. The fingerprint plot of IMS spectra showed the variation in VOCs' composition, indicating the changes of wastewater quality during treatment process. The GC-IMS technique may provide an efficient profiling method for the changes of inlet water and performance of treatment process at WWTPs.

Keywords: VOC; GC-QTOF-MS; GC-IMS; wastewater treatment plant

Citation: Gao, W.; Yang, X.; Zhu, X.; Jiao, R.; Zhao, S.; Yu, J.; Wang, D. Limitations of GC-QTOF-MS Technique in Identification of Odorous Compounds from Wastewater: The Application of GC-IMS as Supplement for Odor Profiling. *Atmosphere* **2021**, *12*, 265. https://doi.org/10.3390/atmos12020265

Received: 25 January 2021

Accepted: 10 February 2021

Published: 17 February 2021

Publisher's Note: MDPI stays neutral with regard to jurisdictional claims in published maps and institutional affiliations.

1. Introduction

In 2019, air pollution accounted for the largest proportion of environmental complaints in China (50.8%), of which malodor problems were up to 41%. Wastewater treatment plants (WWTPs) often received severe reprimand for the sensory nuisances and physiological risks [1,2]. Moreover, it was found that the odor of wastewater effluent may impact the quality of drinking water when released to waters that served as drinking water source [3]. Compared with the industrial odor, the odor of domestic wastewater during treatment in WWTPs is complex and variable, due to its wide sources and various biochemical reactions. As known, odorous emissions from WWTPs are made up of inorganic compound (hydrogen sulfide, ammonia) and organic compound (sulfur organic compounds, nitrogenous organic compounds, alcohols, aldehydes, terpenes, carbonyls, aromatics, fatty acids, alkanes, alkenes, ketones, esters and halogenated hydrocarbons) [4–12]. Among them, volatile sulfur organic compounds (VSOCs) are principal odorants except hydrogen sulfide, and non-sulfur volatile organic compounds (VOCs) are increasingly important to contribute to odorous emission [13]. It is worth noting that, besides the complexity of odor emissions in composition, the concentration difference of compounds is significant. In addition, there are apparent diversities of VOCs emitted from different WWTPs or treatment processes. Thus, in order to assess the pollution impact and develop odor control techniques, it is imperative to obtain the information of compounds contributing to odor and establish the odor emission profile of WWTPs.

Nowadays, odor measurement is usually carried out using sensorial analysis and chemical analysis methods [2,14]. Sensory methods [1] provide information for odor description, odor concentration [4,15,16], odor intensity [3], odor wheels [17–20] and hedonic tones [21]. Chemical analysis can provide accurately and objectively determining information of the chemical composition of odors which mainly apply gas chromatography-mass spectrometry (GC-MS) with different pre-concentration methods. Over the last decade, portable GC-MS instruments have been more widely applied in situ analysis of odor emissions. The development of this technology benefits from the invention of ambient ionization techniques, such as desorption electrospray ionization (DESI) and desorption atmospheric pressure chemical ionization (DAPCI) [22–24]. The portable instruments are especially useful for measuring parameters for mapping of air pollutants. However, the application of laboratory instruments is more widespread due to their advantages in identification ability of analytes. For instance, closed-loop stripping analysis (CLSA) combined with GC-MS has been used for investigating the time and space patterns of VOCs from wastewater treatment plants based on the list of EPA Method [25]. Headspace-solid phase microextraction (HS-SPME) combined with GC-MS has been applied to identify a wide spectrum of VOCs [26]. Meanwhile, the endeavor [27] have been done through combining GC-MS and gas chromatography-flame ionization detection and olfactometry (GC-FID/O) to characterize odorous emissions, and other studies [28–31] have also applied various modalities of GC-MS to characterize odor problems. The summary of methods used for odor determination is listed in Table 1. Most analytic modes of VOCs are using a nontargeted screening approach before determining concentration of selected compounds. In other words, scan mode is used to detect all possible compounds within sample through acquiring total ion chromatogram, then selected ion monitoring (SIM) mode is utilized as a quantitative method for targeted analysis. Although the selectivity and sensitivity of GC-MS applied in organic analysis is excellent and undisputed, there is a shortage of approach for simultaneous target quantification and nontarget screening for VOCs basing on the reported literature data. The ion separation mode of time-of-flight mass spectrometer (TOF-MS) is differed from quadrupole mass spectrometry. The fragments of all ions are pulsed into the flight tube and separated according to their different flight times. Consequently, high-sensitivity and full-spectrum acquisition data are simultaneously obtained in gas chromatography quadrupole-time-of-flight mass spectrometry (GC-QTOF-MS). It is possible to carry out target and nontarget analysis which can acquire more information in one time. Using narrow mass windows to decrease the background noise, the selectivity of QTOF-MS was highly improved and met requirements for detection. However, there is always detection scope restricting the application of analytical method. Thus, a supplementary approach should be considered, and comprehensive detection of VOCs requires multiple methods.

Gas chromatography-ion mobility spectrometry (GC-IMS) is extensively applied to investigate odor measurement/characterization. Generally, GC-IMS spectrum is presented in a three-dimensional form: first dimension is gas chromatography retention time of analytes, second dimension is the ions' drift time, and the third dimension is the signal intensity. Thus, highly resolved fingerprints of VOCs from liquid or solid samples can be obtained through GC-IMS which operated at atmospheric pressure and employed no sample pretreatment [32–34]. GC-IMS analysis has been extensively used in food and environment areas [35–39]. Attributed to its high sensitivity of small molecule compounds, application of GC-IMS may serve as a complement to GC-MS analysis.

Table 1. The number of VOCs (volatile organic compounds) detected using different methods.

References	[25]	[26]	[27]	[28]	[29]	[30]	[31]
Methods	CLSA/ GC-MS	HS-SPME/ GC-MS	SPME/ GC-MS SPME/ GC-FID/O	LGIS/ GC-MS	TSPS/ GC-MS	TD/ GC-MS	Portable GC-MS
Numbers	47	164	56	2	83	35	33

Note: CLSA: Closed-loop stripping analysis, HS-SPME: Headspace-solid phase microextraction, LGIS: Liquid-gas impinger system, TSPS: Three-stage preconcentration system, TD: Thermal desorption, GC-MS: Gas chromatography-mass spectrometry, GC-FID/O: Gas chromatography-flame ionization detection and olfactometry.

In this study, we expect to design an approach base on a different analytical platform for wide screening and characterization of odorous VOCs emitted from wastewaters. The significant differences in VOCs among different wastewater processes were obtained using GC-IMS tool. In view of the advantages of two chromatographic methods, headspace-solid phase microextraction-gas chromatography quadrupole-time-of-flight mass spectrometry (HS-SPME-GC-QTOF-MS) and headspace-gas chromatography-ion mobility spectrometry (HS-GC-IMS) are utilized and combined to provide the comprehensive information of VOCs of wastewater samples. In order to find the optimal way to analyze wastewater odorous VOCs, the experimental design has been previously done and is shown in Figure S1. A subset of VOCs included in the list of EPA Method 524.2 are selected as target compounds. These compounds are hazardous air pollutants (HAPs) that can cause environmental and human health concerns. Among them, some compounds can be perceived as odorants when they reach a certain concentration. In addition, the common odorants appeared in WWTPs are listed in Table 2 referring to the classification of literature [17]. The methodological strategy is validated and then the results are expected to provide information of potential odorants.

Table 2. Main odor substances appeared in WWTPs (wastewater treatment plants) and their descriptors.

Odor Categories	Odor Descriptor	Chemical Compounds	References
Earthy/Musty	Earthy/Musty/Moldy	Geosmin	
		2-Methylisoborneol	
		Trichloro anisole	
Oxidant/Chlorinous	Chlorinous	Monochloramine	
		Dichloramine	
Grassy/Woody	Green/Grassy/ Hay	cis-3-Hexen-1-ol	[27]
		Formaldehyde	
Sulfide/Cabbage/Garlic	Rotten Eggs/	Hydrogen sulfide	[29]
	Decaying Vegetation	Dimethyl trisulfide	[25,26]
		Dimethyl sulfide	[26,29–31]
		Thiophene	
		Methyl mercaptan	[29]
		Dimethyl disulfide	[25,26,29–31]
Rancid/Putrid	Sour Milk/	Methyl thiobutyrate	
	Putrid/	Valeric acid	
	Rancid	Isovaleric acid	
		Butyric acid	
		Heptanal	[26]
		Amyl mercaptan	
Fragrant/Fruity	Soapy/	1-Dodecanal	
	Fruity/	Acetaldehyde	[27]
	Citrusy	Ethyl acetate	[26]
		d-Limonene	[25–27,31]
Ammonia/Fishy	Ammonia/	Ammonia	[29]
	Cat Urine/	Pyridine	[26]
	Fishy	Butylamine	
		Triethylamine	
Solvency/Hydrocarbon	Solvency/ Gasoline	Benzene	[25,26,29,31]
		Toluene	[25–27,29,31]

Table 2. *Cont.*

Odor Categories	Odor Descriptor	Chemical Compounds	References
		m-Xylene	[25–27,29,31]
		Styrene	[25–27,29]
		Heptane	[25,26]
Medicinal/Alcohol	Medicinal/Alcohol	1-Hexanol	[26]
		Phenol	[26,30,31]
		1-Butanol	[26]
Fecal/Sewery	Fecal	Indole	[26,28,30]
		Skatole	[28,30]
Nose feel	Pungent/ Irritating/ Sharp	Ammonia Ozone Chlorine dioxide	

2. Materials and Methods

2.1. Chemicals and Materials

The mixing standard substances and the internal standard (IS) (1,2-dichlorobenzene-d4, 98.5% purity) were purchased from o2si (South Carolina, USA). Dimethyl disulfide (DMDS, 99%) and dimethyl trisulfide (DMTS, 99%) were obtained from J&K Scientific Co., Ltd. (Beijing, China), HPLC-gradient grade methanol and dichloromethane were from Merck (Darmstadt, Germany). Ultrapure deionized water (R $\geq$ 18.2 MΩ cm^{-1}) was produced with a Milli-Q purification system. Standard stock solutions were diluted by methanol or dichloromethane from the pure compounds in different concentrations between 1 and 10 mg/L, and stored at -20 °C. The external calibration solutions were made daily by diluting the standard solutions to the required concentration with ultrapure deionized water.

2.2. Sampling

The samples were taken from a wastewater treatment plant of Beijing (China). The background information of the WWTP was descripted in Supplementary Material and Figure S2. Wastewater samples were collected at nine sampling sites along the treatment process on 2 September 2019, including influent wastewater (IW), rotational flow grit basin inlet (GBI), rotational flow grit basin outlet (GBO), primary sedimentation tank outlet (PSO), anaerobic tank outlet (ANO), aerobic tank outlet (AO), secondary sedimentation tank outlet (SSO), nitrification outlet (NO) and denitrification outlet (DNO). Compared to odor detection on site, the application of solid phase microextraction (SPME) in laboratory was simple and convenient for identifying potential odorants emitted from wastewaters. This approach was considered to determine the odor emission capacity of wastewater, and then to predict the major odor sources in WWTPs. The samples were collected using 25 mL brown glass bottles with minimal headspace and delivered to the laboratory on the day of collection, and stored at 4 °C prior to analysis.

2.3. GC-QTOF-MS Analysis

2.3.1. SPME Conditions

The wastewater samples were subjected to SPME pre-concentration and analyzed by GC-QTOF-MS. The extraction method was carried out using SPME fibers with dual coating of divinylbenzene and carboxene (50/30 μm) suspended in polidimethylsiloxane (DVB/CAR/PDMS) made by Supelco (Bellefonte, PA, USA). First, sample bottles were filled with 10 mL samples. The pre-extraction incubation of the sample was performed in the agitation unit at 500 rpm and at 40 °C for 5 min. Then the SPME device fibers which previously had been conditioned according to manufacturer's recommendations were inserted into the headspace above the sample bottle. The extraction conditions were: stirring at 500 rpm, extraction time 20 min and extraction temperature 40 °C. Final SPME device fibers were instantly desorbed thermally for 5 min at 260 °C in an injector of gas chromatograph.

2.3.2. Chromatographic Conditions

The GC-QTOF-MS instrument comprised of 7890A gas chromatograph and 7200 QTOF mass spectrometer, equipped with an electron ionization (EI) source, obtained from Agilent (Wilmington, DE, USA). Automated recalibration of the mass axis was carried out every 6 injections. Two same columns used for compound separation were DB-5MS capillary columns (15 m $\times$ 0.25 mm $\times$ 0.25 μm), which were also acquired from Agilent. Helium was used as carrier gas at a constant flow of 1.2 mL min^{-1} and the column temperature was programed as follows: 30 °C (4 min), rated at 3 °C min^{-1} to 110 °C (2 min), and finally at 20 °C min^{-1} to 260 °C with a hold time of 1 min. Injections were made in the splitless mode. The transfer line and the EI source were set at 280 and 230 °C, respectively. The solvent delay was fixed at 3 min. The mass range was 40–350 m/z. Perfluorotributylamine was used for the daily mass calibration and as lock mass, and the m/z ion monitored was 218.9856. The TOF mass analyzer was operated in the 2 GHz mode. TOF-MS resolution was approximately 7000 (FWHM). Library search was performed using the commercial NIST library.

2.3.3. Qualitative Screening Protocol

The investigation of nontarget compounds in wastewater, a complex matrix, was hard work because there were huge amounts of peaks which interfered in identification. GC-QTOF-MS always works under full-spectrum acquisition mode at accurate mass, which could obtain more precise m/z information of sample. The potential compounds could be easily identified by using nontarget analysis. The acquired data were processed and carried out by applying the unknown analysis, a module of Agilent MassHunter software, which automatically investigated the presence of nontarget compounds in samples. The identification criterion was the presence of five m/z ions at the expected retention time and at accurate mass (five peaks could be observed in the extracted ion chromatograms). Additionally, contrast mass spectrum between standard and sample were obtained by searching in the library which further confirmed the presence of potential compounds. Next step, the compounds considered as relevant odorant could be added to the list of target analytes, and then these would be quantified by using quantitative method.

2.3.4. Quantitative Method Performance

The presence of nontarget compounds was investigated prior to target method, then the target analysis was subsequently performed, which did not require reanalyzing the samples. The calibration standards of 50, 100, 200, 500 and 1000 ng/L were prepared by dilutions of the mixing standard with ultrapure deionized water and added internal standards (IS) (500 ng/L) to generate the calibration curve. The quantification was based on five-point external calibration curve obtained by plotting the peak areas against the concentration of the corresponding standards. The MassHunter software (Agilent Technologies Inc., Wilmington, DE, USA) automatically processed data and reported quantitative results. The developed method was validated using standard solutions and real samples to evaluate the precision, accuracy, linearity, selectivity, limits of detection (LODs) and limits of quantification (LOQs).

2.4. GC-IMS Analysis

2.4.1. Instrument Parameters

Analyses of GC-IMS were performed on a FlavourSpec GC-IMS from Gesellschaft für Analytische Sensorysteme mbH (G.A.S., Dortmund, Germany). A gas chromatographic column was FS-SE-54-CB-1(5% phenyl, 1%vinyl, 94% methyl polysiloxane) capillary column (15 m $\times$ 0.53 mm, 1 μm film thickness). Nitrogen of 99.99% purity was used as a carrier gas at a programmed flow as follows: 2 mL min^{-1} for 2 min, 20 mL min^{-1} for 8 min, 100 mL min^{-1} for 10 min, 150 mL min^{-1} for 5 min. After injection, the carrier gas was passed through the injector in order to drive the sample into the column, which was kept at 60 °C for timely separation. The total GC runtime was 30 min. Data were acquired in the positive ion mode. Data were viewed and processed by using the software LAV (version 2.2.1) from G.A.S. The identification of VOCs was based on comparing RI and the drift time with the GC-IMS library. For analysis, 2 g of sample was placed in a 20 mL vial that was then closed with magnetic caps. After 20 min incubation at 65 °C, 500 μL of sample headspace was automatically injected into the heated injector (85 °C) of the device by means of a headspace autosampler from CTC Analytics (Zwingen, Switzerland). In order to exclude random errors of the system, each sample was determined twice in parallel.

2.4.2. Data Analysis

Contrary to targeted analysis of GC-QTOF-MS, GC-IMS analysis was similar to nontarget analysis according to plots which represented signal peaks and signal peak intensities of VOCs. Then identification of VOCs was processed by searching in the GC-IMS library. There are two factors influencing the identification in GC-IMS. Firstly, pressure and temperature in the drift tube were mainly factors that could influence the drift time of analyzed ions. In order to avoid deviations between measurements, the drift time of sample spectra was normalized relative to RIP drift time, which proceed automatically in the LAV software. Secondly, due to the difference of measurement conditions, the GC-IMS spectra need be calibrated by the standard substances before library searching. All standard compounds listed in Table S1, and the GC-IMS chromatogram of mixing standard were presented in Figure S3. For each standard, a monomer and a dimer signal (even a trimer) could be observed due to the relatively high concentration. As shown in Figure S3, each component could be detected more than one signal. The strongest signal (rightmost one) was used for calibration.

3. Results and Discussion

3.1. GC-QTOF-MS

3.1.1. Nontarget Analysis

In total, approximately 40 kinds of VOCs were identified using screening protocol, all of which had match factor (MF) greater than 70 (Table S2). It can be seen that the identified compounds mainly included sulfur organic compounds, nitrogenous organic compounds, benzenes, terpenes, carbonyls, aromatics, and halogenated compounds. Some (toluene, chlorobenzene, 1,4-dichlorobenzene, p-cymene, and naphthalene) were assessed as health indicators and already included in list of target compounds. Due to the unpleasant odor description and low odor threshold concentrations (OTCs), VSOCs have previously reported and demonstrated as significant contributor to sewer [40] and sewage sludge composting plants [41]. There were two identified VSOCs (dimethyl disulfide and dimethyl trisulfide) to be selected as target compounds. Figure 1 showed an example of dimethyl disulfide (DMDS) emitted from influent wastewater (IW) detected using the nontarget analysis. At the expected retention time, five representative *m/z* ions were automatically used. Figure 2 showed another example, the identification of dimethyl trisulfide (DMTS) was confirmed by the presence of 5 *m/z* ions also in IW sample. Subsequently, each standard was used to confirm their existence. Then quantitative method of single-point calibration was applied to determine the concentrations of the analytes. DMDS and DMTS were only detected in wastewater samples before PSO with the concentrations ranged from 205.21 ng/L to 689.89 ng/L and from 20.57 ng/L to 70.35 ng/L, respectively. The maximum detected concentrations of DMDS and DMTS were higher than their OTCs, which are 0.0003 mg/kg and 0.00001 mg/kg [42]. DMDS has been recognized as key odorants [43], and it is also found to be the primary odorant in another study [31]. Thus, the presence of DMDS and DMTS in the wastewater during pre-treatment could be smelled by humans and caused odor problems.

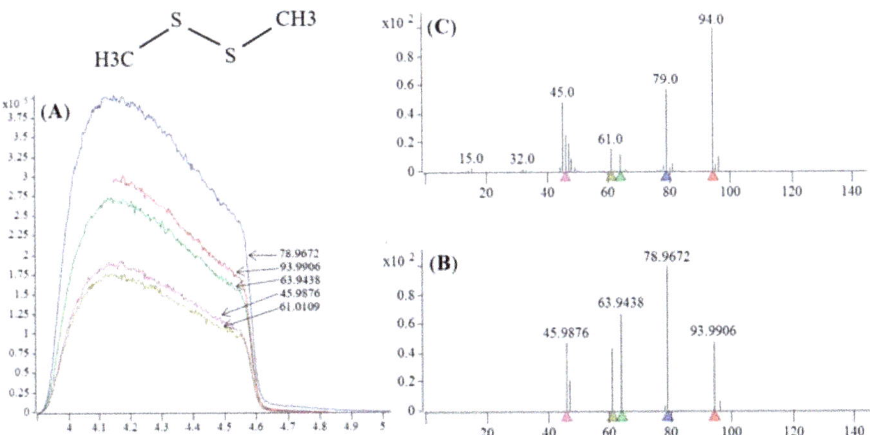

Figure 1. Identification of nontarget DMDS (dimethyl disulfide) by GC-QTOF-MS (gas chromatography quadrupole-time-of-flight mass spectrometry): (**A**) Extracted-ion chromatograms for five m/z ions; (**B**) Deconvoluted accurate mass spectrum in the sample; (**C**) Commercial library mass spectrum of DMDS at nominal mass.

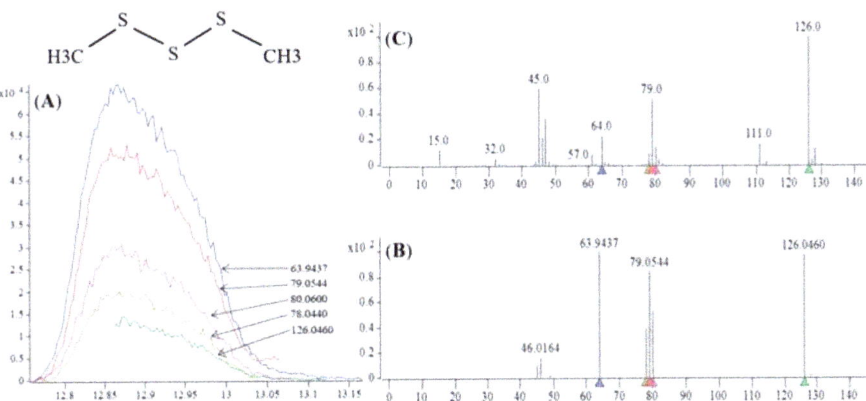

Figure 2. Identification of nontarget DMTS (dimethyl trisulfide) by GC-QTOF-MS: (**A**) Extracted-ion chromatograms for five m/z ions; (**B**) Deconvoluted accurate mass spectrum in the sample; (**C**) Commercial library mass spectrum of DMTS at nominal mass.

The mass accuracy and the ratios between the most abundant ion (Q) and every one of the other measured ions (qi) (Q/qi ratios) are the common problems in nontarget analysis. For example, Q/qi ratios did not accord with that of library mass spectrum, which might be ascribed to the complexity of wastewater matrix. The matrix complexity of environmental samples often leads to the determination errors, thus certain deviations in mass accuracy should be allowed for environmental samples [44].

3.1.2. Target Analysis

The quantitative curve used for target analysis was established using the Agilent MassHunter (QQQ) software. The identification of target analytes in standard substances was carried out by obtaining up to two micro-window extracted ion chromatograms (quantitative and qualitative ions) at selected m/z ions for every compound. Thirty-two VOCs were selected as target compounds in this study, including halogenated hydrocarbons and aromatic hydrocarbons. The accurate m/z of the quantitative and qualitative ions utilized in the method was determined by analyzing the standard substances. The most abundant fragment ion for each compound was selected as the quantitative ion, and typically the second most abundant ion was chosen as the qualitative ion. All the target analytes had unique quantitative and qualitative ions, and all of them were separated during chromatography except for trans- and cis-1,3-dichloropropylene, which co-eluted at 4.024 min, o-, m- and p-xylene, which co-eluted at 8.35 min, and 1,3,5- and 1,2,4-trimethylbenzene, which co-eluted at 12.946 min. These co-eluted compounds were, respectively, constitutional isomers and analyzed together in this method. The information of the target analytes was summarized in Table S3. The results of precision, accuracy, linearity, selectivity, limits of detection (LODs) and limits of quantification (LOQs) are shown in Table S4.

Precision

Intra-day repeatability of the method was assessed using mixing standard at two different levels ($n = 5$, each level) on the same day. Precision was calculated using the mean, standard deviation, and relative standard deviation (RSD, %). The results, expressed as relative standard deviations (RSD, %), were <15%, while RSD values were below 10% for all target VOCs at 50 ng/L level. The method displayed RSDs between 0.81% and 11.11%, confirming its excellent precision.

Accuracy

The accuracy was the percentage recovery of a known amount of target analyte added to the sample, which was presented as the mean recovery of the analyte from the spiked matrix. Both spiked and non-spiked samples were further subjected to the sample preparation procedure described above, and spiked sample was spiked with two different concentrations (50 ng/L and 500 ng/L). The peak areas of the analyte from the spiked and non-spiked samples were used to calculate the concentration of each analyte. Influent wastewater was used in recovery test, in which spiked and non-spiked samples were measured five times in parallel to calculate the recovery rates and RSDs. The results of recovery test were shown in Table S4. The results (recovery rates of 70.11–115.05% and RSDs of 1.08–17.71%) showed accuracy was found to be satisfactory with quantitative requirements.

Selectivity and Linearity

In this study, selectivity was assessed by injecting five blank samples into GC-QTOF-MS by the above SPME procedure. No compound was detectable that could interfere in the identification and quantitation of the target analytes. Due to excellent mass identification power (narrow mass windows) in QTOF-MS, it was possible to identify in case of co-elution and operate trace-level target analysis. Calibration standards (standard concentrations of 50, 100, 200, 500 and 1000 ng/L in triplicate) were prepared to set up calibration equation and evaluate the linearity of the present method. The present method displayed good linearity ($r^2 > 0.99$) over the concentration range of 50–1000 ng/L for each analyte. The linear range of the proposed method was investigated for 32 target VOCs. As seen in Table S4, the linear range was not the same for all VOCs. The reason could be the competition between target compounds in the SPME equilibrium process and the adsorption saturation of SPME.

Limits of Detection (LODs) and Limits of Quantification (LOQs)

The limit of detection (LOD) was estimated by evaluating the signal-to-noise (S/N) ratio for low concentration (50 ng/L) standards injected into the instrument ($n = 5$), and 3 times the baseline noise was used as signal for LOD calculation. The LOD values ranged from 0.2 ng/L to 50 ng/L. The LOQ was calculated by 10 times the baseline noise.

Analysis of Samples

The quantitative method was applied to the wastewater samples which were collected from nine sampling sites of WWTP. There was probability that identified compounds were not detected in target analysis due to their concentrations being lower than the method detection limits. The analytical results are listed in Table S5 in Supplementary Information. Many of the compounds detected in this study have previously been reported in other works [13,29]. The frequently detected compounds included tetrachloroethylene, chlorobenzene, ethylbenzene, 1,4-dichlorobenzene, p-cymene, naphthalene. The most abundant component was naphthalene, and it was present in each wastewater sample with concentrations ranging from 31.16 to 468.16 ng/L. The integral characteristic distribution (quantitative results) of detected compounds had been displayed in heat-map (Figure 3). Other compounds with similar concentration were 1,4-dichlorobenzene and p-cymene. However, the results indicated that tetrachloroethylene, chlorobenzene and ethylbenzene were present only in trace amounts, far lower than the above compounds. The OTCs of tetrachloroethylene (0.24 mg/kg), chlorobenzene (0.08 mg/kg), ethylbenzene (2.21 mg/kg), 1,4-dichlorobenzene (0.018 mg/kg), p-cymene (1 mg/kg), naphthalene (0.006 mg/kg) had been reported [42]. Although the concentration of these frequently detected aromatics in the emission from wastewater treatment were lower than their OTCs and could not cause odor problems, their presence has important influence on the overall odor characteristics due to the synergistic effect [13]. In addition, aromatics such as benzene, m-xylene, p-xylene, dimethylbenzene, dichloromethane, toluene, chlorobenzene, and ethylbenzene are selected as health indicators according to Directory of National Environmental Health Risks of China [29]. Thus, the contribution of these compounds to the odor and health risks should not be ignored. In addition, some common potential odorous compounds, such as aldehydes, ketones and fatty acids were not detected. Given the deficiency of GC-QTOF-MS system, we cannot confirm whether there were other substances that contributed to the wastewater odor in this study. Furthermore, the GC-QTOF-MS analysis was rather time consuming to obtain comprehensive information on odorants, especially for diverse wastewaters. Therefore, to further understand the profiles of wastewater odorous compounds, we performed GC-IMS to qualitatively analyze the wastewater samples.

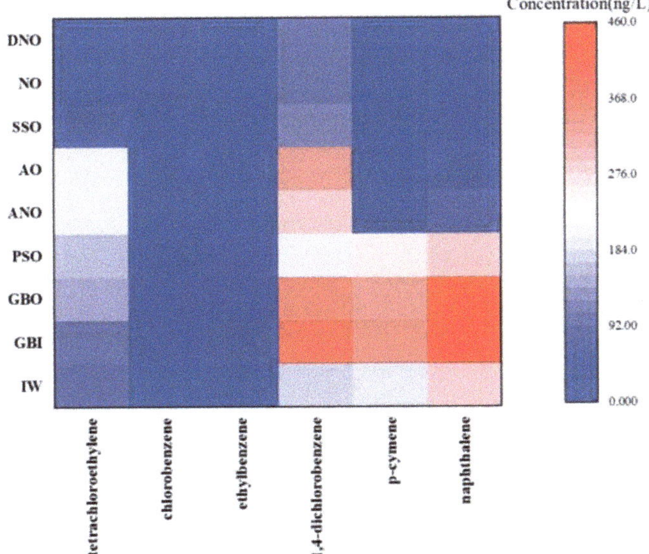

Figure 3. Heat-map visualization of the concentrations for the detected VOCs in each process. (Note: IW: influent wastewater, GBI: rotational flow grit basin inlet, GBO: rotational flow grit basin outlet, PSO: primary sedimentation tank outlet, ANO: anaerobic tank outlet, AO: aerobic tank outlet, SSO: secondary sedimentation tank outlet, NO: nitrification outlet, DNO: denitrification outlet.).

3.2. GC-IMS

The GC-IMS analysis is an analytical method based on the fingerprint in which each composition was represented by a point described by the retention time (measured in seconds, on the y axis), the drift time (measured in milliseconds, on the x axis) and the current signal (measured in millivolts). The signal peaks were represented by colored spots against the blue background in which yellow color to red color represented low intensity peaks to high intensity peaks.

3.2.1. Profile Analysis

As in previous papers, the GC-IMS analysis was generally developed as the fingerprint-based, nontargeted analytical approach for discriminating samples. Therefore, the whole fingerprints were applied to compare the differences in VOCs between samples. The variations of VOCs along wastewater treatment process were assessed by comparing topographic plots of the four groups (IW, PSO, AO and DNO). The profile information was obtained from the peak intensity for all compounds (Figure 4). As seen from Figure 4, the majority of signals located at the zone from the retention time of 100 s–400 s and the drift times of 1 ms–1.5 ms. The retention behavior of analytes was related to boiling point and polarity, so the compounds at the retention time from 100 s to 400 s were considered to be the polar compounds with lower boiling point. In view of signal nature and quantities, the overall VOC's profile of each process stage was different, and the removal efficiency of VOCs was significantly obtained. Comparing two stages (IW and PSO), it was seen that only differences of VOCs in signal intensities were observed. In fact, it was understandable that the stages before PSO were physical processes and would be insufficient to cause distinct transformation of compounds. After PSO, some signals at retention times around 140 s disappeared or the signal intensities were weakened (the area was marked in the red box, Figure 4), which demonstrated those compounds were absorbed or degraded by microorganisms during anaerobic and aerobic conditions in the bioreactors. On the contrary, some signal intensities were enhanced (the area was marked in the green box, Figure 4) which were attributed to the chemical reactions during biochemical process.

3.2.2. Identification Analysis

GC-IMS has already been used for visualizing the composition of odor emissions from wastewater treatment plants, however, the identification and quantification of the compounds have not been carried out [45]. In this work, about 30 signals were found for total samples and identification analysis of odorants was operated. The identification was achieved using IMS library of the LAV software by comparison of corresponding drift time and retention time of the analytes. The results showed that there were substances which could not be identified by IMS library. It was necessary to enrich the GC-IMS database so as to expand qualitative scope. Even so, we could use GC-IMS library to accurately identify 17 substances which were listed in Table S6. The major odorants identified by GC-IMS are alcohols, organic acids, aldehydes and ketones which have been identified in previous study [26]. Some of the compounds such as acetone, acetic acid, butanone, butanol, pentanal, heptanal, and isovaleric acid have also been presented in an odor wheel which is compiled based on characteristics of the odorants and detection reported in the literatures [17,20]. All of them were selected to develop the gallery plot for characterizing odorous profile. Obvious discrepancy among the wastewater samples was clearly observed (Figure 5). Among the identified compounds, five compounds including acetone, 2,3-pentanedione, 3-methyl-1-butanol, ethyl propanoate and methional were not detectable in the samples after PSO treatment unit. Compounds including 2-butanone, 1-propene-3-methylthio, 3-hydroxy-2-butanone and 3-methylbutanal only appeared in the aerobic biological treatment units (ANO and AO). Isovaleric acid and pentanal were present in all the samples, and pentanal showed stronger signal intensity before PSO than after PSO while the case of isovaleric acid was inverse. According to the gallery plot and odor description of compounds, the odor characteristics of the samples at different treatment stages might be inferred. The OTCs of acetone (0.83 mg/kg), 3-methyl-1-butanol (0.004 mg/kg) and isovaleric acid (0.49 mg/kg) had been reported [42], and they were often associated with odor contribution. Acetone (pungent) and 3-methyl-1-butanol (pungent) were only detected in samples before PSO treatment, and isovaleric acid (rancid/acid) was present at higher level after PSO. Thus, their presence might explain the odor character of the corresponding wastewaters. It is worth mentioning that the formation of dimers or trimers of compound occurred in IMS and related to the compounds with high proton affinity or higher concentration. The variation in the signal intensity of acetic acid confirmed the conversion between monomers and dimers. The presence of acetic acid-dimer before PSO manifested that its concentration was probably higher than that after PSO, thus acetic acid was detected as monomer after PSO.

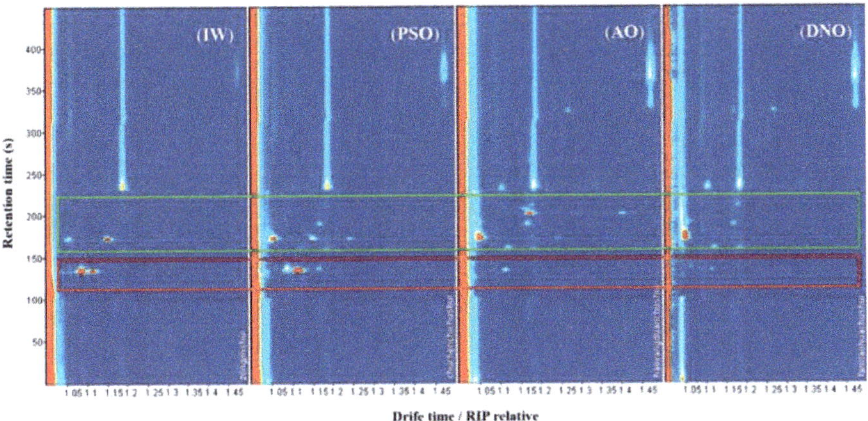

Figure 4. Topographic plots of GC-IMS (gas chromatography-ion mobility spectrometry) spectra of IW (influent wastewater), PSO (primary sedimentation tank outlet), AO (aerobic tank outlet) and DNO (denitrification outlet).

For general domestic WWTPs, the treatment processes mainly include mechanical pre-treatment and biological treatments. The mechanical pre-treatment such as bar screens, grit chamber and primary sedimentation tanks, are typically identified as locations of intense odor emissions. In contrast, the odor concentrations in biological treatments (anaerobic, aerobic tanks and secondary sedimentation) are much lower normally [46,47]. Mechanical treatment will promote the emission of odors by virtue of water flow turbulence at each unit. This is consistent with the analytical results that 1,4-dichlorobenzene, p-cymene, naphthalene, acetone, 2,3-pentanedione, 3-methyl-1-butanol, ethyl propanoate and me-thional are present at higher concentrations before anaerobic tank. New compounds are emitted during the process of biological treatment mostly by means of decomposition of the original components of the sewage or microbial metabolism, such as 2-butanone, 1-propene-3-methylthio, 3-hydroxy-2-butanone and 3-methylbutanal. The above results reinforce the general opinion that the discrepancies of odor emissions are associated with water quality and different treatment units [1].

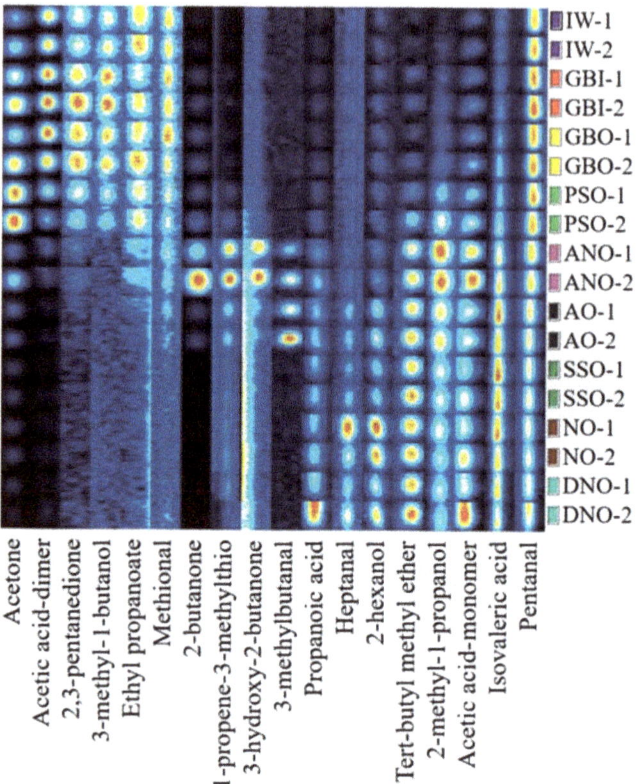

Figure 5. Gallery plot of GC-IMS for VOCs identified in the wastewater samples.

4. Conclusions

A method for wide-scope screening of VOCs in wastewater has been developed, which combine GC-QTOF-MS and GC-IMS. GC-QTOF-MS analysis includes qualitative screening of nontarget compounds and quantitative determination of target compounds, while GC-IMS provides visualized profile information and nontarget identification results. Method validation has been made through analyzing wastewater from nine process units in a municipal wastewater treatment plant. Our results show that 56 substances are identified, which mainly include aromatics (chlorobenzene, toluene, phenol, 2-chloro-phenol, and p-cymene), sulfur organic compounds (DMDS and DMTS), and oxygenated VOCs (alcohols, aldehydes, ketones, and fatty acids). The developed method enables quantification of target compounds in different wastewater samples and recoveries within the range of 70.11–115.05% are obtained with RSDs below or equal to 20%. The determined compounds include tetrachloroethylene, chlorobenzene, ethylbenzene, 1,4-dichlorobenzene, p-cymene, naphthalene, DMDS and DMTS. The presence of these compounds is in various concentration levels. The odorants emitted from wastewater mainly release before primary sedimentation tank and sulfur organic compounds are identified as major odorants. Due to the different response ranges of GC-QTOF-MS and GC-IMS, the results of two methods are not comparable to each other. Instead, GC-IMS can be used as a complementary method for GC-QTOF-MS to achieve more comprehensive analysis. In conclusion, multiple analytical methods should be used for determining odor characteristics in order to assess odor impact and choose odor control techniques.

Supplementary Materials: The following are available online at https://www.mdpi.com/2073-4433/12/2/265/s1, Figure S1: The experimental design of analytical method applied for wastewater samples, Figure S2: Sample sites at municipal wastewater treatment plants, Figure S3: The GC-IMS chromatogram of mixing standards, Table S1: The information of standard compounds used for GC-IMS calibration, Table S2: Information of identified compounds in GC-QTOF-MS, Table S3: Target compounds, CAS number, boiling point, retention index (RI), retention time (RT), quantitative ion (Ti) and qualitative ion (Qi) used for detection of targeted compounds, Table S4: Precision (intra-day) expressed as RSD (%), and accuracy expressed as recovery (%), for the target VOCs. The correlation coefficient, linear range, LOD (ng/L) and LOQ (ng/L) of the method for determining the concentration of VOCs, Table S5: The quantitative results of wastewater samples (ng/L), Table S6: Information of identified compounds in GC-IMS.

Author Contributions: Conceptualization, W.G. and X.Y.; methodology, W.G.; validation, W.G. and X.Z.; resources, R.J. and S.Z.; writing—original draft preparation, W.G.; writing—review and editing, W.G. and X.Y.; supervision, J.Y. and D.W. All authors have read and agreed to the published version of the manuscript.

Funding: This research was funded by the special fund from the State Key Joint Laboratory of Environment Simulation and Pollution Control (Research Center for Eco-environmental Sciences, Chinese Academy of Sciences), project number 19Z01ESPCR and the National Natural Science Foundation of China, grant number 52070186.

Institutional Review Board Statement: Not applicable.

Informed Consent Statement: Not applicable.

Data Availability Statement: Data is contained within the article.

Conflicts of Interest: The authors declare no conflict of interest.

References

1. Gostelow, P.; Parsons, S.A.; Stuetz, R.M. Odour Measurements for Sewage Treatment Works. *Water Res.* **2001**, *35*, 579–597. [CrossRef]
2. Lebrero, R.; Bouchy, L.; Stuetz, R.; Muñoz, R. Odor Assessment and Management in Wastewater Treatment Plants: A Review. *Crit. Rev. Environ. Sci. Technol.* **2011**, *41*, 915–950. [CrossRef]
3. Agus, E.; Zhang, L.; Sedlak, D.L. A framework for identifying characteristic odor compounds in municipal wastewater effluent. *Water Res.* **2012**, *46*, 5970–5980. [CrossRef]
4. Dincer, F.; Muezzinoglu, A. Odor-causing volatile organic compounds in wastewater treatment plant units and sludge management areas. *J. Environ. Sci. Health Part A* **2008**, *43*, 1569–1574. [CrossRef]
5. Hudson, N.; Ayoko, G.A. Odour sampling. 2. Comparison of physical and aerodynamic characteristics of sampling devices: A review. *Bioresour. Technol.* **2008**, *99*, 3993–4007. [CrossRef]
6. Sekyiamah, K.; Kim, H.; McConnell, L.L.; Torrents, A.; Ramirez, M. Identification of Seasonal Variations in Volatile Sulfur Compound Formation and Release from the Secondary Treatment System at a Large Wastewater Treatment Plant. *Water Environ. Res.* **2008**, *80*, 2261–2267. [CrossRef] [PubMed]
7. Zarra, T.; Naddeo, V.; Belgiorno, V.; Reiser, M.; Kranert, M. Odour monitoring of small wastewater treatment plant located in sensitive environment. *Water Sci. Technol.-Wst* **2008**, *58*, 89–94. [CrossRef] [PubMed]
8. Karageorgos, P.; Latos, M.; Kotsifaki, C.; Lazaridis, M.; Kalogerakis, N. Treatment of unpleasant odors in municipal wastewater treatment plants. *Water Sci. Technol.-Wst* **2010**, *61*, 2635–2644. [CrossRef]
9. Lehtinen, J.; Veijanen, A. Odour Monitoring by Combined TD–GC–MS–Sniff Technique and Dynamic Olfactometry at the Wastewater Treatment Plant of Low H_2S Concentration. *Water Air Soil Pollut.* **2010**, *218*, 185–196. [CrossRef]
10. Olujimi, O.O.; Fatoki, O.S.; Odendaal, J.P.; Daso, A.P. Chemical monitoring and temporal variation in levels of endocrine disrupting chemicals (priority phenols and phthalate esters) from selected wastewater treatment plant and freshwater systems in Republic of South Africa. *Microchem. J.* **2012**, *101*, 11–23. [CrossRef]
11. Skoczko, I.; Struk-Sokolowska, J.; Ofman, P. Seasonal Changes in Nirtogen, Phosphorus, BOD AND COD Removal in Bystre Wastewater Treatment Plant. *J. Ecol. Eng.* **2017**, *18*, 185–191. [CrossRef]
12. Skoczko, I.; Struk-Sokolowska, J.; Ofman, P. Modelling Changes in the Parameters of Treated Sewage Using Artificial Neural Networks. *Rocz. Ochr. Srodowiska* **2017**, *19*, 633–650.
13. Fisher, R.M.; Le-Minh, N.; Sivret, E.C.; Alvarez-Gaitan, J.P.; Moore, S.J.; Stuetz, R.M. Distribution and sensorial relevance of volatile organic compounds emitted throughout wastewater biosolids processing. *Sci. Total Environ.* **2017**, *599–600*, 663–670. [CrossRef]
14. Jiang, G.M.; Melder, D.; Keller, J.; Yuan, Z.G. Odor emissions from domestic wastewater: A review. *Crit. Rev. Environ. Sci. Technol.* **2017**, *47*, 1581–1611. [CrossRef]

15. Lambert, S.D.; Beaman, A.L.; Winter, P. Olfactometric characterisation of sludge odours. *Water Sci. Technol.* **2000**, *41*, 49–55. [CrossRef]
16. Dincer, F.; Muezzinoglu, A. Odor Determination at Wastewater Collection Systems: Olfactometry versus H_2S Analyses. *Clean-Soil Air Water* **2007**, *35*, 565–570. [CrossRef]
17. Burlingame, G.A.; Suffet, I.H.; Khiari, D.; Bruchet, A.L. Development of an odor wheel classification scheme for wastewater. *Water Sci. Technol.* **2004**, *49*, 201–209. [CrossRef] [PubMed]
18. Suffet, I.H.; Burlingame, G.A.; Rosenfeld, P.E.; Bruchet, A. The value of an odor-quality-wheel classification scheme for wastewater treatment plants. *Water Sci. Technol.* **2004**, *50*, 25–32. [CrossRef]
19. Decottignies, V.; Huyard, A.; Kelly, R.F.; Barillon, B. Development of a diagnostic tool: The wastewater collection network odour wheel. *Water Sci. Technol.* **2013**, *68*, 839–847. [CrossRef] [PubMed]
20. Fisher, R.M.; Barczak, R.J.; Suffet, I.H.M.; Hayes, J.E.; Stuetz, R.M. Framework for the use of odour wheels to manage odours throughout wastewater biosolids processing. *Sci. Total Environ.* **2018**, *634*, 214–223. [CrossRef]
21. Koch, E.; Winneke, G.; Sucker, K.; Both, R. Odour intensity and hedonic tone - important parameters to describe odour annoyance to residents? *Water Sci. Technol.* **2004**, *50*, 83–92. [CrossRef]
22. Jarmusch, A.K.; Kerian, K.S.; Pirro, V.; Peat, T.; Thompson, C.A.; Ramos-Vara, J.A.; Childress, M.O.; Cooks, R.G. Characteristic lipid profiles of canine non-Hodgkin's lymphoma from surgical biopsy tissue sections and fine needle aspirate smears by desorption electrospray ionization-mass spectrometry. *Analyst* **2015**, *140*, 6321–6329. [CrossRef]
23. Jjunju, F.P.M.; Maher, S.; Li, A.; Badu-Tawiah, A.K.; Taylor, S.; Cooks, R.G. Analysis of Polycyclic Aromatic Hydrocarbons Using Desorption Atmospheric Pressure Chemical Ionization Coupled to a Portable Mass Spectrometer. *J. Am. Soc. Mass Spectrom.* **2015**, *26*, 271–280. [CrossRef]
24. Fedick, P.W.; Pu, F.; Morato, N.M.; Cooks, R.G. Identification and Confirmation of Fentanyls on Paper using Portable Surface Enhanced Raman Spectroscopy and Paper Spray Ionization Mass Spectrometry. *J. Am. Soc. Mass Spectrom.* **2020**, *31*, 735–741. [CrossRef] [PubMed]
25. Escalas, A.; Guadayol, J.M.; Cortina, M.; Rivera, J.; Caixach, J. Time and space patterns of volatile organic compounds in a sewage treatment plant. *Water Res.* **2003**, *37*, 3913–3920. [CrossRef]
26. Kotowska, U.; Zalikowski, M.; Isidorov, V.A. HS-SPME/GC-MS analysis of volatile and semi-volatile organic compounds emitted from municipal sewage sludge. *Environ. Monit. Assess.* **2012**, *184*, 2893–2907. [CrossRef]
27. Kleeberg, K.K.; Liu, Y.; Jans, M.; Schlegelmilch, M.; Streese, J.; Stegmann, R. Development of a simple and sensitive method for the characterization of odorous waste gas emissions by means of solid-phase microextraction (SPME) and GC-MS/olfactometry. *Waste Manag.* **2005**, *25*, 872–879. [CrossRef] [PubMed]
28. Zhou, Y.; Hallis, S.A.; Vitko, T.; Suffet, I.H. Identification, quantification and treatment of fecal odors released into the air at two wastewater treatment plants. *J. Environ. Manag.* **2016**, *180*, 257–263. [CrossRef]
29. Zhu, Y.L.; Zheng, G.D.; Gao, D.; Chen, T.B.; Wu, F.K.; Niu, M.J.; Zou, K.H. Odor composition analysis and odor indicator selection during sewage sludge composting. *J. Air Waste Manag. Assoc.* **2016**, *66*, 930–940. [CrossRef]
30. Gonzalez, D.; Colon, J.; Sanchez, A.; Gabriel, D. A systematic study on the VOCs characterization and odour emissions in a full-scale sewage sludge composting plant. *J. Hazard. Mater.* **2019**, *373*, 733–740. [CrossRef]
31. Nie, E.; Zheng, G.D.; Gao, D.; Chen, T.B.; Yang, J.X.; Wang, Y.W.; Wang, X.K. Emission characteristics of VOCs and potential ozone formation from a full-scale sewage sludge composting plant. *Sci. Total Environ.* **2019**, *659*, 664–672. [CrossRef]
32. Kanu, A.B.; Hill, H.H., Jr. Ion mobility spectrometry detection for gas chromatography. *J. Chromatogr. A* **2008**, *1177*, 12–27. [CrossRef] [PubMed]
33. Armenta, S.; Alcala, M.; Blanco, M. A review of recent, unconventional applications of ion mobility spectrometry (IMS). *Anal. Chim. Acta* **2011**, *703*, 114–123. [CrossRef]
34. Vautz, W.; Franzke, J.; Zampolli, S.; Elmi, I.; Liedtke, S. On the potential of ion mobility spectrometry coupled to GC pre-separation-A tutorial. *Anal. Chim. Acta* **2018**, *1024*, 52–64. [CrossRef] [PubMed]
35. Garrido-Delgado, R.; Mercader-Trejo, F.; Sielemann, S.; de Bruyn, W.; Arce, L.; Valcarcel, M. Direct classification of olive oils by using two types of ion mobility spectrometers. *Anal. Chim. Acta* **2011**, *696*, 108–115. [CrossRef] [PubMed]
36. Garrido-Delgado, R.; Arce, L.; Valcarcel, M. Multi-capillary column-ion mobility spectrometry: A potential screening system to differentiate virgin olive oils. *Anal. Bioanal. Chem.* **2012**, *402*, 489–498. [CrossRef]
37. Garrido-Delgado, R.; Dobao-Prieto Mdel, M.; Arce, L.; Valcarcel, M. Determination of volatile compounds by GC-IMS to assign the quality of virgin olive oil. *Food Chem.* **2015**, *187*, 572–579. [CrossRef]
38. Junger, M.; Bodeker, B.; Baumbach, J.I. Peak assignment in multi-capillary column-ion mobility spectrometry using comparative studies with gas chromatography-mass spectrometry for VOC analysis. *Anal. Bioanal. Chem.* **2010**, *396*, 471–482. [CrossRef] [PubMed]
39. Vautz, W.; Hariharan, C.; Weigend, M. Smell the change: On the potential of gas-chromatographic ion mobility spectrometry in ecosystem monitoring. *Ecol. Evol.* **2018**, *8*, 4370–4377. [CrossRef]
40. Sivret, E.C.; Wang, B.; Parcsi, G.; Stuetz, R.M. Prioritisation of odorants emitted from sewers using odour activity values. *Water Res.* **2016**, *88*, 308–321. [CrossRef]
41. Fisher, R.M.; Le-Minh, N.; Alvarez-Gaitan, J.P.; Moore, S.J.; Stuetz, R.M. Emissions of volatile sulfur compounds (VSCs) throughout wastewater biosolids processing. *Sci. Total Environ.* **2018**, *616–617*, 622–631. [CrossRef]

42. Van Gemert, L.J. *Odour Thresholds: Compilations of Odour Threshold Values in Air, Water and Other Media*; Oliemans Punter & Partners BV: Utrecht, The Netherlands, 2011.
43. Han, Z.L.; Qi, F.; Li, R.Y.; Wang, H.; Sun, D.Z. Health impact of odor from on-situ sewage sludge aerobic composting throughout different seasons and during anaerobic digestion with hydrolysis pretreatment. *Chemosphere* **2020**, *249*, 1–10. [CrossRef]
44. Portoles, T.; Pitarch, E.; Lopez, F.J.; Hernandez, F. Development and validation of a rapid and wide-scope qualitative screening method for detection and identification of organic pollutants in natural water and wastewater by gas chromatography time-of-flight mass spectrometry. *J. Chromatogr. A* **2011**, *1218*, 303–315. [CrossRef] [PubMed]
45. Vera, L.; Companioni, E.; Meacham, A.; Gygax, H. Real Time Monitoring of VOC and Odours Based on GC-IMS at Wastewater Treatment Plants. *Chem. Eng. Trans.* **2016**, *54*, 79–84. [CrossRef]
46. Lewkowska, P.; Cieslik, B.; Dymerski, T.; Konieczka, P.; Namiesnik, J. Characteristics of odors emitted from municipal wastewater treatment plant and methods for their identification and deodorization techniques. *Environ. Res.* **2016**, *151*, 573–586. [CrossRef] [PubMed]
47. Fan, F.; Xu, R.; Wang, D.; Meng, F. Application of activated sludge for odor control in wastewater treatment plants: Approaches, advances and outlooks. *Water Res.* **2020**, *181*, 115915. [CrossRef]

Article

Analysis of Separation Distances under Varying Odour Emission Rates and Meteorology: A WWTP Case Study

Marco Ravina *, Salvatore Bruzzese, Deborah Panepinto and Mariachiara Zanetti

Department of Environment, Land and Infrastructure Engineering, Turin Polytechnic, Corso Duca degli Abruzzi 24, 10129 Turin, Italy; s250847@studenti.polito.it (S.B.); deborah.panepinto@polito.it (D.P.); mariachiara.zanetti@polito.it (M.Z.)
* Correspondence: marco.ravina@polito.it; Tel.: +39-110-907-632; Fax: +39-110-907-699

Received: 23 June 2020; Accepted: 4 September 2020; Published: 10 September 2020

Abstract: A wide variability of odour impact criteria is found around the world. The objective of this research work was to evaluate the influence of the uncertainties related to some individual stages of odour impact assessment in the application of regulatory criteria. The evaluation procedure was established by following the guidelines of the Northern Italian regions. A wastewater treatment plant located in Northern Italy was considered as a case study. Odour dispersion modelling was carried out with the CALPUFF model. The study focused on two phases of the assessment. The first phase was the selection of the meteorology datasets. For low odour concentration thresholds ($C_T = 1$ OU m^{-3}), the results showed that two different years (2018 and 2019) provided similar patterns of the separation distances. The difference between the two years tended to increase by increasing the value of the concentration threshold ($C_T = 3$ OU m^{-3} and $C_T = 5$ OU m^{-3}). The second phase of the assessment was the selection of the open field correction method for wind velocity used in the calculation of odour emission rates (OERs). Three different relationships were considered: the power law, the logarithmic law and the Deaves–Harris (D–H) law. The results showed that OERs and separation distances varied depending on the selected method. Taking the power law as the reference, the average variability of the separation distances was between −7% (D–H law) and +10% (logarithmic law). Higher variability (up to 25%) was found for single transport distances. The present study provides knowledge towards a better alignment of the concept of the odour impact criteria.

Keywords: odour; dispersion modelling; wastewater treatment; odour impact criteria; separation distances

---

1. Introduction

The impact of odour emission sources on sensitive receptors is a hotly debated topic in recent years. For wastewater treatment plants (WWTPs), because of their proximity to sensitive elements and their location in urban and territorial contexts, an olfactory impact evaluation strategy is required, to limit harassments on the surrounding area and to ensure the correct process management. Odour impact assessment presents multiple aspects of complexity. The scientific community agrees in recommending an integrated multi-tool assessment strategy, which supports both qualitative and quantitative analyses, atmospheric dispersion modelling, odour measurement in ambient air, population monitoring as well as the mitigation and control actions of olfactory harassment [1,2]. Odour impact assessment is carried out through the following phases: sampling, characterization, odour emission rate (OER) calculation, atmospheric dispersion modelling and impact evaluation [3].

Recently, a significant contribution to the knowledge of odour sampling methods and tools was brought by research, deepening all odour impact assessment stages and the representativeness requirements that the results must have. Odour sampling and characterization are two critical evaluation phases of the assessment. Their accuracy and representativeness, especially related to measuring instruments and type of analysis, strongly influence the subsequent implementation phases of the assessment [4]. Sensorial analyses are necessary for odour dispersion modelling. The European technique is dynamic olfactometry (DO), regulated by the EN13725: 2004 standard. In this method, the dilution degree necessary to reach the olfactory panel threshold is assessed. Even though the topic of uncertainty relevant to olfactometry is still debated among the scientific community, some studies are proving that the uncertainty of dynamic olfactometry can be estimated between one fourth and fourfold of an actual measurement value [5,6].

The odour emission rate calculation is mainly linked to the emission conditions within the sampling devices. Area sources are commonly sampled with hood methods. Hoods can be static or dynamic, for active and passive source sampling, respectively. The most common dynamic hoods are wind tunnels (WTs) and flow chambers (FCs).

In the atmospheric dispersion modelling phase, the results depend on the model choice and settings, but also on the quality of the input data. The type of model choice is a crucial aspect. This is linked to the features of the simulation domain and the scope of the analysis [7]. To properly evaluate these aspects, it is necessary to conduct a careful study domain analysis, related to the area orography and meteorology, but also emission sources and potentially sensitive receptors. The quality of the incoming meteorological data is also a fundamental factor. The scientific community does not agree on the simulation interval choice: some international jurisdictions prescribe a seasonal or multi-year evaluation duration, believing that it best represents the variability of the emissive and meteorological sources' conditions [8].

Finally, the interpretation of the results is based on standardized assumptions on impact evaluation. The time series of odour concentration provided by dispersion models must be evaluated by odour impact criteria (OIC). The OIC are defined by an odour concentration threshold (C_T), the exceedance probability of this threshold (p_T) and the averaging time used to predict the concentrations (A_T) [9]. If the OIC are specified for an averaging time shorter than 1 h, the peak-to-mean factor (P/M) is commonly applied. The P/M factor allows taking into account the fluctuations of the odour concentrations, which are linked to the atmospheric turbulence and the olfactory sensitivity of the human nose. Applying a P/M factor is simple, but it has a high degree of approximation compared to reality. In many countries, regulations adopt a constant factor. However, concentration fluctuations depend on multiple factors, such as the emissions variability, the source type, the atmospheric stability class and the receptor-source distance.

The results of the application of OIC with odour dispersion modelling are usually reported through the calculation of separation distances. The separation distance is intended to encompass the area within which odour annoyance can be expected, relying on a certain level of protection [10]. The definition of separation distances can be regarded as a practical approach for decision-making on odour pollution because it easily communicates for all stakeholders the area within which odour annoyance can be expected [11].

Previous studies showed that different national and local administrations adopted a wide variety of different parameter combinations [12]. In general, the preferred combinations are either low odour concentration thresholds/high exceedance probabilities or vice versa [13]. Nevertheless, the theme complexity led to different approaches and instruments, resulting in a lack of homogeneity between regulations [12]. Besides, the assessment procedures are often incomplete or lack precise information, generating variability in the results. This variability represents the object of the present study.

The objective of this research work was to evaluate the influence of the uncertainties related to some individual stages of odour impact assessment in the application of the current regulatory criteria. The study focused on two main aspects of the assessment. The first was the meteorology data used in dispersion modelling. The second was the open field correction method for wind velocity used in the calculation of OERs. A WWTP located in Northern Italy, whose odour emissions sources were measured in previous campaigns conducted in 2019, was considered as a case study. The evaluation procedure was established by following the guidelines of the Northern Italian regions, which are based on olfactometry analysis according to the European EN13725 standard [14]. Odour dispersion modelling was carried out using the CALPUFF software.

This study was structured by developing a reference impact assessment and simulation according to the Lombardy Region guidelines [15], and subsequently running alternative simulations with a modified meteorology and odour emission rate (OER) characterization. This paper is structured as follows: the methodology and a description of the reference and alternative simulations are reported in Section 2; the results are reported in Section 3; results are discussed in Section 4; and, finally, some conclusive remarks are reported in Section 5.

2. Methodology

In the present study, the odour impact assessment was based on the maximum impact standard, on which the regulations of many states (including Italy) rely on. In this approach, concentrations deriving from the dispersion modelling analysis are evaluated by applying the following odour impact criteria (OIC) [12]: odour concentration threshold, percentile compliance level, and averaging time for calculating the concentrations.

The study was structured as follows. Firstly, a reference simulation was conducted following Lombardy Region guidelines. Alternative simulations were then made, considering the same modelling domain and emission sources, and alternative simulations aimed at evaluating the impacts on the separation distances of the two factors. The first factor is the meteorology data used in dispersion modelling. Simulations were carried out for two different years (2018 and 2019) to visualize the influence of meteorology on the obtained results. The second factor that was considered is the open field correction method for wind velocity used in the calculation of OERs. To this end, simulations were repeated using three correction methods. The description of the reference and alternative simulations is reported in Table 1. The description of the studied WWTP and the adopted methodology are reported in the following sub-sections.

Table 1. Alternative simulations for the odour impact assessment of the wastewater treatment plant (WWTP) case study.

Simulation	Time Interval	Correction Method for Wind Speed
1 [a]	2019	Power Law
2	2018	Power Law
3	2019	Logarithmic Law
4	2019	Deaves–Harris Correlation

[a] Reference simulation.

2.1. Study Site and Odour Sampling

The WWTP is located in Northern Italy. The site morphology is mainly sub-flat, with a slight slope in the south-east direction towards the river near the plant border. In the eastern part of the domain there are some reliefs. The plant is surrounded by two towns, located NW and SE, respectively. The closest residential area is located 1 km to the plant boundary in the direction NW. The WWTP consists of a line for wastewater treatment and one line for sludge treatment. The wastewater line is made up of the following processes: grid screens, grit and grease removal, primary sedimentation, anoxic and aeration basins, secondary sedimentation and final filtration (Figure 1). The wastewater treatment process generates an amount of primary and secondary sludge with an average TS content of 1%, which is sent to the sludge treatment units. The sludge treatment line consists of the following units: pre-thickening, mesophilic anaerobic digestion, post-thickening and final dewatering.

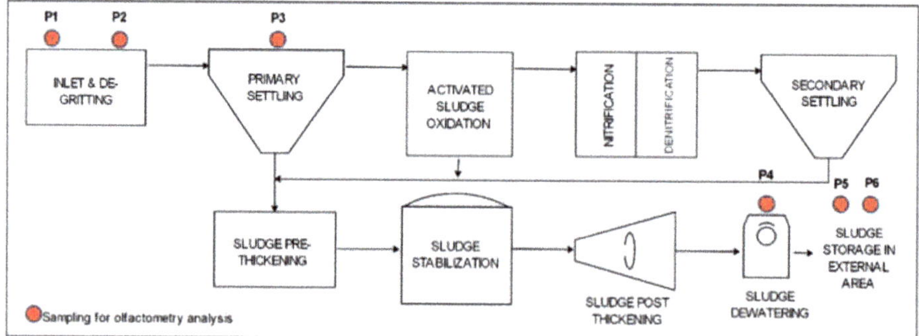

Figure 1. Scheme of the WWTP and position of the sampling for the olfactometry analysis.

A weather monitoring station was installed onsite. The station is composed of the following components:

- an ultra-sonic biaxial anemometer, installed at a height of 10 m above ground;
- a global class 2 radiometer;
- a temperature sensor PT100 1/3 DIN, with a non-vented anti-radiation shield;
- a hygrometer with a non-vented anti-radiation shield;
- a tilt-out tray pluviometer;
- a barometer;
- a Campbell CR800 data logging system.

The data logging system provides average values of the weather variables over 10-min intervals.

2.2. Odour Measurements

Odour sampling was carried out at the plant in January 2019. This work used the olfactometry analysis results to calculate the emission rates, as required by EN13725: 2004 standard [14]. Air samples from passive area sources (referred to as P2, P3, P5 and P6 in Figure 1) were collected employing a wind tunnel (WT) and following the standard procedure. The WT sampling flow was 2.5 m^3 h^{-1}, corresponding to an average velocity of 0.035 m s^{-1}. In addition to the area sources, two volumetric sources were monitored (referred to as P1 and P4 in Figure 1). The first was the plant inlet, a closed channel where the incoming wastewater is conveyed to the preliminary treatments. Some openings are present on this channel that are potential odour sources. The second was in correspondence of the sludge dehydration section, a closed building in which all the air collected in the sludge treatment line is treated with a wet scrubber. Sampling was done by placing small tubes that correspond to the building openings, and collecting the air in Nalophan bags with the use of a pump. Single samples and a replicate were collected from each source. Odour concentrations were determined in an ODOURNET TO8 olfactometer according to standard EN 13725: 2004.

2.3. OER Calculation

In the reference simulation of the present study, the OER was calculated following the indications provided by the Italian regional technical guidelines. The OER inside the WT (OER_{WT}, OU s^{-1}) was calculated from the specific odour emission rate (SOER), as follows (Equations (1) and (2)):

$$SOER = \frac{Q_{effl} \cdot C_{od}}{A_{base}} \tag{1}$$

$$OER_{WT} = SOER \cdot A_{emiss} \tag{2}$$

where $SOER$ is the specific odour emission rate (OU m^{-2} s^{-1}), Q_{effl} is the effluent volumetric flow rate leaving the hood (m^3 s^{-1}), A_{base} is the instrument base area (m^2) and A_{emiss} is the emission source area (m^2). The values of the parameters of the WT are reported in Table 2. Values of the OER_{WT} for each of the area sources are reported in Table 3.

Table 2. Parameters of the wind tunnel (WT).

Parameter	Description	Value	Unit of Measure
Q_{effl}	Air flow rate	2.5	m$^3 \cdot$h^{-1}
A_{base}	WT area	0.125	m^2
v_0	Sweep air velocity	0.035	m$\cdot$s^{-1}

Table 3. Values of the odour emission rate (OER) inside the WT (OER_{WT}) for the passive area sources.

ID	Plant Area	C_{od} (OU m^{-3})	A_{emiss} (m^2)	OER_{WT} (OU s^{-1})
P2	Grit removal	180	2880	2880
P3	Primary settler	540	16,989 (4 modules of 4247 m^2 each)	12,742
P5	Sand external storage	1100	945	5768
P6	Stabilized sludge external storage	3600	3168	63,360

To calculate the hourly values of the OER, OER_{WT} was corrected to account for open field conditions [16]. The main uncertainty of this phase is linked to the significance of the relationship that allows obtaining the emissive flow in the open field starting from that recorded within the dynamic hoods [17]. In the Italian regulations, a correlation dependent on the relationship between the actual wind speed at the dynamic hood height (v_1) and that of the air flowed inside (v_0) is proposed according to Equation (3):

$$OER = OER_{WT} \cdot \left(\frac{v_1}{v_0}\right)^{0.5} \tag{3}$$

The dependence on the square root is a simplification since it relates to the laminar flow condition on a flat surface. Some authors [18] proposed a modification to the equation to improve how the dependence between the emission speed and the wind speed in turbulent conditions is expressed (Equation (4)):

$$OER = OER_{WT} \cdot \left(\frac{v_1}{v_0^*}\right)^{0.78} \tag{4}$$

In addition to changing the proportionality factor, the correlation contained a corrected value of the wind speed inside the hood (v_0^*), taking into account the geometric characteristics of the device.

The OER estimation is closely related to the wind speed profile adopted to obtain v_1 at the flow hood height. In this study, three different wind profile models were evaluated: the power law, the logarithmic law (log law) and the Deaves–Harris correlation (D–H law). The power law does not require the knowledge of complex meteorological data, but the stability class and the prevalent type of land-use class of the surface must be known, expressed through the α parameter [19] (Equation (5)):

$$v_1 = v_2 \cdot \left(\frac{h_1}{h_2}\right)^{\alpha} \tag{5}$$

where v_1 is the wind speed at the flow hood height (m s^{-1}), v_2 is the wind speed at the meteorological station height (m s^{-1}), h_1 is the flow hood height (1 m), h_2 is the meteorological station height (10 m) and α is Hellman's parameter (-). Values of α proposed by Hanna et al. [20] for rural areas were considered. These values, which depend on the atmospheric stability class, are reported in Table 4. The power law is typically valid in the 30–300 m range, but not for the upper and lower limits of the PBL. Although it is the most used, it does not provide a detailed estimate of the speed at low heights (1–10 m).

Table 4. Values of Hellman's parameter α used in Equation (5) [20].

Stability Class	$\alpha[-]$
A	0.07
B	0.07
C	0.10
D	0.15
E	0.35
F	0.55

The logarithmic law (Equation (6)) was observed to be more suitable for the velocity profile evaluation close to the ground level since it accounts for friction velocity u^* and surface roughness z_0. Furthermore, unlike the power law, it does not constitute an empirical expression as it is derived from similarity theory, according to

$$v_1 = \frac{u^*}{K_v}\left[ln\frac{h_1}{z_0} + \psi\left(\frac{h_1}{L_m}\right)\right]$$

(6)

where u^* is the friction velocity (m s^{-1}), K_v is Von Kármán's constant (0.41), z_0 is the surface roughness length (0.625 m), L_m is the Monin–Obukhov length (m) and ψ is a stability factor, related to the atmospheric stability class (-). The values of the parameters in Equation (6) were extracted by the output of the CALMET simulation. The CALMET model follows the approach introduced by Holtslag and van Ulden [21], where the calculation of u^*, L_m and ψ are differentiated depending on stable and unstable atmospheric conditions. More details can be found in the CALMET user's manual [22].

The D–H correlation, also known as the logarithmic with parabolic defect model equation, is defined as reported in Equation (7):

$$v_1 = \frac{u^*}{K_v}\left[ln\frac{h_1}{z_0} + 5.75\, ln\left(\frac{h_1}{H}\right) - 1.88ln\left(\frac{h_1}{H}\right)^2 - 1.33ln\left(\frac{h_1}{H}\right)^3 + 0.25ln\left(\frac{h_1}{H}\right)^4 \right]$$

(7)

where H is the equilibrium boundary layer height, equal to $\frac{u^*}{6f_c}$ (m); f_c is the Coriolis' parameter, equal to $2\Omega\sin(\varphi)$ (s^{-1}); Ω is the Earth rotation rate, equal to $7.2921\cdot10^{-5}$ (rad s^{-1}); and φ is the latitude (rad). The D–H law is an extension of the previously analysed laws since it includes both scale parameters u^* and z_0 (inherited from the logarithmic profile) and the PBL height parameter. For these reasons [19], the D–H law can accurately describe the entire PBL and also represent its upper and lower boundary conditions. However, the applicability of this correlation has still to be studied. Cook [19] pointed out that, in the wind speeds design range, the correspondence between the D–H and power law models is to be considered excellent.

For the volumetric sources, since it was not possible to measure the airflow rate from the building openings, the OER was calculated starting with the hourly air exchange rate (Equation (8)). According to the information provided by the plant operator, an exchange rate equal to 10 h^{-1} was assumed.

$$OER = C_{od} \cdot AER \cdot V$$

(8)

where C_{od} is the odour concentration (OU m^{-3}), AER is the air exchange rate (h^{-1}) and V is the source volume (m^3).

2.4. Odour Impact Assessment

In Italy, as well as other countries worldwide, compliance assessment takes the form of modelling. By analysing the impact maps and the extension of overcoming isopleths, the assessment must understand the measures to be taken to avoid that smell significantly impacts on the receptors.

In this study, the dispersion modelling phase was carried out using CALPUFF [23]. CALPUFF contains algorithms for modelling the following aspects: puff splitting and merging, building, stack-tip downwash effects, dry and wet deposition, wind shear, chemical transformations, partial penetration into the inversion layers and interaction with complex areas. A detailed plume rise schematization and different puff sampling ways are provided; they can adapt to wind conditions and the presence of buildings.

Technical guidelines require simulations of at least one-year duration and recommend a domain geometry choice that allows to include all potential receptors. Simulations were conducted on a square domain of 16.2 km × 16.2 km, with 10 vertical layers and a 200 m grid step. Surface meteorological data were collected by the weather monitoring station installed onsite. Wind calm conditions were not excluded for the dispersion calculation. Upper air data were collected from radiosounding measurements located at the Milano Linate Airport. Weather observations were first processed with the CALMET model. The area is characterized by mainly agricultural land use. Depletion processes (dry and wet deposition, chemical transformations) were not set up since their effect on atmospheric removal may be considered negligible at this scale of analysis [24]. The same regulatory indications suggest this choice for olfactory impact studies.

Concerning the OIC, the levels of protection are respected through three different approaches: adjust the odour concentration threshold (C_T), adjust the exceedance probability of this threshold (p_T) and introduce correction factors related to the hedonic tone of the emissions [11,25]. In some countries, C_T is assumed as a constant value, whereas p_T is varied depending on location and emission offensiveness. Other countries adopted a constant p_T and modify the C_T for adjusting the criteria to the required level of protection. Previous studies analysing the international regulatory framework indicated three general different groups: a high C_T combined with a low p_T (e.g., $C_T = 10$ OU m^{-3}; $p_T = 1\%$); a low C_T combined with a high p_T (e.g., $C_T = 1$ OU m^{-3}; $p_T = 10\%$); and a low C_T combined with a low p_T (e.g., $C_T = 1$ OU m^{-3}; $p_T = 1\%$).

In the Northern Italian regions, the maximum impact standard is based on the frequency with which a given C_T is exceeded. Three odour impact criteria must be reported in the concentration maps [15]: 1 OU m^{-3}, 3 OU m^{-3} and 5 OU m^{-3}. It is defined that 50% of the population perceives the odour at 1 OU m^{-3}; 85% of the population perceives the odour at 3 OU m^{-3}; and 90–95% of the population perceives the odour at 5 OU m^{-3}. These values were derived from the study of Nicell [26]. Analogously to many other member states of the European Union, the Italian guidelines set the 98th percentile ($p_T = 2\%$) for odour modelling [12]. The work of Sommer-Quabach et al. [13] showed that for low exceedance probabilities, such as $p_T \leq 2\%$, the separation distance has the potential to be driven by a few distinct, uncommon meteorological conditions. To convert from hourly concentrations to short-term odour peaks, a P/M factor of 2.3 was applied. An element of uncertainty [12] of the Italian standards is the fact that the exceeding criteria do not provide any indication of the average time to which the peak concentrations is referred. Previous works [27] demonstrated that the P/M factor depends on several parameters, like the stability of the atmosphere, intermittency, travel time or distance from the source. Countries like Australia [28] or Austria [29] introduced variable P/M values. Variable P/M were also implemented in the Austrian Odour Dispersion Model (AODM), the regulatory Austrian Gaussian model and in the German Lagrangian model LASAT [10].

3. Results

As reported in Table 1, Simulation 1 and Simulation 2 differed only in the weather input (2019 and 2018, respectively). Seasonal wind distributions of the two years are reported in Figures 2 and 3. These figures show the typical wind distribution of this region, which is regulated by prevailing NE and

SE directions. NE winds generally have higher speeds, especially in spring and autumn. However, low wind speeds (<3 m s^{-1}) have higher occurrence frequencies due to the presence of the Alps that surround the entire region, acting as a barrier for continental winds. A share of 29.8% and 22.4% of wind calms was registered for the years 2018 and 2019, respectively. Daily wind variations show that the evening and night hours registered the lowest intensities, with a morning increase and maximum values in the early afternoon. Even though wind distribution is similar in the two years, wind presence was higher in 2019 than in 2018. If the atmospheric stability class distribution is considered, in both years stable conditions prevailed (47%), followed by unstable (37%) and neutral conditions (16%). The Pasquill–Gifford atmospheric stability class distribution and wind directions for the years 2018 and 2019 are reported in Figures 4 and 5, respectively. These figures also show a similar trend between the two years. In case of unstable conditions (Class A and B), NE is the prevailing wind direction, although the frequency of other directions is not negligible. This distribution reflects daytime conditions, where stability is mainly driven by thermal convection and higher-speed winds. Conversely, in case of stable conditions (Class E and F), SE is the prevailing wind direction. In this area of study, SE winds typically have a low speed (Figure 3) and occur after sunset due to the balancing of the residual thermal energy between the valley and the surrounding reliefs. This latter situation of atmospheric stability and low winds may, in principle, favour the dispersion of odours to considerable distances.

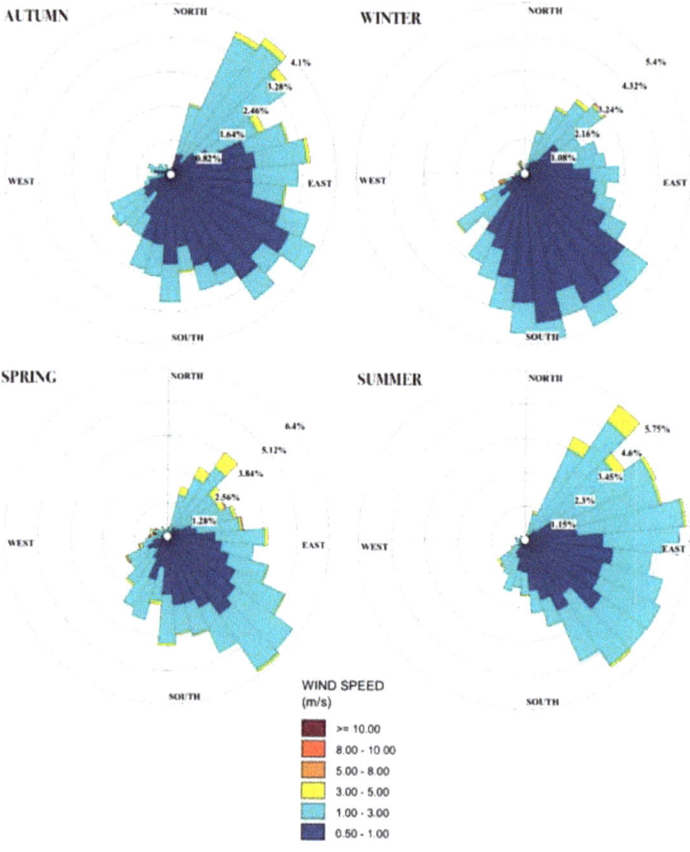

Figure 2. Seasonal wind distribution of the year 2018.

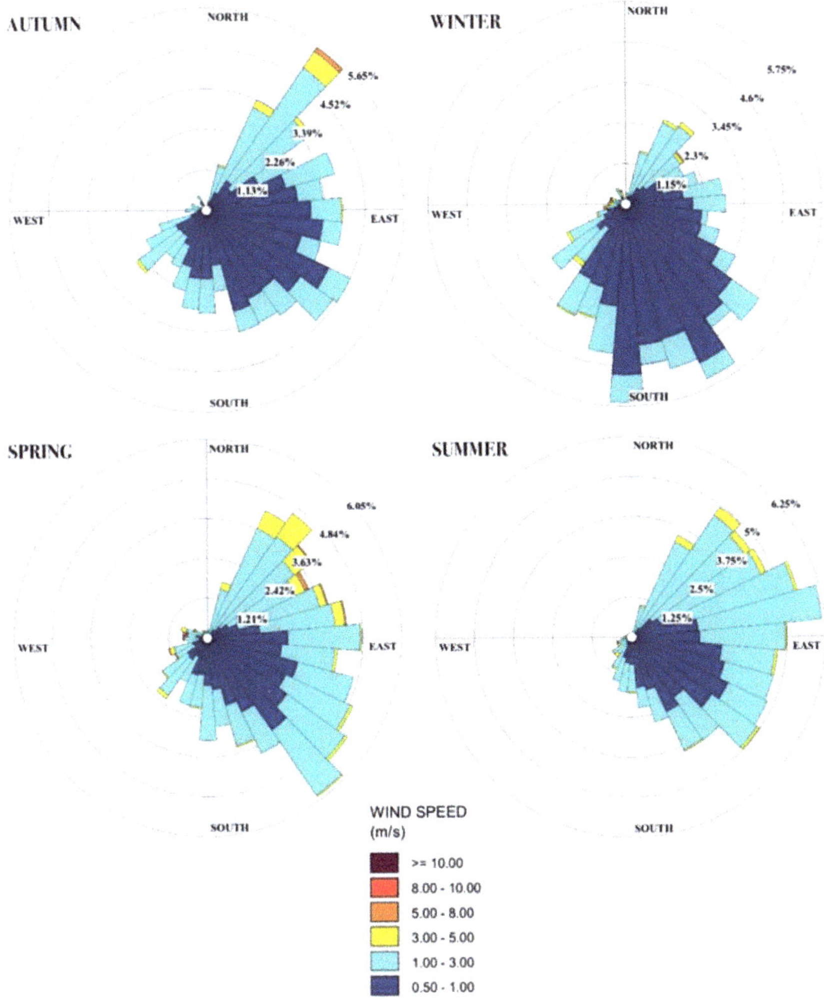

Figure 3. Seasonal wind distribution of the year 2019.

The OERs were calculated following the methodology reported in Section 2. Following Equation (5) (power law), Equation (6) (log law) and Equation (7) (D–H law), the hourly values of the OERs were obtained depending on the atmospheric conditions. To obtain a comparison, the hourly OERs were divided into classes, and their probability distributions were considered. In Figure 6, the distribution of the OER of the primary settler (P3 in Figure 1) is reported. Since OER_{WT} is constant for each source, the trend of a single source is also representative of other sources. Figure 6 indicates how the OER changed depending on the correction method adopted. This comparison shows that the application of different wind speed correction methods (power law, log law and D–H law) provided different OER values. The distributions of both the power law and the log law indicate a peak density. For the power law, the most frequent value of the OER is around 15,000 OU/s. For the log law, the most frequent value of the OER is around 30,000 OU/s. The application of the D–H law provided two peak densities: the first is in correspondence of the values of the OER around zero, and the second is around 22,000

OU/s. For higher values of the OER, the distributions of the log and the D–H laws perform similarly. For the power law, the distribution is shifted towards higher OER values.

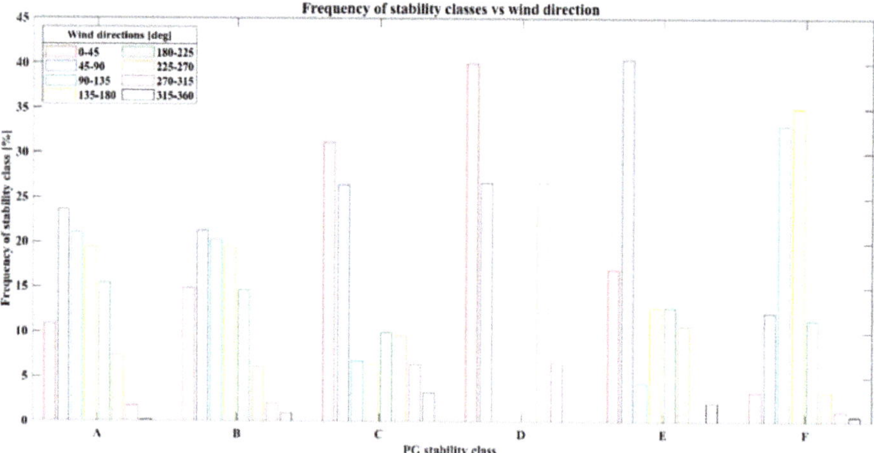

Figure 4. The Pasquill–Gifford (PG) atmospheric stability class distribution and wind directions for the year 2018.

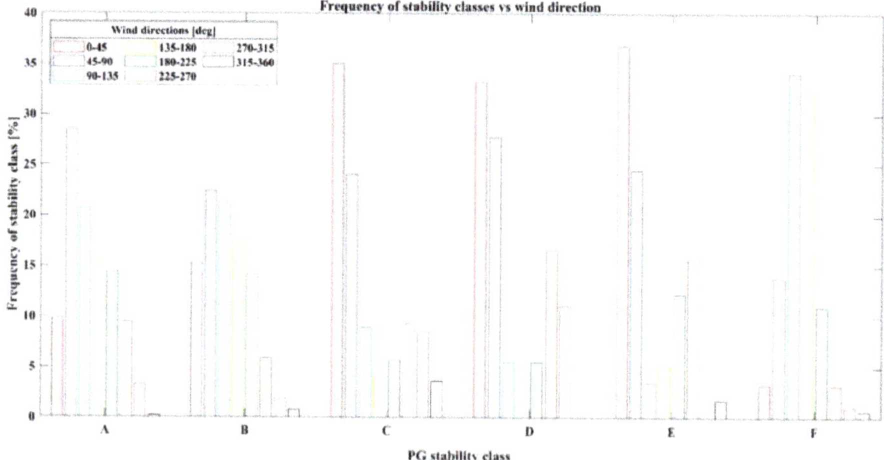

Figure 5. The PG atmospheric stability class distribution and wind directions for the year 2019.

The separation distances corresponding to Simulation 1 (year 2019, power law) and Simulation 2 (year 2018, power law) for C_T equal to 1 OU m^{-3}, 3 OU m^{-3} and 5 OU m^{-3} are reported in Figure 7. This figure shows that the odour impact area of the WWTP may be significant. The odour impact area is extended in the SW direction, in accordance with the anaemological data. Furthermore, the separation distances are higher in the NE direction, which is not fully in agreement with the prevailing wind directions reported. The 1 OU m^{-3} isopleth reached towns up to 6 km away from the plant boundary, far beyond the 3 km limit set by the guidelines as a radius within which to verify the olfactory harassment extent. Contour lines referred to $C_T = 3$ OU m^{-3} and $C_T = 5$ OU m^{-3} (which indicate a greater frequency of harassment perception by the population) extend beyond the plant borders, including the nearby residential areas. Figure 7 also shows the effect of the different meteorological data compared based on the same emission scenario. Contour lines show that the separation distances for the 1 OU m^{-3} isopleths of the two years are similar, while some difference is reported for the 3 OU m^{-3} and 5 OU m^{-3} isopleths. Simulation 2 generated lower separation distances than Simulation 1. The shape of the impacted areas shows that the prevailing NE winds contribute to odour dispersion in the area. The contribution of SE winds is instead less evident. This graph also shows how the presence of the reliefs in the southern area contributes to limiting the odour dispersion in this direction. The maximum separation distances for $C_T = 1$ OU m^{-3}, $C_T = 3$ OU m^{-3} and $C_T = 5$ OU m^{-3}, respectively, were

- for simulation 17,000 m in direction NE; 4300 m in direction SW; 3510 in direction SW;
- for simulation 26,870 m in direction NE; 3580 m in direction SW; 2680 in direction SW.

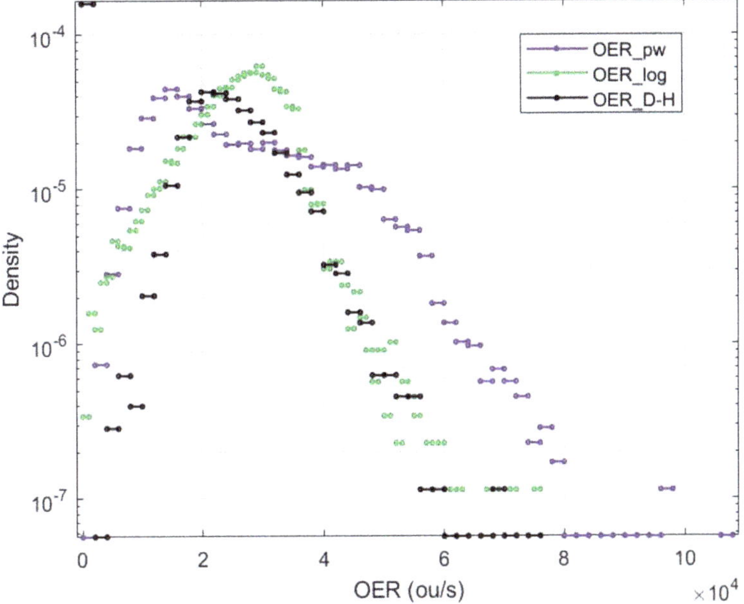

Figure 6. Frequency distribution of the OER values. OER_pw refers to Simulation 1. OER_pw refers to Simulation 3. OER_D-H refers to Simulation 4.

Taking Simulation 1 as the reference, the average difference on the eight main cardinal positions was −2%, −9% and −17% for $C_T = 1$ OU m^{-3}, $C_T = 3$ OU m^{-3} and $C_T = 5$ OU m^{-3}, respectively. A maximum difference of −16% (SW direction), −20% (SW direction) and −32% (S direction) was found for the three values of C_T.

The separation distances corresponding to Simulation 3 and Simulation 4, compared to Simulation 1, are reported in Figure 8. This figure shows how the odour concentration changed by changing the correction method for wind speed in the OER calculation. The same meteorological data (year 2019) were used in Simulations 1, 3 and 4. As reported in Figure 8, the application of the D–H law generated lower separation distances. Conversely, separation distances were higher in case the log law was applied. The maximum separation distances for $C_T = 1$ OU m^{-3}, $C_T = 3$ OU m^{-3} and $C_T = 5$ OU m^{-3}, respectively, were

- for simulation 37,800 m in direction NE; 4550 m in direction SW; 3720 in direction SW;
- for simulation 46,500 m in direction SW; 4120 m in direction SW; 3300 in direction SW.

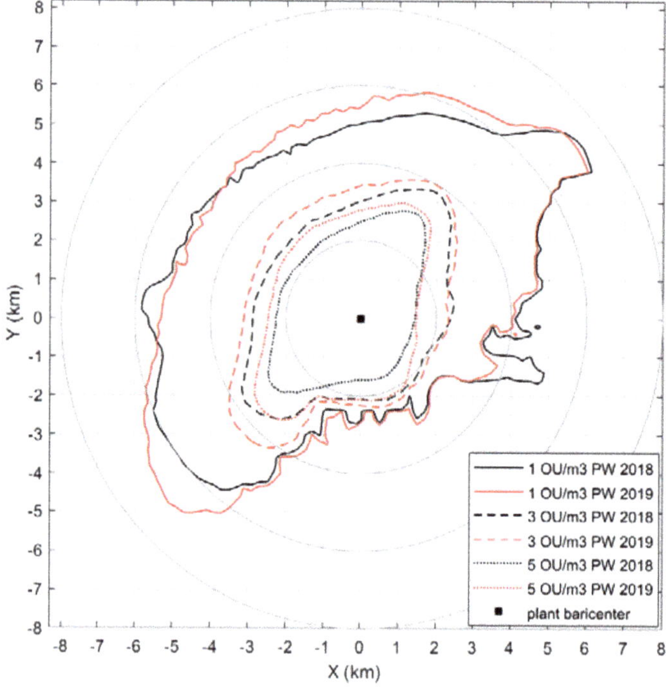

Figure 7. Separation distances generated by Simulation 1 (year 2019, power law) and Simulation 2 (year 2018, power law).

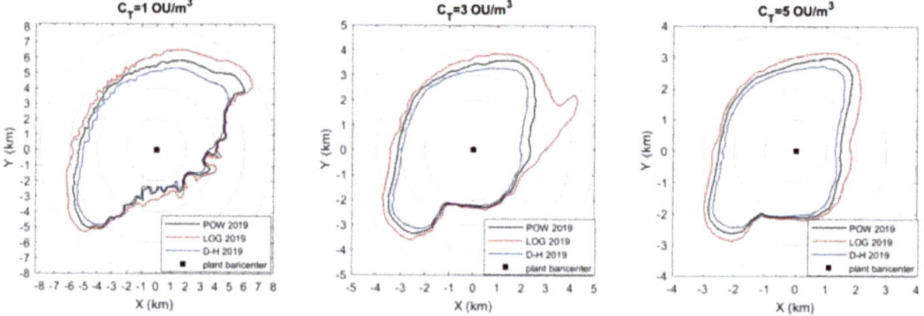

Figure 8. Separation distances generated by Simulation 1 (year 2019, power law), Simulation 3 (year 2019, log law) and Simulation 4 (year 2019, D–H law), for $C_T = 1$ OU m^{-3} (left), $C_T = 3$ OU m^{-3} (center) and $C_T = 5$ OU m^{-3} (right).

Taking Simulation 1 as a reference, the average variation of the separation distances on the eight main cardinal positions with the application of the log-law was +8%, +10% and +10% for $C_T = 1$ OU m^{-3}, $C_T = 3$ OU m^{-3} and $C_T = 5$ OU m^{-3}, respectively. The average variation of the separation distances with the application of the D–H law was −7% for all values of C_T. A maximum difference of +19% (log law, SW direction), +25% (log law, NE direction) and +20% (log law, E direction) was found for the three values of C_T. In the S and SE directions, the difference between the three simulations is less visible. This is probably due to the presence of the topographic reliefs, which act as a barrier to odour dispersion. Increasing the level of C_T, the difference between the separation distances tended to be lower. Contour lines of $C_T = 3$ OU m^{-3} also show a different shape in the ENE direction if the log law is applied. A further comparison of the separation distances in the main transport directions for the three applied relationships is reported in Figure 9. This figure shows that the application of the log-law generally yields higher separation distances in all directions. Except for the E, SE and S directions, where the effect of the orography is evident, the difference maintains the proportionality D–H law < power law < log law.

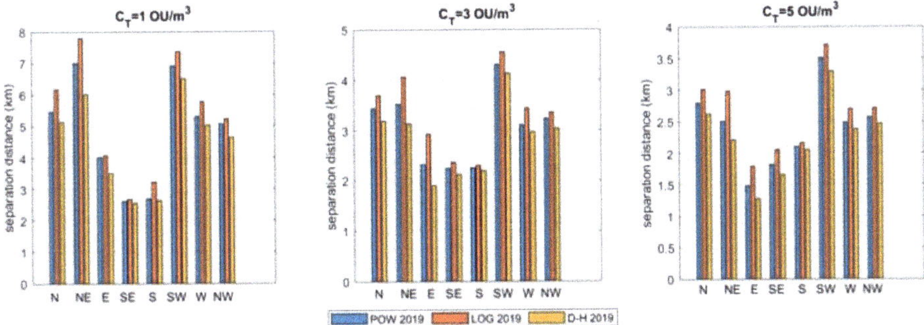

Figure 9. Direction-dependent separation distances generated by Simulation 1 (year 2019, power law), Simulation 3 (year 2019, log law) and Simulation 4 (year 2019, D–H law) for $C_T = 1$ OU m^{-3} (left), $C_T = 3$ OU m^{-3} (center) and $C_T = 5$ OU m^{-3} (right).

4. Discussion

This study focused on the aspects of variability related to the application of regulatory OIC criteria to a WWTP located in Northern Italy. Two factors were investigated. The first factor was the use of different meteorology datasets as input to odour dispersion modelling. The second factor was the adoption of different correction methods for the wind speed profile used in the calculation of OERs.

Simulations 1 and 2 showed that the odour impact area is extended in the NE–SW direction, partially in accordance with the anaemological data. Figure 7 showed that the extent of the distances was probably a combination of many factors, in particular the frequency distribution of atmospheric stability and wind speeds per wind direction sector. A similar trend was found by Brancher et al. [11].

Compared to 2019, the year 2018 showed similar wind distribution patterns. The distribution and frequency of the atmospheric stability classes were also similar in the two years. As reported in Figure 7, for $C_T = 1$ OU m^{-3}, the average difference in the resulting separation distances was around 7%. Conversely, for $C_T = 3$ OU m^{-3} and mostly for $C_T = 5$ OU m^{-3}, an average 17% difference (maximum 32%) was reported. These variations may be attributed to the differences in wind distribution in the two years, which is more visible closer to the source. At higher distances, the results reflect the combined effect of the different factors that regulate odour dispersion (multiple wind components and convective turbulence in particular).

In another study, Brancher et al. [9] analysed the variation of the separation distances over a 5-year meteorological dataset. For an OIC with a $C_T = 1$ OU m^{-3}, according to the present study, the results showed some inter-annual variability, especially in the main wind directions. The authors also found that, at the two investigated sites, the mean direction-dependent separation distances over the individual meteorological years were largely in agreement with the distances determined for the five years of meteorology data. They concluded that a one-year dataset of hourly meteorological observations is enough to be taken as a plausible length of time to attain reliable distances. Even though the present study is partly in agreement with these findings, the results show that, for higher values of C_T, higher inter-annual variability of separation distances could be expected.

The second aspect analysed in the present study showed that the application of different correction methods for wind speed calculation affects the resulting separation distances. Compared to the power law, the log law provides higher distances (8–10%), while the D–H law provides lower distances (7%). The variation is higher along with the prevailing wind directions. Maximum variations are recorded for the log law and can be up to 19%, 25% and 20% for $C_T = 1$ OU m^{-3}, $C_T = 3$ OU m^{-3} and $C_T = 5$ OU m^{-3}, respectively. The results reflect the distribution of the OER values reported in Figure 6. The reason for such a discrepancy must be investigated by analysing the form of Equations (5)–(7), which describe the wind speed correction. If the power law is applied (Equation (5)), the resulting value of the wind speed is highly dependent on the Hellman's parameter (α) assigned to each observation (Table 4). Taking $h_2 = 10$ m and $h_1 = 1$ m, for stable conditions ($\alpha = 0.07$), a slight correction is applied, i.e., the corrected value v_1 is close to the value v_2 registered by the anemometer. Conversely, for unstable conditions ($\alpha = 0.55$), the value of v_2 is significantly reduced compared to v_1. If the log law is considered, the second term in Equation (6) represents the contribution of stability-induced turbulence. At the height h_1, the contribution of this term is generally low. This means that the term $\ln (h_1/z_0)$ is close to the value of K_v; thus, the value of v_1 approximates the value of u^*. Considering that, at h_2, the u^*/u ratio is close to unity; it comes out that the value of v_1 is close to v_2 in most cases. These aspects could be the main reasons for the differences in the OER values of Figure 6 and, consequently, on the results of the separation distances reported in Figure 8. Finally, if the D–H law is considered, Equation (7) shows that this relationship is regulated by the term H, which is the equilibrium boundary layer height. Equation (7) approximates to the log law when the term h_1/H equals unity, i.e., $H = 1$. For low values of u^*, as in the present case study, the entire second term of Equation (7) tends to be null; thus, at the height h_1, v_2 reaches values close to zero. For this reason, the application of the D–H law resulted in a higher frequency of OER values close to zero (Figure 6). This approach can be considered interesting and is worthy of further investigations as, in principle, it seems to provide an approximation of the wind speed conditions at lower heights that is closer to the real conditions, especially when the height of the source is close to the value of z_0.

Calculating the OER has been considered in previous studies. However, there are no studies in which the different OER calculation methods are matched to the dispersion modelling and compared in terms of separation distances. Previous works analysed (i) the correlation between emissive flows in the open field and within the dynamic hoods and (ii) the law used to describe the wind speed profile. Lucernoni [30] addressed the issue of dependence on the velocity law by comparing the three correlations presented in this study, finding differences consistent with the results of this article. The same authors also underlined the greater reliability of the logarithmic profile for reduced elevations (0–100 m), while highlighting its complexity regarding application. Furthermore, they recognized the possibility of adopting the D–H relationship, citing it in a subsequent study concerning an OER correction of the passive area sources in Northern Italy [18].

It must be pointed out that the present study was based on constant values of OER_{WT}. Odour emissions from WWTPs are known to fluctuate over time [31,32]. The use of time-varying OERs, however, is not yet easily feasible in odour modelling, as multiple monitoring campaigns are expensive in terms of time and costs. Repeating the investigation with alternative WWTP case studies and/or OER_{WT} values, e.g., those deriving from regulatory emission factors, could provide important information in the view of discussing the impact of different OIC on separation distances.

5. Conclusions

The objective of the present study was to investigate the variability of two factors (meteorology and OER calculation) related to the regulatory odour impact assessment, applying a modelling analysis to a wastewater treatment plant located in Northern Italy. The odour impact criteria of the Northern Italian regions were considered as the benchmark. Possible alternative technical choices were analysed to understand their influence on the resulting separation distances.

Regarding the influence of different meteorological years on the separation distances, this study was partially in agreement with the existing literature. For low odour concentration thresholds ($C_T = 1$ OU m^{-3}), the results showed that the two different years (2018 and 2019) provided similar patterns of the separation distances. The difference between the two years tended to increase by increasing the value of the concentration threshold (C_T = 3 OU m^{-3} and C_T = 5 OU m^{-3}). Overall, differences in the separation distances were greater for the prevailing wind directions compared to non-prevailing wind directions.

The comparative analysis of the wind speed correction methods in the calculation of the OERs considered three different relationships: the power law, the logarithmic law and the Deaves–Harris law. The results showed that the OERs and separation distances varied depending on the selected method. Taking the power law as a reference, the average variability of the separation distances was between −7% (D–H correlation) and +10% (logarithmic law). Higher variability (up to 25%) was found for single transport distances. The main reason of such variability may be attributed to the different parameterization of the PBL of the three methods in relationship with the atmospheric stability class. From this study, it cannot be concluded if one method is more representative of another. However, it is highlighted that the application of the Deaves–Harris law is interesting but needs further investigation.

The present study confirmed that the representativeness of the odour impact assessment depends not only on the evaluator's choices but also on the application of the current regulatory provisions on odour emissions. Populations and administrations are increasingly concerned with environmental odour problems. Although the odour impact criteria are the result of both technical and political considerations, it seems plausible that the different assessment methods should provide similar separation distances. The present study provided knowledge towards a better alignment of the concept of the odour impact criterion.

Author Contributions: Conceptualization, M.Z., D.P. and M.R.; Methodology, M.Z., D.P. and M.R.; Software, M.R. and S.B.; Validation, M.Z., D.P. and M.R.; Formal Analysis, M.R. and S.B.; Investigation, M.R. and S.B.; Resources, M.Z., D.P., M.R. and S.B.; Data Curation, M.R. and S.B.; Writing—Original Draft Preparation, M.R. and S.B.; Writing—Review & Editing, M.R. and S.B.; Visualization, M.R., D.P. and M.Z.; Supervision, D.P. and M.Z.; Project Administration, D.P. and M.Z.; Funding Acquisition, M.Z. All authors have read and agreed to the published version of the manuscript.

Funding: This research did not receive any specific grant from funding agencies in the public, commercial, or not-for-profit sectors.

Conflicts of Interest: The authors declare no conflict of interest.

Abbreviations

CALMET	California Meteorological model
CALPUFF	California Puff model
C_T	threshold concentration
DO	dynamic olfactometry
D–H	Deaves–Harris
EN	electronic nose
FC	flux chamber
GC-MS	gas chromatography-mass spectrometry
OER	odour emission rate
OIC	odour impact criteria
OU	odour unit
P/M	peak to mean
PBL	planet boundary layer
p_T	exceedance probability of the concentration threshold
U.S. EPA	United States Environmental Protection Agency
WT	wind tunnel
WWTP	wastewater treatment plant

References

1. Byliński, H.; Kolasińska, P.; Dymerski, T.; Gębicki, J.; Namieśnik, J. Determination of odour concentration by TD-GC×GC–TOF-MS and field olfactometry techniques. *Mon. Chem.* **2017**, *148*, 1651–1659. [CrossRef] [PubMed]
2. Barczak, R.J.; Kulig, A. Comparison of different measurement methods of odour and odorants used in the odour impact assessment of wastewater treatment plants in Poland. *Water Sci. Technol.* **2017**, *75*, 944–951. [CrossRef] [PubMed]
3. Belgiorno, V.; Naddeo, V.; Zarra, T. *Odour Impact Assessment Handbook*; John Wiley & Sons, Ltd.: Hoboken, NJ, USA, 2013; ISBN 978-1-119-96928-0.
4. Jiang, G.; Melder, D.; Keller, J.; Yuan, Z. Odor emissions from domestic wastewater: A review. *Crit. Rev. Environ. Sci. Technol.* **2017**, *47*, 1581–1611. [CrossRef]
5. Boeker, P.; Haas, T. A Monte-Carlo simulation of the measurement uncertainty of olfactometry. *Chem. Eng. Trans.* **2008**, *15*, 109–114.
6. Capelli, L.; Sironi, S.; Del Rosso, R.; Guillot, J.-M. Measuring odours in the environment vs. dispersion modelling: A review. *Atmos. Environ.* **2013**, *79*, 731–743. [CrossRef]
7. Carrera-Chapela, F.; Donoso-Bravo, A.; Souto, J.A.; Ruiz-Filippi, G. Modeling the Odor Generation in WWTP: An Integrated Approach Review. *Water Air Soil Pollut.* **2014**, *225*, 1932. [CrossRef]
8. Luciano, A.; Torretta, V.; Mancini, G.; Eleuteri, A.; Raboni, M.; Viotti, P. The modelling of odour dispersion as a support tool for the improvements of high odours impact plants. *Environ. Technol.* **2017**, *38*, 588–597. [CrossRef]
9. Brancher, M.; Piringer, M.; Franco, D.; Belli Filho, P.; De Melo Lisboa, H.; Schauberger, G. Assessing the inter-annual variability of separation distances around odour sources to protect the residents from odour annoyance. *J. Environ. Sci.* **2019**, *79*, 11–24. [CrossRef]
10. Piringer, M.; Knauder, W.; Petz, E.; Schauberger, G. Factors influencing separation distances against odour annoyance calculated by Gaussian and Lagrangian dispersion models. *Atmos. Environ.* **2016**, *140*, 69–83. [CrossRef]
11. Brancher, M.; Piringer, M.; Grauer, A.F.; Schauberger, G. Do odour impact criteria of different jurisdictions ensure analogous separation distances for an equivalent level of protection? *J. Environ. Manag.* **2019**, *240*, 394–403. [CrossRef]
12. Brancher, M.; Griffiths, K.; Franco, D.; Lisboa, H. A review of odour impact criteria in selected countries around the world. *Chemosphere* **2017**, *168*, 1531–1570. [CrossRef] [PubMed]
13. Sommer-Quabach, E.; Piringer, M.; Petz, E.; Schauberger, G. Comparability of separation distances between odour sources and residential areas determined by various national odour impact criteria. *Atmos. Environ.* **2014**, *95*, 20–28. [CrossRef]
14. EN 13725:2004. Air Quality—Determination of Odour Concentration by Dynamic Olfactometry. Available online: https://bit.ly/3m0t21F (accessed on 2 September 2020).
15. Regione Lombardia. Determinazioni Generali in Merito Alla Caratterizzazione Delle Emissioni Gassose in Atmosfera Derivanti da Attività a Forte Impatto Odorigeno. Serie Ordinaria n. 8-Lunedì 20 febbraio 2012. Available online: https://www.regione.lombardia.it/wps/wcm/connect/e7464d04-b9c5-496f-8546-257170f9e658/DGR+3018_2012_Linee+guida+odori.pdf?MOD=AJPERES&CACHEID=ROOTWORKSPACE-e7464d04-b9c5-496f-8546-257170f9e658-m6YF2gp (accessed on 2 September 2020).
16. Barclay, J.; Scire, J. *Generic Guidance and Optimum Model Settings for the CALPUFF Modelling System for Inclusion into the 'Approved Methods for the Modelling and Assessments of Air Pollutants in NSW, Australia'*; NSW Office of Environment and Heritage: Sydney, Australia, 2011.
17. Bliss, P.; Jiang, K.; Schulz, T. The Development of a Sampling System for the Determination of Odour Emission Rates from Areal Surfaces: Part II. *Mathematical Model. Journal of the Air & Waste Management Association* **1995**, *45*. [CrossRef]
18. Lucernoni, F.; Capelli, L.; Busini, V.; Sironi, S. A model to relate wind tunnel measurements to open field odorant emissions from liquid area sources. *Atmos. Environ.* **2017**, *157*, 10–17. [CrossRef]
19. Cook, N.J. The Deaves and Harris ABL model applied to heterogeneous terrain. *J. Wind Eng. Ind. Aerodyn.* **1997**, *66*, 197–214. [CrossRef]

20. Hanna, S.R.; Briggs, G.A.; Hosker, R.P., Jr. *Handbook on Atmospheric Diffusion*; DOE/TIC-11223; Atmospheric Turbulence and Diffusion Lab National Oceanic and Atmospheric Administration: Oak Ridge, TN, USA, 1982; p. 5591108.

21. Holtslag, A.A.M.; van Hulten, P. A Simple Scheme for Daytime Estimates of the Surface Fluxes from Routine Weather Data. *J. Clim. Appl. Meteorol.* **1983**, *22*, 517–529. [CrossRef]

22. Scire, J.S.; Robe, F.R.; Fernau, M.E.; Yamartino, R.J. *A User's Guide for the CALMET Meteorological Model*; Earth Tech, Inc.: Concord, MA, USA, 2000.

23. Scire, J.; Strimaitis, D.; Yamartino, R. *A User's Guide for the CALPUFF Dispersion Model*; Earth Tech, Inc.: Concord, MA, USA, 2000; p. 521.

24. Zannetti, P. Dry and Wet Deposition. In *Air Pollution Modeling*; Springer: Boston, MA, USA, 1990; pp. 249–262. ISBN 978-1-4757-4467-5.

25. Griffiths, K.D. Disentangling the frequency and intensity dimensions of nuisance odour, and implications for jurisdictional odour impact criteria. *Atmos. Environ.* **2014**, *90*, 125–132. [CrossRef]

26. Nicell, J.A. Development of the odour impact model as a regulatory strategy. *Int. J. Environ. Pollut.* **1994**, *4*, 124–138.

27. Schauberger, G.; Piringer, M.; Schmitzer, R.; Kamp, M.; Sowa, A.; Koch, R.; Eckhof, W.; Grimm, E.; Kypke, J.; Hartung, E. Concept to assess the human perception of odour by estimating short-time peak concentrations from one-hour mean values. Reply to a comment by Janicke et al. *Atmos. Environ.* **2012**, *54*, 624–628. [CrossRef]

28. Freeman, T.; Cudmore, R. *Review of Odour Management in New Zealand*; Air Quality Technical Report No. 24; New Zealand Ministry of Environment: Wellington, New Zealand, 2002.

29. Piringer, M.; Knauder, W.; Petz, E.; Schauberger, G. A comparison of separation distances against odour annoyance calculated with two models. *Atmos. Environ.* **2015**, *116*, 22–35. [CrossRef]

30. Lucernoni, F. Politecnico di Milano. The Evaluation of the Odour Emission Rate for passive area sources: A new approach. *Chem. Eng. Trans.* **2015**, *43*, 2015.

31. Lebrero, R.; Bouchy, L.; Stuetz, R.; Muñoz, R. Odor Assessment and Management in Wastewater Treatment Plants: A Review. *Crit. Rev. Environ. Sci. Technol.* **2011**, *41*, 915–950. [CrossRef]

32. Morosini, C.; Favaron, M.; Cogo, D.; Basilico, D.I. Gestione e trattamento degli odori in Regione Lombardia: Indagine su 35 impianti di depurazione delle acque reflue. *Ing. dell'Ambiente* **2016**, *3*. (In Italian) [CrossRef]

Article

Are Empirical Equations an Appropriate Tool to Assess Separation Distances to Avoid Odour Annoyance?

Marlon Brancher [1], Martin Piringer [2], Werner Knauder [2], Chuandong Wu [3], K. David Griffiths [4] and Günther Schauberger [1,*]

[1] WG Environmental Health, Department of Biomedical Sciences, University of Veterinary Medicine Vienna, Veterinärplatz 1, A-1210 Vienna, Austria; Marlon.Brancher@vetmeduni.ac.at
[2] Department of Environmental Meteorology, Central Institute for Meteorology and Geodynamics, Hohe Warte 38, A-1190 Vienna, Austria; Martin.Piringer@zamg.ac.at (M.P.); Werner.Knauder@zamg.ac.at (W.K.)
[3] School of Chemistry and Biological Engineering, University of Science and Technology Beijing, Beijing 100083, China; wuchuandong@ustb.edu.cn
[4] Environmental and Conservational Sciences, Murdoch University, 90 South Street, Murdoch, Western Australia 6150, Australia; dgr_2001@yahoo.com.au
* Correspondence: Gunther.Schauberger@vetmeduni.ac.at

Received: 29 April 2020; Accepted: 24 June 2020; Published: 27 June 2020

Abstract: Annoyance due to environmental odour exposure is in many jurisdictions evaluated by a yes/no decision. Such a binary decision has been typically achieved via odour impact criteria (OIC) and, when applicable, the resultant separation distances between emission sources and residential areas. If the receptors lie inside the required separation distance, odour exposure is characterised with the potential of causing excessive annoyance. The state-of-the-art methodology to determine separation distances is based on two general steps: (i) calculation of the odour exposure (time series of ambient odour concentrations) using dispersion models and (ii) determination of separation distances through the evaluation of this odour exposure by OIC. Regarding meteorological input data, dispersion models need standard meteorological observations and/or atmospheric stability typically on an hourly basis, which requires expertise in this field. In the planning phase, and as a screening tool, an educated guess of the necessary separation distances to avoid annoyance is in some cases sufficient. Therefore, empirical equations (EQs) are in use to substitute the more time-consuming and costly application of dispersion models. Because the separation distance shape often resembles the wind distribution of a site, wind data should be included in such approaches. Otherwise, the resultant separation distance shape is simply given by a circle around the emission source. Here, an outline of selected empirical equations is given, and it is shown that only a few of them properly reflect the meteorological situation of a site. Furthermore, for three case studies, separation distances as calculated from empirical equations were compared against those from Gaussian plume and Lagrangian particle dispersion models. Overall, our results suggest that some empirical equations reach their limitation in the sense that they are not successful in capturing the inherent complexity of dispersion models. However, empirical equations, developed for Germany and Austria, have the potential to deliver reasonable results, especially if used within the conditions for which they were designed. The main advantage of empirical equations lies in the simplification of the meteorological input data and their use in a fast and straightforward approach.

Keywords: environmental odour; emission; annoyance; separation distance; dispersion models; empirical equations

---

1. Introduction

Odours from industrial, municipal, and agricultural activities are the most common causes of public complaints to authorities, besides noise. Effectively tackling such complaints is an essential part of environmental odour management practices. Separation distances between odour-emitting facilities and residential areas can be imposed to restrict annoyance within acceptable levels. Doing so has the potential to prevent complaints from being generated in the first place. These separation distances divide the circumjacent area around a source into a zone which is protected from annoyance and a zone closer than the separation distance where annoyance is likely to be expected. The term setback distance has also been commonly used in this context.

Governments around the world have set jurisdictional limit values for environmental odours, called odour impact criteria (OIC), to orientate compliance demonstration procedures. By this means, separation distances can be calculated on a case-by-case basis and in a direction-dependent manner. Dispersion modelling is a method extensively used for such a purpose. A variety of dispersion models are in use, differing mainly by their mathematical formulation to the physics driving the transport and dispersion of pollutants in the atmosphere. Currently, the most frequently used dispersion models are based on the Gaussian, Lagrangian, and Eulerian descriptions [1–7].

Estimates of downwind ambient concentrations of air pollutants, including odours, can be computed using the emission of the source, geophysical and meteorological data as the primary inputs. Even for cases in which the source emission is constant over time, the ambient concentrations will vary according to the meteorological conditions at the time of the release. Accordingly, meteorology is widely acknowledged as the principal factor in this context. Today's advanced dispersion models can also treat building downwash or buoyancy effects, which is not possible when applying empirical equations (EQs). The importance of acquiring good quality meteorological input data that is representative of the site being modelled is palpable as it is the meteorological expertise needed to obtain consistent results from dispersion models. Meteorological parameters such as wind direction and velocity, a measure of the degree of turbulence (atmospheric stability) and mixing height, typically on an hourly basis, form the basis of such datasets.

Hourly time series of ambient odour concentrations predicted by dispersion models are used to quantify odour exposure. The evaluation of the odour exposure is performed by OIC, commonly relying on a protection level. The level of protection is differentiated, for instance, by the zoning/land use of the area and facility status (new or existing installations). A review of OIC in 28 countries has been presented elsewhere [8]. Due to the application of such jurisdictional OIC, the annoyance potential of odours at a particular site can be assessed by direction-dependent separation distances [9–11].

The OIC are defined by an odour concentration threshold C_T, the exceedance probability P of this threshold (or percentile $1 - P$) and the model averaging time. There are basically two alternatives to calculate direction-dependent separation distances by OIC. The first alternative uses a constant C_T so that the protection level is adopted by P. This tactic has been defined in the German national guideline [12], which is from time to time adopted for compliance demonstrations in Austria and Switzerland too. The second alternative, which is established in the majority of the countries [8,13], works the other way round. First, P is set, then C_T is used to cover the targeted protection levels.

Odour-related separation distances do not apply to all kinds of sources. This approach is more appropriate for ground-level or low-height sources. These sources are typical of, for example, wastewater treatment plants and concentrated animal feeding operations. It is difficult to define the two zones given by separation distances for relatively high point sources unaffected by adjacent structures. The separation distance approach is part of an integrated multi-tool strategy to manage environmental odours [8].

Simplified tools have also been developed to enable a simpler determination of odour-related separation distances. Such tools are often called empirical equations (EQs). In this paper, we present and discuss various EQs used in some jurisdictions. These EQs were in some cases incorporated in regulations or national guidelines. Although a number of EQs are listed (see next section, Table 1), they do not represent an exhaustive review of available EQs, or even just those that use meteorological data as predictors. However, some background context is provided regarding the subset of EQs that do include meteorological predictors. These EQs have been well described individually in the published literature. However, little work has been done comparing the performance of EQs that include meteorology as a predictor with each other and dispersion modelling calculations. This has been done to demonstrate the advantages and shortcomings of such parsimonious methods.

2. Selected Empirical Equations to Assess the Separation Distance

There is a wide spectrum of empirical methods to assess separation distances. The presented EQs fill the gap between dispersion models, which are the gold standard and benchmark and guesswork on the other side. Simple EQs deliver a unique, fixed distance, thereby shaping a circle around the source. Consequently, this procedure does not take into account the meteorological conditions of a site. More elaborate EQs include meteorological predictors such as wind frequencies and mean wind velocities within direction sectors. In these cases, the EQ coefficients have been derived from regression analyses of dispersion model calculations.

Table 1 presents selected international EQs and their main input factors. This list is given with the intention of facilitating the calculation of separation distances in future works. The EQs are sorted descending by the complexity of the input parameters and the shape of the separation distance.

- Ideally, the following objectives should be fulfilled by EQs [5]:
- Odour emission rate: it should be quantified in the same way as it is done for dispersion models, using the odour emission rate (ou_E s^{-1}) even if the emission geometry cannot be taken into account;
- Odour impact criteria: the separation distance is calculated in reference to a certain protection level via the same odour impact criterion which are used for dispersion models;
- Meteorology: the meteorological situation of the site should be defined by wind statistics. These refer to at least the relative frequency of the wind direction for 10° sectors. Such meteorological datasets are accessible, for example, from national weather services;
- "Paper and pencil": the method should be fast and easy to use to be appropriate as a screening tool.

The first two EQs in Table 1 are the German VDI (Verein Deutscher Ingenieure) equation [5,14] and the Austrian equation [15]. The German VDI EQ uses only one meteorological predictor (wind direction frequency), whereas the Austrian EQ uses, in addition, the mean wind velocity for wind direction sectors of 10°. The mean wind velocity was added as a predictor with the objective of considering atmospheric stability, pragmatically. The reason this was considered is centred on the greater variability in atmospheric stability conditions and higher calm frequencies of Austrian sites as compared to most German sites.

Both EQs are based on a power function $E = a\ S^b$. Such a power function calculates the separation distance E (m) as a function of the odour source strength, given as the odour emission rate S (ou_E s^{-1}).

In Germany, the two coefficients a and b of the power function were fitted to dispersion model results obtained with AUSTAL2000, a Lagrangian particle dispersion model. The coefficient a depends on two predictors. The first is the relative frequency of the wind direction sectors F (‰) of 10°. The second is the odour exceedance probability P (%) of the odour impact criterion. The exponent b depends on P only. The exceedance probability is set depending on the required protection level. The equation for the separation distance E_G by the German VDI regression model reads

$$E_G = [(-0.0137\,P + 0.689) \times F + 0.251\,P + 0.0590]\,S^{\frac{1}{1.79 + 0.204\,P}} \tag{1}$$

Table 1. List of selected empirical equations (EQs) for the determination of the separation distance E used in different countries with indication to the predictors (independent variables) of the power function $E = a\, S^b$ with the factor a, the exponent b and the odour emission rate S. The protection level is related to odour impact criteria (OIC) (C_T concentration threshold in $\mathrm{ou_E\, m^{-3}}$ and P exceedance probability in %).

Name and Reference	Factor a	Emission S	Exponent b	OIC C_T/p_T	Separation Distance Shape
German VDI [1,2] [5,14]	wind frequency, exceedance probability	odour emission rate ($\mathrm{ou_E\ s^{-1}}$)	exceedance probability	$0.25\ \mathrm{ou_E\ m^{-3}}$ / 7–40%	Shape corresponding to the wind frequency of 10° sectors
Austria [1,2] [15]	wind frequency, wind velocity, exceedance probability	odour emission rate ($\mathrm{ou_E\ s^{-1}}$)	wind frequency, exceedance probability	$0.25\ \mathrm{ou_E\ m^{-3}}$ / 3–24%	Shape corresponding to the wind frequency of 10° sectors
Belgium [2] [16]	surface roughness, protection level (zoning)	number of animals, species, ventilation system, manure, feeding	$b = 0.5$	$10\ \mathrm{ou_E\ m^{-3}}$ / 2%	Ellipse by 1.2 E and 0.8 E, orientated in the prevailing wind direction
Old Austrian guideline [17,18]	wind frequency, zoning	number of animals, species, technical factor (ventilation system, manure and feeding)	$b = 0.5$	–	Shape parametrised by the frequency of the wind direction for 45° sectors (factor between 0.6 and 1.0)
Purdue setback model [2] [19]	wind frequency, zoning, topography, orientation and shape of the building	odour emission rate ($\mathrm{ou_E\ s^{-1}}$)	$b = 0.5$	–	Shape parametrised by the frequency of the wind direction for 45° sectors (factor between 0.75 and 1.0)
Ontario M (cf. Guo, et al. [20])	$a = 1$	species, number of animals, zoning, manure	$b = 1$	–	Circle
W-T [1,2] [21]	$a = 1.60$	odour emission rate ($\mathrm{ou_E\ s^{-1}}$)	$b = 0.6$	–	Circle

[1] Compared against dispersion models. [2] Used for an intercomparison of empirical equations.

For example, for an odour emission rate S = 14,000 ou$_E$ s^{-1} (which can correspond to about 1860 fattening pigs [22]), wind frequency F = 20‰ for a 10° sector and odour exceedance probabilities for residential areas (P = 10%) and rural areas (P = 15%), the separation distance E_G (m) returns 207 m and 152 m, respectively.

A minimum distance of 50 m was used for the regression coefficients estimation. It is thus recommended to reset to 50 m separation distances smaller than this minimum.

The German VDI EQ has been derived from dispersion model calculations at 23 sites. The German regulatory Lagrangian particle dispersion model (AUSTAL2000) was used. The exponent b and the multiplicative factor a of the power function ($E_G = a\,S^b$) have been obtained by three input parameters which were restricted in the range of an improved fit [5]. The basis of the power function is the odour emission rate S (ou$_E$ s^{-1}) in the range between $500 \le S \le 50{,}000$ ou$_E$ s^{-1}. The two other predictors are the frequency of the wind direction F (‰) of a 10° sector ($10 \le F \le 60$‰) and the odour exceedance probability P (%) ($7 \le P \le 40$%) of the odour impact criterion [12]. A single-point source with vertical release at a height of 5 m was considered for running the dispersion model. To apply the German VDI EQ also for wide-stretched odour sources (e.g., area sources), a focal point must be calculated by using the coordinates of individual odour sources, which are weighted by individual odour emission rates. An additional distance, depending on the extension of the wide-stretched source, will be added to the baseline separation distances calculated from the EQ [14]. Due to this determination, the requirement for a conservative approach was satisfied. To reduce the influence of the livestock building on the dispersion, the lowest calculable separation distance was set to 50 m. A guidance for the application of the German VDI EQ can be found in [23] discussing how the elongation of an odour source can be taken into account, how the wind statistics can be adapted for sites that are influenced by a valley wind system, how the hedonic tone of the species can be taken into account, how abatement measures to reduce odour emission can be considered, and case studies.

For the Austrian EQ, the regression analysis was done on the basis of dispersion calculations. The power function $E = a\,S^b$ is defined by the factor a and the exponent b, both of which depend on two meteorological parameters (the relative frequency of the wind direction F and the mean wind velocity W of the wind direction for 10° sectors), as well as P. The equation for the separation distance E_A by the Austrian regression model reads

$$E_A = P^{-0.389}\left(165\,F^{0.0289} - 3.63\,W - 150\right) S^{\frac{1}{-0.0381F+0.0191P+2.31}} \tag{2}$$

For the same inputs as in the example given above, but now considering a mean wind velocity of W = 2.5 m s^{-1} for a 10° sector, the separation distance E_A (m) returns 165 m and 97 m, respectively.

A minimum distance of 100 m was used for the regression coefficients estimation. This means that separation distances which are smaller than this minimum have to be reset to 100 m.

The Austrian EQ [15] has been derived based on dispersion model calculation at 6 sites. The model used was the Austrian Odour Dispersion Model (AODM) [2,24,25], a Gaussian plume dispersion model with a calculation scheme for the peak-to-mean factor, which depends on the travel distance and the stability of the atmosphere [26]. The exponent b and the multiplicative factor a of the power function ($E_A = a\,S^b$) are determined by four input parameters, which were also restricted in the range for an improved fit. The odour emission rate S (ou$_E$ s^{-1}) lies in the range between $400 \le S \le 24{,}000$ ou$_E$ s^{-1}. The two meteorological predictors for 10° sectors are the relative frequency of the wind direction $F \le 160$‰ and the mean wind velocity $W < 4$ m s^{-1}. The odour exceedance probability P (%) lies in the range of $3 \le P \le 24$%. The source geometry was a single-point source with a height of 6 m. The lowest calculable separation distance was set to 100 m, according to the limitations of the Gaussian dispersion model. The main difference between the two referred EQs is the best-fit approach for the Austrian equation, in contrast to a worst-case calculation for Germany.

The two EQs quantify the odour emission rate S (ou$_E$ s^{-1}) in the same way as it has to be done for dispersion modelling studies in their respective jurisdictions. This is a major advantage because the discrepancy between the results of the EQs and dispersion models is then related to effects other than the odour emission rate. Based on the simplification, the geometry of the emission cannot be included in the calculation by the EQs. The inclusion of such predictors is reserved to dispersion models. While the exponent b of the power function depends on the wind direction frequency and/or the odour exceedance probability in these equations, this parameter has a constant value for all other EQs in the range between 0.5 and 1.0.

The separation distance E_P of the Purdue EQ [19] reads as

$$E_P = a_P\, S^{0.5} \tag{3}$$

with the factor $a_P = 6.19\, F\, L\, T\, V$, which includes the impact of the wind frequency for 45° sector F, in the range of $0.75 < F < 1.0$, the land use factor L, describing the protection level for agricultural areas and pure residential areas between $0.5 < L < 1.0$, the topography T, describing good ventilated areas (flat terrain) and narrow valleys between $0.8 < L < 1.0$, and the orientation and the shape factor, describing the length to width (L/W) ratio of the livestock building V in relation to the wind direction in the range between $1.0 < V < 1.15$. The source strength S is given by the product of the specific odour emission factor per animal place ($ou_E\ s^{-1}$) and the number of animals. Calculations are supported by a spread sheet [27].

The Williams and Thompson model (W-T) [21] is a worst-case approach giving the distance $E_{W\text{-}T}$ from the source, within which complaints are likely, to the odour emission rate S ($ou_E\ s^{-1}$).

$$E_{W-T} = 1.60\, S^{0.6} \tag{4}$$

The other three EQs (Belgium [16], Old Austrian [17,18], and the Ontario (cf. Guo, Jacobson, Schmidt, Nicolai and Janni [20])) were not included in the comparison, because the odour emission rate S is parametrised by several empirical factors, which exclude an intuitive comparison with the other EQs.

A previous work [20] compared four EQs, Ontario (cf. [20]), the Williams and Thompson (W-T) [21], the old Austrian [18] and Purdue [19] EQs, against the Minnesota OFFSET model [28,29]. The last EQ does not deliver a pure separation distance. For several stability classes, wind velocities and protection levels (related to OIC), the factor a and the exponent b are given. In [20], the Minnesota OFFSET model was used as a reference for the other four EQs. All these EQs are based on a power function. The Purdue EQ [19] was a further development of the old Austrian EQ [18], with the advantage that the odour source S is given as odour rate and not parametrised by several empirical factors as it was previously suggested by the old Austrian EQ.

For some geographical areas, local environmental agencies have recommended locally adapted dispersion-based tools in combination with meteorological data for calculating separation distances. This simplifies the application of such tools because no specific meteorological knowledge is compulsory to run them.

For example, the German province North Rhine-Westphalia has developed a model called Screening Model for Odour Dispersion (SMOD) for planning and informative purposes of licensing procedures [30,31]. In The Netherlands, the V-Stacks model was developed for the entire area of the country [32]. For the Canadian province Manitoba, look-up tables have recently been developed based on AERMOD simulations [33]. Due to the fact that the meteorological data are an integral part of locally adapted solutions, these models cannot be transferred directly to other regions.

An EQ primarily developed for use in Switzerland [34] was not included in the intercomparison (Table 1) due to several reasons: (i) the odour emission rate is quantified by the odour intensity, (ii) the dilution is quantified by an exponential function, (iii) the description of the input parameters is incomplete, which means that a comparison with other EQs is not straightforward. Because no meteorological predictors are included for the calculations with the Swiss model, the separation distance is represented by a circle around the source. Similarly, an EQ derived with a view to assessing broiler farms in Western Australia [35] was not included as it did not to reach the point of having policy status.

Figure 1 depicts the separation distances calculated by using five EQs for a livestock building with 3000 fattening pigs (22,500 ou$_E$ s^{-1}, [14]) for the meteorological situation for Wels, an Austrian site. The prevailing wind directions at this site were from WSW and ENE. The comparison shows the impact of the meteorological situation (exposed by the wind rose) on the separation distances, calculated by the EQs. The five EQs were selected from Table 1, the German and the Austrian EQs, which include the wind statistics of the site, the Belgian EQ with a rough parametrisation for the prevailing wind direction, the Purdue EQ, which uses the wind frequency of 45° sectors and the W-T EQ which does not consider the wind frequency as a predictor. For Easterly winds, the impact of the width of the wind sectors can be seen in comparison to the German and Austrian EQs with 10° sectors and the Purdue EQ with 45° sectors. The impact of the wind frequency is very similar for the Belgium and the Purdue EQs. The Belgian EQ shows only a weak influence of directional wind frequencies and forms an ellipse orientated towards the prevailing wind directions with 20% longer major axes. However, the wind frequency is not taken into account in detail. The range for the wind frequency factor F of the Purdue EQ is $0.75 < F < 1.0$, which results in a maximum of 33% greater separation distance in the prevailing wind direction, compared to 20% for the Belgian EQ. The German VDI and the Austrian EQs show the highest sensitivity to the wind frequency. The W-T EQ shows that a circle with a constant separation distance for all directions is unsuitable to describe the meteorological situation of the dilution process in the atmosphere. Even if the claim of this EQ is a worst-case assessment, the enormous overestimation of separation distances for several directions will not help in some cases to find an appropriate location for an odour source concerning the wind situation of the site and residential areas.

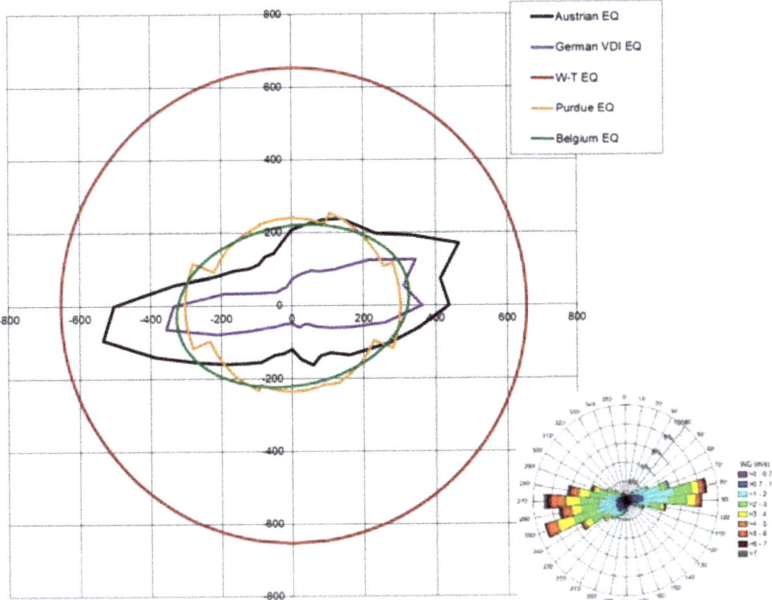

Figure 1. Separation distances calculated using the German VDI [5,14], Austrian [15], Belgian [16], Purdue [19] and Williams and Thompson (W-T) [21] empirical equations (EQs) for a protection level of rural areas and a 3000 head fattening pig livestock building. The wind statistics (frequency of wind directions and wind velocity per 10° sectors) from a site in Wels, Austria, are shown separately.

3. Case Studies

In this section, separation distances due to three selected EQs are qualitatively compared to those obtained from dispersion models. This is undertaken to explore the advantages and shortcomings of EQs. Moreover, the potential of EQs to be used as easy-to-use screening tools is tested. This comparison is carried out using available model data sets and thus is non-systematic. It is not intended to compare the output of the different dispersion models but the differences in separation distances between EQs and models. The following EQs have been selected: the two EQs from Germany and Austria, which include the wind statistics per 10° wind sectors, and the W-T EQ, which delivers a circle.

Dispersion models used for the comparison are AODM (used in the derivation of the Austrian EQ), LASAT (similar to AUSTAL2000, which was used to derive the German EQ) and the U.S. regulatory air quality model AERMOD. The latter has no link to any of the three EQs used for the comparison. Both an empirical-based peak-to-mean (p2m) procedure and a constant factor of 4 (f4) to account for short-term concentrations are considered for LASAT and AODM, as these models have in-house implementations of both schemes. As the p2m scheme could not promptly be incorporated for AERMOD, consequently, only the f4 scheme was used. The selected sites are located in Brazil, China, and Austria. Throughout the comparison, an odour impact criterion currently enforced in Germany [12] is applied. The German criterion is defined by an ambient concentration of 1 ou_E m^{-3} as a short-term value. The constant factor of 4, f4, is in use to account for the human perception of odours, which translates into an hourly mean of $C_T = 0.25$ ou_E m^{-3}.

3.1. São José dos Pinhais, Brazil

The Brazilian site (25.555° S, 49.132° W) is located in São José dos Pinhais, a city relatively close to Curitiba, the capital of the state of Paraná. This site lies in a plateau (known as "Planalto de Curitiba") shaped by gently rolling terrain. Different land uses can be found scattered around the site. The AERMOD Modelling System was used for the present investigation [36,37]. The extent to which AERMOD is suitable for assessing source impacts in this site has been shown in a previous study [38]. AERMOD is acknowledged as an advanced steady-state Gaussian plume model. It incorporates boundary layer turbulence theory and scaling concepts. The modelling system consists of three core modules: (i) the AERMOD dispersion model itself, (ii) the AERMET meteorological processor, and (iii) the AERMAP terrain processor. Versions 18,081 of these modules were used.

Here, a single-point source was selected for the prediction of ambient odour concentrations. The odour emission rate was $S = 15{,}000$ ou_E s^{-1}. The source geometry was assumed circular, with a height of 10 m from the ground, a diameter of 1.0 m, and vertical release. The effluent was released with an exit velocity and temperature of 3.0 m s^{-1} and 20 °C, respectively. A total of 2736 receptors were placed on a polar grid with a minimum distance from the source of 50 m. The receptors were set at breathing height (1.5 m). A digital elevation model was built using AERMAP on the basis of terrain data in SRTM1 (resolution of ~30 m). Elevations from near 880–918 m above sea level were observed within the model domain. The elevated terrain option was selected to consider that terrain heights can be above or below the stack base elevation.

Surface and upper air meteorological observations for 2015 were used as input data to AERMOD. The surface dataset, given by hourly values of air temperature, atmospheric pressure, cloud cover, wind direction and velocity, was obtained from the NOAA Integrated Surface Database [39] for Afonso Pena International Airport (25.531° S, 49.167° W, ~4.5 km from the site). The frequency of calm winds (<0.5 m s^{-1}) amounted to ~3.4% during 2015. Calms were not disregarded but reset to 0.5 m s^{-1}. Upper air soundings were obtained from the NOAA/ESRL Radiosonde Database [40] for the same airport and used as collected. The surface and upper air meteorological datasets were processed using AERMET, which in turn estimates the required boundary layer parameters for use by AERMOD. The Brazilian site has wind directions (at 10 m height) primarily from east to southeast (E–SE) together with secondary maxima from northeast to east-northeast (NE–ENE). The annual mean wind velocity was ~3.0 m s^{-1}. High velocities were observed from almost all quadrants, with a maximum of 19.0 m s^{-1} for the period.

Finally, separation distances were calculated using AERMOD and the two previously described EQs. The tolerated exceedance probability of the concentration threshold was exemplarily taken to be $P = 10\%$.

3.2. Beijing, People's Republic of China

The Chinese site (40.10° N, 116.16° E) is located ~30 km northwest from central Beijing. The city, with an average elevation of 43.5 m, is located at the northern part of the North China Plain. Beijing is enclosed by the Yanshan Mountains (average elevation 600–1500 m) to the north and the Taihang Mountains (average elevation 1500–2000 m) to the west [41]. Beijing is characterised by a monsoon-influenced, temperate continental climate, with humid and hot summers and dry, cold, and windy winters [42].

Odour emissions are due to a 300-head commercial dairy farm [43]. The terrain near the site is mostly flat, typically farmland. The farm has 0.67 km^2 of area, including feedlot pens, a feed mill, a slurry treatment workshop, and administrative offices. The farm contains three feedlot pens with a total area of 42,000 m^2 for raising the cows. The identified and quantified odour sources in the dairy farm include three barns for 100 cows each and the feed storage of corn silage.

Here, the results obtained in our previous work [43] are considered to augment the comparative analysis with the other two sites. Details on the use of the EQs, the AERMOD modelling approach and quantification of odour emissions at the Chinese site can be found in the referred previous work [43]. For convenience, a summary is presented below.

AERMOD, AERMET and AERMAP (versions 18081) were applied. The three barns were treated as area sources, for which a release height of 0.05 m was assumed. The feed storage was treated as a stack (release height of 2.5 m, exit velocity of 0.5 m s^{-1}, exit temperature of 20 °C and diameter of 0.8 m). The total odour emission rate of the dairy farm was $S = 5850$ ou$_E$ s^{-1}. This value was attained based on emission factors given in the VDI 3894 Part 2 [14]. A polar receptor network with a minimum distance from the source of 50 m was defined, totalling 1656 receptors set at breathing height (1.5 m). A digital elevation model was created using AERMAP with terrain data in SRTM1. Elevations from near 40–55 m above sea level occurred within the model domain. The elevated terrain option was selected.

Surface and upper-air meteorological observations for 2017 were used as input data. The surface dataset mainly corresponds to measurements at a station called Haidian (39.98° N, 116.28° E, ~17 km from the site). Calm winds (~6.2% of the observations) were reset to 0.5 m s^{-1}. Upper air soundings were acquired from the NOAA/ESRL Radiosonde Database for the Beijing Capital International Airport (40.08° N, 116.60° E, ~36 km from the site). The Chinese site has prevailing wind directions (at 10 m height) from northeast (NE) and southwest (SW). The annual mean wind velocity was ~1.5 m s^{-1}.

The application of the two EQs is limited to a single odour source. According to the German guideline VDI 3894 Part 2 [14], in the case of a single source, there will be also only one emission focal point. The present case, however, comprises four sources (i.e., three barns and one feed storage). An overall emission focal point was thus determined on the basis of the coordinates of the individual sources, which are weighted by their respective odour emission rates. Subsequently, the emission focal point of the dairy farm was used as the reference point (origin of the coordinate system) for the separation distance determination.

The German odour impact criterion, as for the Brazilian site, was selected. However, the tolerated exceedance probability of the concentration threshold was taken to be $P = 15\%$ because the vicinity of the dairy farm is mostly rural.

3.3. Kittsee, Austria

Separation distances have also been calculated for Kittsee (48.109° N, 17.070° E,), east of Vienna in Austria, near Bratislava [11]. Kittsee is situated within flat terrain, mainly farmland. For this site, separation distances were determined using two dispersion models and then compared to those obtained from the two investigated EQs. The dispersion models used are AODM, a steady-state

Gaussian plume model adapted for odour assessments, and the Lagrangian particle model LASAT. For both dispersion models, an empirical-based peak to mean (p2m) approach, giving the separation distance as a function of the travel distance and atmospheric stability [2,24–26], has been applied as a post-processing tool [44]. This procedure gives two sets of separation distances, namely, one for each dispersion model.

As described before, in Germany, f4 is in use, which is independent of the distance from the source and meteorological conditions [45]. Also, f4 was considered for the two dispersion model outputs, by this means returning two additional sets of separation distance results. Accordingly, for Kittsee, four different model-related sets of separation distances, besides other two related to the investigated EQs, are attained.

A dataset of more than one year of half-hourly ultrasonic anemometer measurements (03.03.2006–31.05.2007) was used. At this site, high wind velocities occurred mainly from NW, often associated with frontal systems and storms. The secondary prevailing winds are from ENE, showing on average lower wind velocities as they are mainly observed in anti-cyclonic conditions. Calm winds frequency was ~0.75% during the period. Calms were discarded. The mean wind velocity of the period was 4.1 m s^{-1}. A single-point source was selected with an odour emission rate of S = 5200 ou$_E$ s^{-1}, 8 m high and vertical exit velocity of 3.0 m s^{-1}.

The German odour impact criterion was used for a protection level of residential areas ($P = 10\%$).

3.4. Summary of Input Parameters

The parameters which describe the odour source and the selected dispersion model are summarised in Table 2.

Table 2. Summary of source and site features and input parameters of the three selected case studies in São José dos Pinhais, Brazil; Beijing, People's Republic of China, and Kittsee, Austria, for the selected EQs and dispersion models.

Case Study	Source and Site Features	Input Parameters	Dispersion Model
São José dos Pinhais, Brazil [38]	Single-point source with vertical release Mean wind velocity = 2.9 m s^{-1} Elevations from 880–918 m within the model domain	Odour emission rate S = 15000 ou$_E$ s^{-1} Emission height = 10 m Exit velocity = 3 m s^{-1} Protection level $P = 10\%$	AERMOD + f4
Beijing, People's Republic of China [43]	Single-point source with vertical release and three area sources Mean wind velocity = 1.5 m s^{-1} Elevations from 40–55 m within the model domain	Total odour emission rate S = 5850 ou$_E$ s^{-1} Area sources emission height = 0.05 m Point source emission height = 2.5 m Point source exit velocity = 0.5 m s^{-1} Protection level $P = 15\%$	AERMOD + f4
Kittsee, Austria [11]	Single-point source with vertical release Mean wind velocity = 4.1 m s^{-1} Flat terrain	Odour emission rate S = 5200 ou$_E$ s^{-1} Emission height = 8 m Exit velocity = 3 m s^{-1} Protection level $P = 10\%$	LASAT + f4 LASAT + p2m AODM + f4 AODM + p2m

3.5. Comparison of the Separation Distance Calculated by Dispersion Models and EQs

For the three case studies, Figure 2 depicts the calculated separation distances as contour plots (left panels), and the site-dependent meteorological input data (right panels) as wind roses. The wind roses summarise the wind distribution at the sites, displaying their strength, direction and frequency per 10° sectors over the specified period of data collection. The separation distance shapes resemble to a great extent the wind distribution of the sites. The largest distances tended to occur along with the prevailing wind directions. In other words, the calculated separation distances were significantly influenced by meteorological conditions.

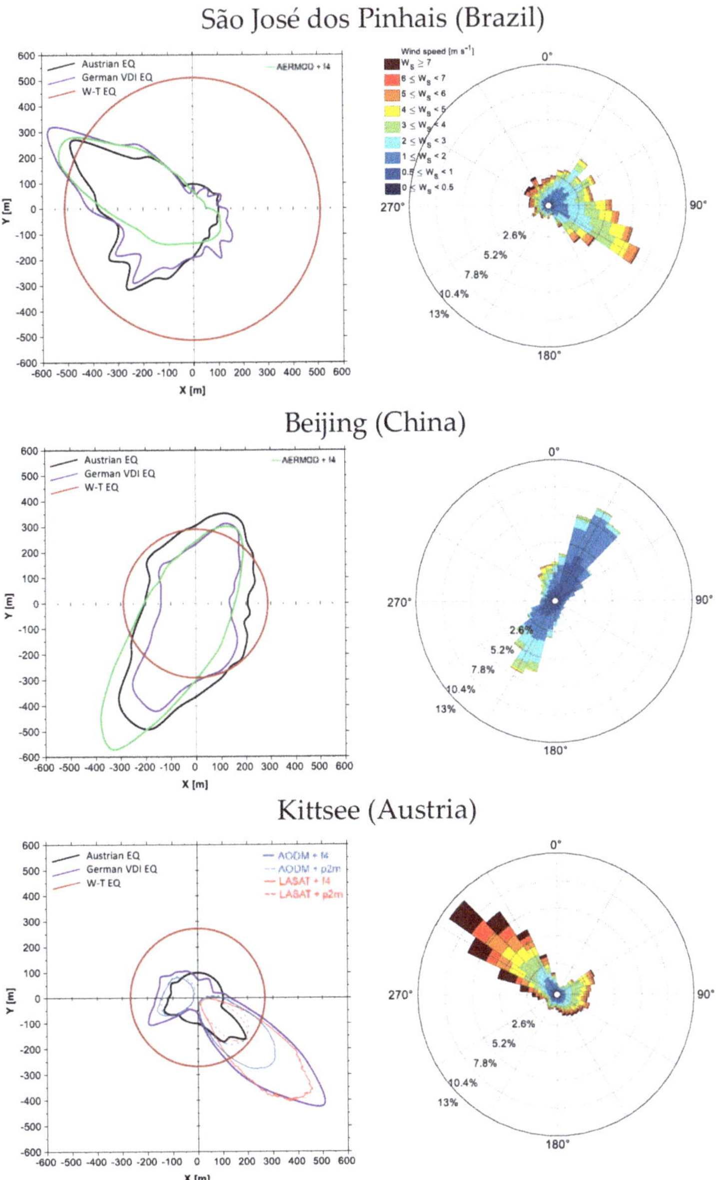

Figure 2. Separation distances calculated using the German VDI, the Austrian, and the W-T EQs against dispersion models for the three case studies under investigation and their respective wind roses.

The first case study is from a site in São José dos Pinhais (Brazil) for which the comparison of separation distances between the AERMOD dispersion model against the two EQs is shown. The conformity of the direction-dependent separation distances derived from these three methods is satisfactory. In this case, the separation distances show a maximum for the two prevailing wind directions, SE and NE, meaning more elongated distances towards NW and SW. For SE winds with a portion of higher wind velocities, the agreement is closer compared to the secondary prevailing wind direction blowing from NE. Moreover, for these NE winds, AERMOD delivers lower distances than the two EQs. The Austrian EQ separation distance pattern towards the 45° sector is due to the limitation to a minimum distance given by this equation. That is, all those separation distances between North and East were reset to 100 m because the Austrian EQ returned smaller values than this minimum, as explained in Section 2. Even if AERMOD was not used to derive the parameters of the two EQs, the shape of the separation distances reflects the meteorological situation of the site in a comparable way.

The second case study is from Beijing (China). Overall, the separation distances show a good correspondence between the two EQs and AERMOD for this case. The Austrian EQ gives generally higher values for the separation distances compared to the German VDI EQ. For the secondary prevailing winds from SW with higher wind velocities, the separation distances of the two EQs and AERMOD show a better agreement compared to the distances related to the north-easterly prevailing wind direction. The wind rose of the site shows high wind velocities for wind directions from N and NW, which results in separation distances towards SE comparable to those in the main wind directions.

The third case study is from Kittsee (Austria). This site features separation distance results from two dispersion models, LASAT and AODM, using the p2m approach, in comparison to the two investigated EQs. While the parameters of the German VDI EQ have been derived using the German regulatory dispersion model AUSTAL2000, the parameters of the Austrian EQ have been derived from AODM calculations. Results applying f4 from [12] are also shown. The first couple (LASAT + f4 and the German VDI EQ) shows a good agreement for the largest separation distances, especially for the NW winds. These large separation distances in the far-field are a well-known effect due to the use of f4. This overestimation for separation distances of several hundred meters was discussed in detail in a previous study [25]. The second couple (AODM + p2m and the Austrian EQ) shows in general a good agreement for the calculated separation distances as well. For the secondary prevailing wind direction from ENE, thus with a transport direction towards WSW, the agreement between the separation distances of the two EQs is much closer. The inconsistency for the NW winds (transport direction towards SE) could be explained by the uncertainty intrinsically related to the estimation of instantaneous ambient concentrations by the p2m approach. For Kittsee, the wind velocity is distinctly higher as compared to the Beijing site. This could be a reason for the higher deviation between the dispersion models and the two EQs in Kittsee.

In accordance with the German OIC [12], the protection level in the Austrian and German EQs is taken into account for a constant odour threshold $C_T = 0.25$ ou$_E$ m^{-3} and an adjustable exceedance probability P (%). In the German guideline, for "commercial, industrial and agricultural areas" and "villages", the exceedance probability is taken as $P = 15\%$. For "residential and mixed areas" as $P = 10\%$. These benchmark exposure limits can be reduced for unpleasant odours and increased for pleasant odours through weighting factors. However, such weighting factors are not considered here. The main condition that was necessary to be followed for the present investigation was to match the exceedance probability P of the odour impact criterion for the two methods (EQ and dispersion models). If a weighting factor for hedonic tone adjustment had been considered, this would return in the end the same P for both methods, thus allowing the separation distances to be compared.

As the emission sources assumed for São José dos Pinhais (Brazil) and Kittsee (Austria) are hypothetical, a reason for selection of stacks and its parameters has to be given. For these cases, we intended to approximate emissions from livestock buildings (mechanically ventilated). In turn, a single-point source was set for each case because it is well known that dispersion models can potentially provide more accurate results for this source typology.

In general, the three case studies demonstrate that the two EQs from Germany and Austria, which include sufficient meteorological data by a 10° wind rose, provide a first guess of the direction-dependent separation distances. Other EQs, which are less elaborated, seem inappropriate for such applications. The consideration of appropriate EQs can be particularly important in the context of tiered regulatory frameworks. If needed, a more detailed investigation using state-of-the-art dispersion models has to be applied.

The case study of Kittsee shows (Figure 2c) a closer relationship between the EQ and the underlying dispersion model with their corresponding peak-to-mean approaches. The dispersion models with f4 show a better agreement with the German EQ. The Austrian EQ, which was derived by using the variable p2m, shows a better agreement between dispersion models using this scheme. In short, it can be seen that the best match is achieved when the peak-to-mean approach that was used to derive the EQ matches with the peak-to-mean approach of the dispersion model.

The W-T EQ was selected for the case studies because it delivers a constant separation distance for all directions. The circle given by this EQ shows that the meteorologically-driven dispersion and transport in the atmosphere cannot be explained by such an oversimplification. Even if this EQ was designed as a worst-case scenario, the separation distances towards the prevailing winds are underestimated for all case studies. Contrary, all the separation distances for wind directions with a low frequency are overestimated. This emphasises that the concept of EQs needs at least a meteorological input based on the wind direction frequency. As expected, the use of 10° wind sectors compared to 45° sectors (as for the old Austrian and the Purdue EQs) shows that this feature improves the separation distance resolution considerably (Figure 2).

4. Discussion

From the list of EQs given in Table 1 and discussed throughout Section 2, the range of empirical methods to give a first guess of separation distances becomes apparent. Simple EQs that do not take into account the meteorological conditions of a site deliver a unique, fixed distance, thereby shaping a circle around the source. More elaborated EQs, especially the Austrian and German EQs, include meteorological predictors (wind direction frequency and average wind velocity per wind direction sector) in their formulations (Figure 1).

Only for the latter case study, the separation distances are compared against those of two different dispersion models using two different peak-to-mean approaches (Figure 2). These EQs employ the highest resolution of the wind direction (thirty six 10° sectors) compared to eight 45° sectors as a separation distance predictor and are consequently considered by the authors to be state-of-the-art. The Austrian EQ also includes the mean wind velocity per wind direction sector as a predictor with the objective of pragmatically considering atmospheric stability. The reason this was considered is centred on the greater variability in atmospheric stability conditions and higher calm frequencies of Austrian sites as compared to most German sites. A simple circular EQ (W-T) was also included in the comparisons shown in Figure 2 to show the remarkable differences in separation distances between EQs considering or not considering meteorology. The impact of the peak to mean approach on the separation distance is here shown only for Kittsee (Figure 2), but was recently discussed more elaborately [46].

The high level of simplifications when using EQs becomes obvious by comparing the indispensable input data for dispersion models and EQs. The characterisation of the emission is reduced to either a constant emission rate or to the number of animals and an odour emission factor to quantify the emission rate. The second approach by the number of animals reduces the applicability of such EQs solely to livestock houses. The German and the Austrian EQs use the odour emission rate, which means that they can also be applied to non-agricultural odour sources. Most of the EQs cannot include the geometry of the emission source (height, vertical velocity, outlet air temperature, area vs. point sources). The highest reduction of input data can be seen for the characterisation of the meteorological situation of a site. For dispersion models, the minimum meteorological input data are usually characterised by hourly values of wind velocity, wind direction, and stability of the atmosphere (with $N = 8760 \times 3 = 26{,}280$ data points), whereas the meteorological input required by the EQs is reduced to the wind frequency distribution for 10° sectors ($N = 36$, German VDI EQ) and mean wind direction for 10° sectors ($N = 72$, Austrian EQ) or even less information (45° sectors with $N = 8$, old Austrian and Purdue EQs). Most of the remaining EQs do not incorporate any meteorological input (Table 1).

The OIC are country-specific; either the odour concentration threshold C_T is held constant and the exceedance probability is adapted to the protection level P, or the other way round with a constant exceedance probability P and a variable odour concentration threshold C_T for a certain protection level. It seems difficult to compare the two approaches but [13] could show that a similarity for the separation distance could be found for various OIC. E.g., the German OIC for pigs in a rural area with $P = 15\%$ and $C_T = 1$ ou$_E$ m^{-3} corresponds to an OIC with $P = 2\%$ and $C_T = 5.4$ ou$_E$ m^{-3}, which is roughly the Irish OIC. This means that the Austrian and the German EQs can potentially be used as a substitute for those countries as well, where the exceedance probability P is hold constant and the odour concentration threshold C_T is adapted to the protection level.

Figure 3 shows a schematic diagram comparing the state-of-the-art procedure to calculate separation distances using a dispersion model against the structure of EQs. It is acknowledged that dispersion models can simulate many more physical processes than EQs and thus often have to be run with more complex input data and parameterisations. In the simplified schematic shown here, both procedures start with the odour emission rate S and end with the direction-dependent separation distance E. The red arrows show the simplifications of the input parameters (meteorology and OIC) in the EQs.

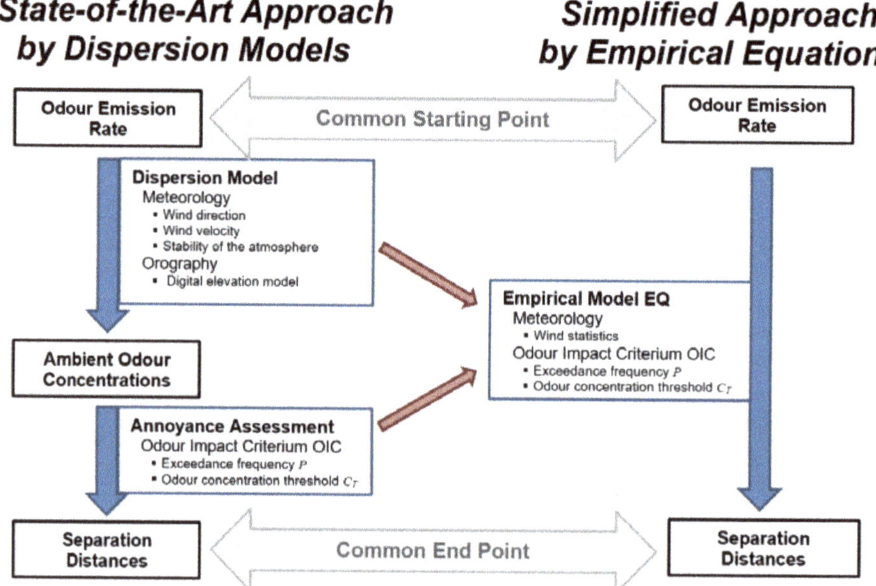

Figure 3. Schematic diagram of the state-of-the-art determination of separation distances by the use of a dispersion model (left side) and the simplified approach shown by the Austrian and German EQs (right side).

A major advantage of using EQs lies in their simplified handling of the influence of meteorological input on separation distances. Dispersion models are typically inputted with annual hourly time series of, for example, wind direction and velocity, atmospheric stability and mixing height. A meteorological dataset with lower temporal resolution (e.g., mean daily data) is not possible, because the OIC, which are needed for the calculation of separation distances, are typically derived for hourly data. Furthermore, atmospheric stability is not routinely measured at standard meteorological stations. Three-axis ultrasonic anemometers, in particular, offer the possibility of estimating atmospheric stability via the Obukhov length, based on wind and turbulence measurements [11,47]. For most of the stability classification schemes, additional non-standard meteorological parameters are necessary to determine stability estimates in the form of classes. The scheme of Golder [48] developed for Turner stability classes (the scheme developed by Reuter [49] is very similar to it) has been used here for AODM. Such a scheme considers sun elevation angle, cloud base height and cloud cover; alternatively, the radiation balance (net radiation) or the vertical temperature gradient, each in combination with the wind velocity, can be used. The details of the schemes are given in Section 4.6 of [50] and in [11].

Conversely, separation distances from EQs are determined by the equation coefficient values, which are derived from a statistical analysis of the time series of modelled ambient odour concentrations by OIC. The EQ procedure includes implicit input of the exceedance probability P to the empirical equations. This means that the two-step procedure of the state-of-the-art modelling methodology is reduced to a single step for EQs.

The German and Austrian EQs are based on model calculations for flat terrain. This means that complex topographical features cannot be considered. The adaptation of meteorological data (wind statistics) to the orographic situation of a site is described in a handbook about the German VDI EQ [23].

Furthermore, most of the existing EQs have been derived for livestock buildings. The odour emission rate related to this sector can be estimated, for example, using emission factors related to the animal species, numbers and body mass, manure handling system, and laying area, among other parameters. These parameters are subsequently used as activity values to scale the odour emission rate. Only those EQs can be applied for other types of odour sources such as wastewater treatment plants and municipal solid waste disposals, which use the odour emission rate (ou$_E$ s^{-1}) as input parameter.

5. Conclusions

In this work, empirical equations were selected for investigation on the basis that, each in a different and unique manner, they greatly simplify the process of determining the direction-dependent influence of meteorological data on calculated separation distances when compared to dispersion modelling. These EQs are fast and low cost to use. They also meet the requirements that the odour emission rate is quantified in the same way as it is done for dispersion models and separation distances are determined for odour impact criteria as for dispersion models. These constraints allow for a meaningful comparison of empirical equation separation distances against modelled separation distances. In practice, the required meteorological information for the utilisation of the German and Austrian empirical equations are wind statistics in the form of the wind direction frequency distribution and mean wind velocity for 10° sectors (the latter for Austria only).

The investigated empirical equations, developed from simple linear regression models, can potentially calculate distances that are representative of modelling, particularly towards prevailing winds, if the conditions for which they were developed are observed. Otherwise, the empirical equations may give results very different from those provided by dispersion models.

The results suggest that some of the investigated empirical equations can be usefully incorporated as screening-level analysis tools in tiered regulatory odour assessment frameworks. A tiered framework recognises that tools such as simple power function-based equations may be sufficient to demonstrate that a proposal presents a low risk of impacting on amenities at nearby sensitive receptors in some cases. If compliance is demonstrated using such simple screening-level tools, more complex and costly investigations using dispersion modelling might be avoided. In contrast, an assessment using more refined tools such as dispersion models may be required if screening level assessments are not passed.

Author Contributions: G.S.: conceptualisation, writing—original draft preparation; M.B., G.S. and M.P.: methodology; W.K., M.B., G.S., M.P. and C.W.: preparation of input data and modelling; G.S., M.P., K.D.G. and

M.B.: writing—review and editing; M.B., G.S. and M.P.: projects' administration; M.B., G.S., M.P. and C.W.: funding acquisition; G.S. and M.P.: supervision. All authors have read and agreed to the published version of the manuscript.

Funding: Marlon Brancher is supported by the Austrian Science Fund (FWF) in the framework of the Lise Meitner Programme [project number M 2548-N29]. We also thank the National Key R&D Program of China (Nos. 2016YFC0700603, 2016YFC0700600 and 2016YFE0115500), the National Natural Science Foundation of China (21576023), and the Fundamental Research Funds for the Central Universities (No. FRF-TP-17-047A1). The cooperation between Austria and China was funded by the OeAD-GmbH (No. CN 10/2016).

Conflicts of Interest: The authors declare no conflict of interest. The funders had no role in the design of the study; in the collection, analyses, or interpretation of data; in the writing of the manuscript, or in the decision to publish the results.

References

1. Hayes, E.; Curran, T.P.; Dodd, V. A dispersion modelling approach to determine the odour impact of intensive poultry production units in Ireland. *Bioresour. Technol.* **2006**, *97*, 1773–1779. [CrossRef] [PubMed]
2. Schauberger, G.; Piringer, M.; Petz, E. Diurnal and annual variation of the sensation distance of odour emitted by livestock buildings calculated by the Austrian odour dispersion model (AODM). *Atmos. Environ.* **2000**, *34*, 4839–4851. [CrossRef]
3. Douglas, P.; Hayes, E.; Williams, W.; Tyrrel, S.; Kinnersley, R.; Walsh, K.; O'Driscoll, M.; Longhurst, P.; Pollard, S.; Drew, G. Use of dispersion modelling for Environmental Impact Assessment of biological air pollution from composting: Progress, problems and prospects. *Waste Manag.* **2017**, *70*, 22–29. [CrossRef] [PubMed]
4. Pandey, G.; Sharan, M. Performance evaluation of dispersion parameterization schemes in the plume simulation of FFT-07 diffusion experiment. *Atmos. Environ.* **2018**, *172*, 32–46. [CrossRef]
5. Schauberger, G.; Schmitzer, R.; Kamp, M.; Sowa, A.; Koch, R.; Eckhof, W.; Eichler, F.; Grimm, E.; Kypke, J.; Hartung, E. Empirical model derived from dispersion calculations to determine separation distances between livestock buildings and residential areas to avoid odour nuisance. *Atmos. Environ.* **2012**, *46*, 508–515. [CrossRef]
6. Sironi, S.; Capelli, L.; Centola, P.; Del Rosso, R.; Pierucci, S. Odour impact assessment by means of dynamic olfactometry, dispersion modelling and social participation. *Atmos. Environ.* **2010**, *44*, 354–360. [CrossRef]
7. Capelli, L.; Sironi, S.; Del Rosso, R.; Guillot, J.-M. Measuring odours in the environment vs. dispersion modelling: A review. *Atmos. Environ.* **2013**, *79*, 731–743. [CrossRef]
8. Brancher, M.; Griffiths, K.D.; Franco, D.; Lisboa, H.D.M. A review of odour impact criteria in selected countries around the world. *Chemosphere* **2017**, *168*, 1531–1570. [CrossRef]
9. Piringer, M.; Knauder, W.; Petz, E.; Schauberger, G. Factors influencing separation distances against odour annoyance calculated by Gaussian and Lagrangian dispersion models. *Atmos. Environ.* **2016**, *140*, 69–83. [CrossRef]
10. Griffiths, K.D. Disentangling the frequency and intensity dimensions of nuisance odour, and implications for jurisdictional odour impact criteria. *Atmos. Environ.* **2014**, *90*, 125–132. [CrossRef]
11. Piringer, M.; Knauder, W.; Petz, E.; Schauberger, G. A comparison of separation distances against odour annoyance calculated with two models. *Atmos. Environ.* **2015**, *116*, 22–35. [CrossRef]
12. GOAA. *Guideline on Odour in Ambient Air (GOAA). Detection and Assessment of Odour in Ambient Air*; LAI (Bund/Länderarbeitsgemeinschaft für Immissionsschutz): Berlin, Germany, 2008.
13. Sommer-Quabach, E.; Piringer, M.; Petz, E.; Schauberger, G. Comparability of separation distances between odour sources and residential areas determined by various national odour impact criteria. *Atmos. Environ.* **2014**, *95*, 20–28. [CrossRef]
14. VDI 3894 Part 2. *Emissions and Their Impact from Livestock Operations. Method to Determine the Separation Distance for Odour*; Verein Deutscher Ingenieure, VDI: Düsseldorf, Germany, 2012.
15. Schauberger, G.; Piringer, M.; Jovanovic, O.; Petz, E. A new empirical model to calculate separation distances between livestock buildings and residential areas applied to the Austrian guideline to avoid odour nuisance. *Atmos. Environ.* **2012**, *47*, 341–347. [CrossRef]

16. Nicolas, J.; Delva, J.; Cobut, P.; Romain, A.-C. Development and validating procedure of a formula to calculate a minimum separation distance from piggeries and poultry facilities to sensitive receptors. *Atmos. Environ.* **2008**, *42*, 7087–7095. [CrossRef]

17. Schauberger, G.; Piringer, M. Guideline to assess the protection distance to avoid annoyance by odour sensation caused by livestock husbandry. In Proceedings of the 5th International Livestock Environment Symposium ILES V, St. Joseph, MI, USA, 29–31 May 1997.

18. Schauberger, G.; Piringer, M. Assessment of the protection distance to avoid annoyance by odour sensation caused by livestock husbandry by the Austrian guideline. In Proceedings of the International Symposium, Vinkeloord, The Netherlands, 6–10 October 1997; pp. 677–684.

19. Lim, T.T.; Heber, A.J.; Ni, J.-Q.; Grant, R.; Sutton, A.L. ODOR IMPACT DISTANCE GUIDELINE FOR SWINE PRODUCTION SYSTEMS. *Proc. Water Environ. Fed.* **2000**, *2000*, 773–788. [CrossRef]

20. Guo, H.; Jacobson, L.D.; Schmidt, D.R.; Nicolai, R.E.; Janni, K.A. Comparison of five models for setback distance determination from livestock sites. *Can. Biosyst. Eng.* **2004**, *46*, 6.17–16.25.

21. Williams, M.L.; Thomson, N. The effect of weather on odour dispersion from livestock buildings and from fields. In *Odour Prevention and Control of Organic Sludge and Livestock Farming*; Nielsen, V.C., Voorburg, J.H., L'Hermite, P.L., Eds.; Elsevier Applied Science: New York, NY, USA, 1985; pp. 227–233.

22. VDI 3894 Part 1. *Emissions and Immissions from Animal Husbandry—Housing Systems and Emissions—Pigs, Cattle, Poultry, Horses*; Verein Deutscher Ingenieure, Ed.; Verein Deutscher Ingenieure, VDI: Düsseldorf, Germany, 2011.

23. Eckhof, W.; Gallmann, E.; Grimm, E.; Hartung, E.; Kamp, M.; Koch, R.; Lang, M.; Schauberger, G.; Schmitzer, R.; Sowa, A. *Emissionen und Immissionen von Tierhaltungsanlagen—Handhabung der Richtlinie VDI 3894*; KTBL-Schrift 494: Darmstadt, Germany, 2012; p. 216.

24. Piringer, M.; Petz, E.; Groehn, I.; Schauberger, G. A sensitivity study of separation distances calculated with the Austrian Odour Dispersion Model (AODM). *Atmos. Environ.* **2007**, *41*, 1725–1735. [CrossRef]

25. Schauberger, G.; Piringer, M.; Petz, E. Calculating direction-dependent separation distance by a dispersion model to avoid livestock odour annoyance. *Biosyst. Eng.* **2002**, *82*, 25–37. [CrossRef]

26. Piringer, M.; Knauder, W.; Petz, E.; Schauberger, G. Use of ultrasonic anemometer data to derive local odourrelated peak-to-mean concentration ratios. *Chem. Eng. Trans.* **2014**, *40*, 103–108.

27. Available online: https://engineering.purdue.edu/~{}odor/setback.htm (accessed on 25 June 2020).

28. Jacobson, L.D.; Guo, H.; Schmidt, D.R.; Nicolai, R.E.; Zhu, J.; Janni, K.A. Development of the OFFSET model for determination of odor-annoyance-free setback distances from animal production sites: Part I. Review and experiment. *Trans. ASAE* **2005**, *48*, 2259–2268. [CrossRef]

29. Nimmermark, S.A.; Jacobson, L.D.; Schmidt, D.R.; Gay, S.W. Predictions by the Odor From Feedlots, Setback Estimation Tool (OFFSET) compared with observations by neighborhood monitors. *J. Air Waste Manag. Assoc.* **2005**, *55*, 1306–1314. [CrossRef] [PubMed]

30. Hartmann, U.; Janicke, L.; Janicke, U.; Höscher, M. A Screening Model for Odor Dispersions (SMOD). In *Gerüche in der Umwelt: Innenraum- und Außenluft; Tagung, Bad Kissingen, 13. und 14. November 2007*; VDI Verl.: Düsseldorf, Germany; pp. 302–312.

31. Janicke, L. *SMOD—Erstellung Eines Screening-Modells Für Geruchsimmissionen*; Landesamtes für Natur, Umwelt und Verbraucherschutz NRW: Essen, Germany, 2007.

32. Available online: https://www.infomil.nl/onderwerpen/landbouw/geur/model-v-stacks/ (accessed on 25 June 2020).

33. La, A.; Zhang, Q.; Gao, Z.; Guo, H. A Dispersion-Based Tool for Assessing Odor Impact of Hog Operations. In Proceedings of the 2019 ASABE Annual International Meeting, Boston, MA, USA, 7–10 July 2019; p. 1.

34. Steiner, B.; Keck, M. *Grundlagen zu Geruch und dessen Ausbreitung für die Bestimmung von Abständen bei Tierhaltungsanlagen*; Agroscope: Tänikon, Switzerland, 2018; p. 44.

35. Griffiths, D. A risk-based procedure for broiler farm separation distance calculations. In Proceedings of the 21st International Clean Air and Environment Conference, Sydney, Australia, 7–11 September 2013.

36. Perry, S.; Cimorelli, A.; Paine, R.J.; Brode, R.W.; Weil, J.C.; Venkatram, A.; Wilson, R.B.; Lee, R.F.; Peters, W.D. AERMOD: A Dispersion Model for Industrial Source Applications. Part II: Model Performance against 17 Field Study Databases. *J. Appl. Meteorol.* **2005**, *44*, 694–708. [CrossRef]

37. Cimorelli, A.; Perry, S.; Venkatram, A.; Weil, J.C.; Paine, R.J.; Wilson, R.B.; Lee, R.F.; Peters, W.D.; Brode, R.W. AERMOD: A Dispersion Model for Industrial Source Applications. Part I: General Model Formulation and Boundary Layer Characterization. *J. Appl. Meteorol.* **2005**, *44*, 682–693. [CrossRef]
38. Brancher, M.; Piringer, M.; Franco, D.; Filho, P.B.; Lisboa, H.D.M.; Schauberger, G. Assessing the inter-annual variability of separation distances around odour sources to protect the residents from odour annoyance. *J. Environ. Sci.* **2019**, *79*, 11–24. [CrossRef] [PubMed]
39. Available online: https://www.ncdc.noaa.gov/isd (accessed on 25 June 2020).
40. Available online: https://ruc.noaa.gov/raobs (accessed on 25 June 2020).
41. Xiang, Y.; Zhang, T.; Liu, J.; Lv, L.; Dong, Y.; Chen, Z. Atmosphere boundary layer height and its effect on air pollutants in Beijing during winter heavy pollution. *Atmos. Res.* **2019**, *215*, 305–316. [CrossRef]
42. Xu, M.; Sbihi, H.; Pan, X.; Brauer, M. Local variation of PM2.5 and NO2 concentrations within metropolitan Beijing. *Atmos. Environ.* **2019**, *200*, 254–263. [CrossRef]
43. Wu, C.; Brancher, M.; Yang, F.; Liu, J.; Qu, C.; Schauberger, G.; Piringer, M. A Comparative Analysis of Methods for Determining Odour-Related Separation Distances around a Dairy Farm in Beijing, China. *Atmosphere* **2019**, *10*, 231. [CrossRef]
44. Schauberger, G.; Piringer, M.; Schmitzer, R.; Kamp, M.; Sowa, A.; Koch, R.; Eckhof, W.; Grimm, E.; Kypke, J.; Hartung, E. Concept to assess the human perception of odour by estimating short-time peak concentrations from one-hour mean values. Reply to a comment by Janicke et al. *Atmos. Environ.* **2012**, *54*, 624–628. [CrossRef]
45. Janicke, L.; Janicke, U.; Ahrens, D.; Hartmann, U.; Müller, W.J. Development of the odour dispersion model AUSTAL2000G in Germany. In Proceedings of the Environmental Odour Management, VDI-Berichte 1850, Cologne, Germany, 17–21 November 2004.
46. Brancher, M.; Hieden, A.; Baumann-Stanzer, K.; Schauberger, G.; Piringer, M. Performance evaluation of approaches to predict sub-hourly peak odour concentrations. *Atmos. Environ. X* **2020**, 100076. [CrossRef]
47. Piringer, M.; Knauder, W.; Petz, E.; Schauberger, G. Determining separation distances to avoid odour annoyance with two models: A comparison of two sites. *Chem. Eng. Trans.* **2016**, *54*, 7–12. [CrossRef]
48. Golder, D. Relations among stability parameters in the surface layer. *Boundary Layer Meteorol.* **1972**, *3*, 47–58. [CrossRef]
49. Reuter, H. Die Ausbreitungsbedingungen von Luftverunreinigungen in Abhängigkeit von meteorologischen Parametern. *Theor. Appl. Clim.* **1970**, *19*, 173–186. [CrossRef]
50. Piringer, M.; Joffre, S. *The Urban Surface Energy Budget and the Mixing Height in European Cities: Data, Models and Challenges for Urban Meteorology and Air Quality—Final Report of Working Group 2 of COST-715 Action*; Piringer, M., Joffre, S., Eds.; Demetra Ltd.: Sofia, Bulgaria, 2005.

Review

Summary and Overview of the Odour Regulations Worldwide

Anna Bokowa [1], Carlos Diaz [2], Jacek A. Koziel [3,*], Michael McGinley [4], Jennifer Barclay [5], Günther Schauberger [6], Jean-Michel Guillot [7], Robert Sneath [8], Laura Capelli [9], Vania Zorich [10], Cyntia Izquierdo [2], Ilse Bilsen [11], Anne-Claude Romain [12], Maria del Carmen Cabeza [13], Dezhao Liu [14], Ralf Both [15], Hugo Van Belois [16], Takaya Higuchi [17] and Landon Wahe [3]

[1] EOC Environmental Odour Consulting Corp, Oakville, ON L6J 2Y2, Canada; bokowa.anna@environmentalodourconsulting.com

[2] Ambiente et Odora S.L., 48001 Bilbao, Spain; carlosdiaz@olores.org (C.D.); cyntiaizquierdo@olores.org (C.I.)

[3] Department of Agricultural and Biosystems Engineering, Iowa State University, Ames, IA 50011, USA; lwahe@iastate.edu

[4] St. Croix Sensory Inc., Stillwater, MN 55082, USA; mike@fivesenses.com

[5] Atmospheric Science Global Ltd., Auckland 0600, New Zealand; jenniferbarclay@xtra.co.nz

[6] WG Environmental Health, Department for Biomedical Sciences, University of Veterinary Medicine, 1210 Vienna, Austria; gunther.schauberger@vetmeduni.ac.at

[7] Environmental Engineering, LSR, IMT Mines Alès, 30319 Alès, France; jean-michel.guillot@mines-ales.fr

[8] Silsoe Odours Ltd., Silsoe, Bedford, Bedfordshire MK45 4HP, UK; robert.sneath@silsoeodours.co.uk

[9] Politecnico di Milano, Department of Chemistry, Materials and Chemical Engineering "Giulio Natta", 20133 Milano, Italy; laura.capelli@polimi.it

[10] Ecometrika, The Synergy Group, Santiago 1030000, Chile; vzorich@ecometrika.com

[11] VITO, Flemish Institute for Technological Research, 2400 Mol, Belgium; ilse.bilsen@vito.be

[12] University of Liege, 8362 Arlon, Belgium; acromain@ulg.ac.be

[13] Directorate of Sectorial and Urban Environmental Affairs, Ministry of Environment and Sustainable Development, Bogota 110311, Colombia; mcabeza@minambiente.gov.co

[14] College of Biosystems Engineering and Food Science, Zhejiang University, Hangzhou 310058, China; dezhaoliu@zju.edu.cn

[15] North Rhine—WestphaliaOffice for Nature, Environment and Consumer Protection (LANUV), 45133 Essen, Germany; Ralf.Both@lanuv.nrw.de

[16] Van Belois Environmental Services, 6812 DM Arnhem, The Netherlands; vanbelois@luchtenleefomgeving.nl

[17] Graduate School of Sciences and Technology for Innovation, Yamaguchi University, Yamaguchi 755-8611, Japan; takaya@yamaguchi-u.ac.jp

* Correspondence: koziel@iastate.edu; Tel.: +1-515-294-4206

Citation: Bokowa, A.; Diaz, C.; Koziel, J.A.; McGinley, M.; Barclay, J.; Schauberger, G.; Guillot, J.-M.; Sneath, R.; Capelli, L.; Zorich, V.; et al. Summary and Overview of the Odour Regulations Worldwide. *Atmosphere* **2021**, *12*, 206. https://doi.org/10.3390/atmos12020206

Academic Editor: Prashant Kumar

Received: 5 December 2020
Accepted: 25 January 2021
Published: 3 February 2021

Publisher's Note: MDPI stays neutral with regard to jurisdictional claims in published maps and institutional affiliations.

Abstract: When it comes to air pollution complaints, odours are often the most significant contributor. Sources of odour emissions range from natural to anthropogenic. Mitigation of odour can be challenging, multifaceted, site-specific, and is often confounded by its complexity—defined by existing (or non-existing) environmental laws, public ordinances, and socio-economic considerations. The objective of this paper is to review and summarise odour legislation in selected European countries (France, Germany, Austria, Hungary, the UK, Spain, the Netherlands, Italy, Belgium), North America (the USA and Canada), and South America (Chile and Colombia), as well as Oceania (Australia and New Zealand) and Asia (Japan, China). Many countries have incorporated odour controls into their legislation. However, odour-related assessment criteria tend to be highly variable between countries, individual states, provinces, and even counties and towns. Legislation ranges from (1) no specific mention in environmental legislation that regulates pollutants which are known to have an odour impact to (2) extensive details about odour source testing, odour dispersion modelling, ambient odour monitoring, (3) setback distances, (4) process operations, and (5) odour control technologies and procedures. Agricultural operations are one specific source of odour emissions in rural and suburban areas and a model example of such complexities. Management of agricultural odour emissions is important because of the dense consolidation of animal feeding operations and the advance of housing development into rural areas. Overall, there is a need for continued survey, review, development, and adjustment of odour legislation that considers sustainable development, environmental stewardship, and socio-economic realities, all of which are amenable to a just, site-specific, and sector-specific application.

Keywords: odour legislation; air quality; air pollution; odor; smell; odour units; dispersion modelling; agriculture; environmental regulations; policy

1. Introduction

This paper is a collaborative work by seventeen international odour experts sharing comprehensive summaries and evaluations of odour policy and legislation from seventeen countries/regions: Europe (Austria, Belgium, France, Germany, Hungary, Italy, the Netherlands, Spain, the UK), Asia (China, including Hong Kong, Japan), Australasia (Australia, New Zealand), North America (the USA, Canada), and South America (Chile, Colombia).

While the authors acknowledge that this paper is only a snapshot in time of current worldwide odour policy, the content of the paper will always maintain historical value (i.e., the status of odour regulatory approaches as of 2019) and will likely remain relevant as a gauge for changes made to regulations in the future and which tend to evolve slowly.

Odour issues are currently one of the major causes of environmental grievances around the world and, in some countries, are routinely the cause of most environmental complaints to regulatory authorities. There continue to be multiple reasons for the prominence of odour complaints, including an unrelenting urban expansion of residential areas into land use areas once predominantly agricultural with few largely isolated facilities; increases in facility operations and their size; increasingly higher aesthetic, environmental expectations of citizens, who are less familiar and tolerant of odours than in the past; and concerns over potential health risks from airborne odourous substances.

In most countries, environmental legislation covers most types of common air pollutants, and there is little variation between jurisdictions with such legislation. However, odour legislation tends to be much more varied and varies across a wide spectrum: from having little to no specific mentioning in environmental legislation to extensive and rigid detailing in odour source testing, odour dispersion modelling, ambient odour monitoring, setback distances, process operations, and odour control procedures. Odour legislation can be highly variable from one jurisdiction to the next.

Odour issues are very complex, and, therefore, an excellent understanding of the formation of odour released into the atmosphere and exposure is important. The exposure of individuals living in odour-prone areas may lead to immediate annoyance, which in the long term may lead to it being defined as a nuisance. In some countries, odour policies are based solely on odour nuisance criteria, and so the question arises on how to determine odour nuisance. There are several guidelines for nuisance such as use and loss of enjoyment of the property, interference with the normal conduct of business, damage to animal and plant life, human health and safety, or property damage. Some countries, provinces, or states have also defined odour concentrations at which the odour nuisance could occur, taking into consideration several factors such as frequency and duration of odour episodes. Therefore, a common use of the FIDOL factors (frequency, intensity, duration, offensiveness, and location/receptor) is often used by some jurisdictions to determine the likelihood of odour annoyance in the area.

Nuisance can also be determined based on the validity of odour complaints and odour measurements. The odour measurements are either performed at the sources [1,2] or at locations where odour may be present by conducting direct odour monitoring [3,4]. Measurements conducted at the sources include estimating odour emission rates at each potential odour source in $ou_E \cdot s^{-1}$ and the use of dispersion modelling to establish odour concentrations (in ou or in some countries recorded as in $ou_E \cdot m^{-3}$) at sensitive receptors, at the property line, or at any other affected areas. Some countries set the limit for odour, either based on dispersion modelling criteria at the nearest sensitive receptor, or property boundary (for example, in New Zealand and some states of Australia (Tasmania), or in certain Canadian provinces such as Ontario province), or based on direct odour monitoring at the affected areas (for example, in Germany and some American states). The limits are either called the odour impact criteria (OIC) or odour concentration or detection thresholds. The OIC are based on odour concentrations and the accepted probability of exceeding the concentration (i.e., percentile) to define compliance. In some countries where there is no odour control, the odour limit may be determined by some specific and relatively easy-to-measure compounds such as hydrogen sulfide or ammonia. In some European countries such as France, odour exposure limits are also set as emission limit values (ELV) in $ou_E \cdot s^{-1}$ or $ou_E \cdot h^{-1}$. On the other hand, in several U.S. states, the dilution-to-threshold (D/T) field olfactometry approach is used to set the limits.

Odour nuisance depends on various predictors of odour, which are often summarised with the acronym FIDO (frequency, intensity, duration, and offensiveness), with factors not presented in any prioritised order [5]. In New Zealand and Australia, a fifth factor, "L", as in FIDOL, refers to the odour location [6]. This additional factor refers to the sensitivity of the surrounding residential area. For example, odours near a school may increase concerns for citizens.

Almost all odour policies specify criteria or otherwise reference the intensity component of FIDO: either through a measure of odour concentration as odour units per cubic metre ($ou_E \cdot m^{-3}$) from laboratory olfactometry [1]; as an odour index threshold value through the triangle bag method; as perceived odour intensity [7,8]; or as offensiveness [9] (or as D/T) through field olfactometry measurements [10].

The frequency and duration of odour episodes are often taken into consideration through dispersion modelling of odour emission rates to determine odour exposure to receptors and the number of hours in a year with odours present. The OIC limit the number of odour hours or provide a requirement for per cent of the hours in a year without odours (e.g., 98%). Secondly, frequency and duration are assessed through field inspection and documentation of the odours present.

In any investigation of odours, the character of the offending odours is documented to identify their source. Some policies have different criteria or even different approaches for specific odour sources.

Currently, odour policies are highly variable between countries, individual states or provinces, and even between counties and towns. These policies include (1) no specific mention in environmental legislation, (2) regulation of pollutants which are known to have an odour impact, (3) consideration of odour perception as a nuisance, (4) setting standards for specific odourants or other contaminants such as hydrogen sulfide; and (5) extensive detail for odour assessments, including odour source testing, dispersion modelling, ambient odour monitoring, setback distances, process operations, and odour control technologies and procedures, and (6) other approaches. While there are differences in the details of these policies, all policies outlined in this paper include one or more of these FIDO factors of odour nuisance. This paper outlines these varying approaches and discusses the advantages and disadvantages of the systems.

2. Europe—A Common Approach

In twenty-eight (28) European Union countries, odour is regulated through the Directive 2010/75/EU of the European Parliament and of the Council of 24 November 2010 on Industrial Emissions—in short, the Industrial Emission Directive or IED. The IED establishes a general framework for determining limits, including odour limits for many industrial activities/processes intending to (among others) control odour emissions.

The covered sectors include, for example, the energy industry, metals production and processing, waste management, chemical and mineral industry, and agriculture sectors such as animal production.

A complete list of sectors can be found in Annex 1 of the IED.

This European IED rules that installations should operate only if they hold a written permit or, in certain cases, if they are just registered. The permit conditions are defined to achieve a high level of protection for the entire environment. These conditions are commonly based on the concept of the best available technique (BAT). To determine BATs and limit imbalances in the EU with regard to the level of emissions from industrial activities, reference documents for BAT (named BREF) are drawn up [11].

There are over 30 BAT reference (BREF) documents published, which are related to different sectors. The new BREF documents also include the new figure of best available technology associated emission levels (BAT AELs), defining a range for the emission limits for any installation pursuing a permit [12]. As of 2019, the only BREF in Europe that set an odour limit is the recently published Waste Treatment BREF that establishes a range of 200 to 1000 $ou_E \cdot m^{-3}$ as the maximum allowed odour concentration for some BATs related to the biological treatment of waste [11].

It is important to note that BREFs are neither prescriptive nor exhaustive. The BREFs do not consider local conditions, so their application does not relieve the countries' permitting authorities from an obligation to make site-specific judgments. That means that during the permitting procedure, the responsible authority has to take into consideration all information provided by the BREF, including the operator's application and the local conditions to set an odour limit.

Some other legally binding documents and guidelines related to odours are available in a few European countries. Those odour regulations are used when no specific criteria are set in a BREF, or when the IED does not cover the odour-emitting activity.

Specific odour regulations and policies in France, Germany, Austria, Hungary, the United Kingdom, Spain, the Netherlands, Italy, and Belgium are introduced below.

2.1. France

France has an overall odour regulation based on the IED for any activity included in this regulation. In addition, France has specific regulations regarding odour control for two special activities: animal by-product processing plants and composting plants. Further, there are some common emission limit values (ELV) for the food and beverage processing industry.

2.1.1. Animal by-Product Processing Plants

The order from 12 February 2003 related to animal by-product processing plants is still in force despite several revisions [13]. Article 28 of that order lists different, relevant OIC, depending on the facility status. For a new plant, the OIC are set to 5 $ou_E \cdot m^{-3}$ in a radius of 3 km from the fence of the installation less than 44 h per year (99.5th percentile). This calculation is based on emission factors.

For an existing plant, the OIC are set to 5 $ou_E \cdot m^{-3}$ in a radius of 3 km from the fence of the installation less than 175 h per year (98th percentile). This calculation has to be made from on-site odour measurements, followed by air dispersion modelling. If dispersion modelling is not performed, the odour concentration should not exceed 1000 $ou_E \cdot m^{-3}$ for any source, no matter the stack height. However, if there are any odour complaints, the inspector may require an odour dispersion modelling or may ask for an increase in the frequency of odour measurements.

According to point 10 of the same order, if the odour concentration at the existing plant stack exceeds an ELV of 100,000 $ou_E \cdot m^{-3}$, then an olfactometric measurement according to the EN 13725 [1] must be performed every three months (Table 1). The frequency of the odour concentration measurement can be reduced to once per year if the plant is equipped with a representative and permanent electronic sensing device.

Table 1. Frequency of odour concentration checks for animal by-products processing plants [13].

Odour Concentration ($ou_E \cdot m^{-3}$)	Frequency of Odour Concentration Checks	Frequency of Odour Concentration Checks (with an Electronic-Sensor)
>100,000	quarterly	annual
5000–100,000	biannual	biennial
<5000	annual	triennial

If the plant's odour concentration is between 5000 and 100,000 $ou_E \cdot m^{-3}$, then an olfactometric measurement must be performed every six months according to the EN 13725 [1]. This measurement frequency can be reduced to once every two years if the plant has an electronic sensor for an odour monitoring system installed. If the odour concentration at the plant is less than 5000 $ou_E \cdot m^{-3}$, then an olfactometric measurement according to the EN 13725 must be performed every year. This measurement frequency could be reduced to once every three years if the plant has an electronic sensor for an odour monitoring system installed.

2.1.2. Composting Plants

The order of 22 April 2008 related to composting plants is currently in force despite several revisions [14,15]. According to Article 26, there are different OIC regulating composting plants. For both existing plants and new plants, the OIC are set to 5 $ou_E \cdot m^{-3}$ in a radius of 3 km from the fence of the installation less than 175 h per year (98th percentile). This calculation has to be made from on-site odour measurements in the case of existing plants and based on estimations in the case of new plants. In the case of existing plants, all the odour sources should be identified. If the sum of all the odour emissions is less than 20,000,000 $ou_E \cdot h^{-1}$ or if the plant is located in an area with a low risk of odour impact, there is no need to do anything else. If any of these two criteria are not met, an odour dispersion model should be performed in order to verify that the existing plant complies with the OIC of 5 $ou_E \cdot m^{-3}$ in a radius of 3 km from the fence of the installation less than 175 h per year (98th percentile). If the OIC are exceeded, the existing plant has to send an Odour Management Plan to reduce its impact to meet the previously outlined criteria.

2.1.3. Food and Beverage Industries

In the case of the food and beverage industry, there are some odour ELVs in $ou_E.h^{-1}$ which depend on the emission point's height according to the following (Table 2):

Table 2. Odour emission limit values (ELVs) for food and beverage industries [16–18].

Height of Point Source Emission (m)	Odour Emission Limit ($ou_E \cdot h^{-1}$)
0	1000×10^3
5	3600×10^3
10	$21,000 \times 10^3$
20	$180,000 \times 10^3$
30	$720,000 \times 10^3$
50	3600×10^6
80	$18,000 \times 10^6$
100	$36,000 \times 10^6$

One of the main points about the legislation regarding the food and beverage industry is that the minimum stack height is fixed and it is a function of the odour emission limits. These limits were previously mentioned for other industries, but it was often a better choice to treat the effluent and decrease the emitted concentration than to build very high stacks.

2.2. Germany

According to § 3 (1) of the Federal Immission Control Act [19], harmful effects on the environment are caused by many substances present in ambient air. According to their nature, extent or duration, they are liable to cause hazards, considerable disadvantages, or considerable nuisance to the general public or the neighbourhood. In the case of odours, the type of ambient odour is considered by the description of the smell, and the ambient odour extent or level is quantified by odour detection above the recognition threshold and uses the concept of the odour hour. The duration is expressed by the odour frequency (odour hours per year). If the odour frequency exceeds the specific exposure limit values given in the German Guideline on Odour in Ambient Air (GOAA) [20], the odour exposure is classified as a "considerable nuisance" according to the BImSchG [19].

In Germany, odour regulation for livestock farms and industrial installations has a long-lasting history. After several attempts to regulate odour exposure, e.g., by setback distances for livestock farms and industrial installations, a concept based on odour frequencies as detailed in the first GOAA in 1993 was developed. The GOAA was developed further in 2008, considering odour intensity, hedonic tone, and annoyance potential of specific odours [20]. The concept given in Figure 1 has been approved in many cases and is generally accepted at court. This concept has the outstanding advantage that the results of grid measurements and dispersion modelling can be directly compared because both methods aim to determine recognisable odours in terms of odour frequencies.

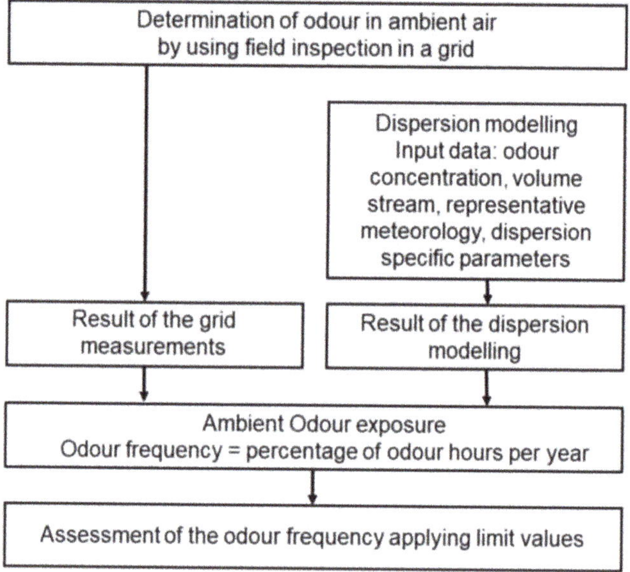

Figure 1. The concept of the GOAA [21].

Determination of odour in ambient air by using field inspections in a grid has to be conducted according to guideline VDI 3940-1:2006 [22]. However, in 2017, EN 16841 Part 1 [3] superseded this guideline. This method allows the standardised measurement of recognisable odours (in terms of odour hours) in the field by panel members. An odour hour is obtained when the percentage odour time of a single measurement reaches or exceeds 10% by convention. It is a statistical survey, which considers different times of the day, week, and year (for details, see EN 16841 Part 1) [3]. The grid method is the only method to determine the perceived odour in the field over periods of six months or a whole year. Therefore, the method is mainly applied in cases of complaints in the neighbourhood of odour sources or for determining the odour frequency. The disadvantage of this method is, among others, that the duration of the survey (at least six months) and that the representativeness of the results depend on the meteorology and the emission variation during that time interval.

Dispersion modelling with either the measured or estimated odour emission is the method that is used in most of the cases in Germany. The Lagrangian dispersion model AUSTAL2000G (G (Geruch) stands for odour) is used to calculate odour frequencies. In the

AUSTAL2000G model, the spread of the particles in the atmosphere is simulated depending on the wind speed and direction. The mean odourant concentration is calculated as the average hourly value. If the average hourly value is above an assessment threshold of cthr = 0.25 $ou_E \cdot m^{-3}$, the relevant hour is counted as an odour hour [23]. On this basis, the annual mean odour hour frequency is calculated [23]. For odour emission measurements, EN 13725:2003 [1] is applied in combination with guideline VDI 3880:2011 [2] on static sampling and VDI 3884-1:2015 [24] with supplementary instructions for application of EN 13725 [1].

One of the disadvantages of dispersion modelling is that the input data (Figure 1) are often vague. Odour emissions cannot be quantified sufficiently for some sources, such as diffuse sources or area sources. Additionally, the representativeness of the meteorology for the location is often limited. Looking at court cases, the recoverable claim is that the results of the calculation of odour frequencies by dispersion modelling have to be conservative. If they are compared with grid measurement results as a measure for the existing odour frequency, they need to be at least equal or higher [25].

Finally, the odour hour frequency is assessed by applying limit values. These limit values are the outcome of several investigations where the odour frequency was correlated with the annoyance degree of residents (e.g., [26,27]). The limit values expressed as relative odour frequencies per year are 0.10 (10%) for residential and mixed areas, 0.15 (15%) for commercial and industrial areas, and 0.15 (15%) only for livestock odours in villages with a mixture of houses and farms.

Another finding of these investigations was the lower annoyance potential of clearly pleasant odours [28]. A definition of clearly pleasant odours is given by the GOAA as well as the method. The method to be applied is the polarity profile method [29]. For clearly pleasant odours, a weighting factor of f = 0.5 can be used before applying the limit value.

In an investigation, especially on livestock farming, the odour frequency caused by cattle, pigs, and poultry was correlated with the annoyance degree of residents. A lower annoyance potential was found for dairy cows, including young cattle (f = 0.5), and for fattening pigs and sows (f = 0.75), whereas a higher annoyance potential was found for poultry (f = 1.5). Investigations in 2017 [30] combining plume measurements according to VDI 3940-2:2006 [31] (superseded by EN 16841–2:2017 [4]) and the polarity profile method showed a lower annoyance potential for horses and fattening bulls (f = 0.5). A follow-up study in 2019 showed similar results for sheep and nanny goats [32].

In practice, the GOAA is used by responsible authorities all over Germany. It is applied in the licensing and surveillance of installations, in cases of odour complaints, and in urban land use planning. The measurements are carried out by accredited laboratories based on the standards EN 13725 [1] and EN 16841 Part 1 [3] (former VDI 3940–1:2006 [22]). Currently, it is planned to include the GOAA in the Technical Instructions on Air Quality Control [33], which would further increase their legal bindingness for local authorities.

2.3. Austria

Austria's regulations distinguish between limit values that have a legal basis and guiding or target values, which are only part of guidelines without a legal basis. In general, there are no legal limit values for odours in Austria. For spa areas [34], a target value for the exceedance probability p_T = 3% for an odour concentration of C_T = 1 $ou_E \cdot m^{-3}$ (similar to Germany) is suggested. The Austrian Academy of Sciences published a guideline (without legal relevance) with two limit values (both have to be taken into account to fulfil the criteria) [35]. The odour concentration threshold C_T is only given verbally as odour intensity:

- An exceedance probability of p_T = 8% for a "weak" odour intensity;
- An exceedance probability of p_T = 3% for a "strong" odour intensity.

To apply these two limit values for dispersion modelling, the verbally given odour intensity needs to be converted to odour concentration. For the "weak" intensity, an odour concentration of $C_T = 1$ $ou_E \cdot m^{-3}$ is used, and for the "strong" intensity, an odour concentration in the range $ou_E \cdot m^{-3} < C_T < 8$ $ou_E \cdot m^{-3}$ is used.

The local government of Styria issued a guideline [36], which suggests new odour impact criteria taking into account the annoyance potential (hedonic tone) for four categories, characterised by typical odour sources: (1) small (e.g., biofilter, silage, horses, sheep, goat), (2) medium (brewery oil mill, domestic fuel, pig), (3) high (e.g., bitumen, refinery, kitchen, poultry), and (4) very high (e.g., nauseating smell, tannery, composting facility, some parts of wastewater treatment plants). For continuous emitting sources, the exceedance probability p_T (%) of the guideline is related to the German odour hour definition, which means the odour concentration threshold is $C_T = 1$ $ou_E \cdot m^{-3}$ (Table 3). For discontinuous emitting sources, the OIC are defined by a certain exceedance probability of $p_T = 2\%$ and an odour concentration C_T (Table 4).

Table 3. Odour impact criteria (OIC) based on the guideline of the local government of Styria, Austria, for continuous sources [36], defined by a certain threshold concentration of $C_T = 1$ ou_E m^{-3} and the exceedance frequency p_T (%).

Annoyance Potential	Exceedance Probability p_T (%)	Exceedance Probability p_T (%)
	Non-Livestock Sources	Livestock Sources (Pure Residential Areas/Agricultural Dominated Villages/Other Utilization)
Small	40	40/50/-
Medium	15	15/20/30
High	10	10/15/20
Very high	2	-/-/-

Table 4. Odour impact criteria (OIC) based on the guideline of the local government of Styria, Austria, for discontinuous sources [36], defined by a certain exceedance probability of $p_T = 2\%$ and the threshold concentration C_T.

Annoyance Potential	Threshold Concentration C_T ($ou_E \cdot m^{-3}$)	Threshold Concentration C_T ($ou_E \cdot m^{-3}$)
	Non-Livestock Sources	Livestock Sources (Pure Residential Areas/Agricultural Dominated Villages/Other Utilization)
Small	15	15/20/-
Medium	5	5/7/10
High	4	4/5/7
Very high	1	-/-/-

This odour interval is related to short-term concentrations to mimic the odour perception of the human nose [37]. In summary, Austria uses a variable peak-to-mean (the relevant short-term peak odour concentrations are calculated with a stability-dependent peak-to-mean algorithm) approach [38,39], but in many cases, the German peak-to-mean factor (F) equal to 4 is used as a constant value [40]. Therefore, the concentration for a 1 h mean value of $C_T^* = 0.25$ $ou_E \cdot m^{-3}$ is used. A new approach to calculating the short-term odour concentration based on one-hour mean values is used beside the two other methods. The "concentration variance" method [41] was implemented in the Lagrangian dispersion models GRAL and LASAT [42,43]. A critical review of the three methods can be found in Brancher et al. (2020) [42].

Austria also has a guideline related to livestock buildings (guideline for the evaluation of ambient odour emitted by livestock buildings [44]) that offers two alternatives to evaluate the emission of farm animals. The first alternative uses a qualitative comparison of the odour emission rate, the impact of the ventilation system on the emission characteristics, and the local conditions at the livestock farm. This results in a dimensionless odour number, which is then used to assess if this farm size is common in this area. The second alternative is the application of a dispersion model, where an empirical equation is derived to simplify the calculation of the separation distance [45]. In the Austrian guideline, the

German GOAA [20] and the Austrian Academy of Sciences [35] are mentioned for odour impact criteria, which can be selected to assess odour annoyance and calculate separation distances.

2.4. Hungary

Hungary does not have legal national odour impact criteria in use. However, to avoid any odour annoyance, it is suggested that the exceedance probability (pT) = 2% (98th Percentile) is used for an odour concentration threshold range of $3\ ou_E \cdot m^{-3} < C_T < 5\ ou_E \cdot m^{-3}$ [46].

2.5. United Kingdom

The regulators are slightly different in the four countries, England, Wales, Northern Ireland, and Scotland, which make up the United Kingdom, but the regulations are substantially the same.

Numerous individual local authorities and four environment agencies are responsible for regulating the impact of odourous emissions from industrial and commercial premises in England, Wales, Scotland, and Northern Ireland [47]. Waste activities, larger industrial processes, and intensive livestock farms are regulated by the environment agencies under the IPPC directive through Environmental Permitting Regulations (EPR), and smaller enterprises, as well as those below the size thresholds for the EPR, are regulated by local authorities. Local authorities use three regimes for odour control: (i) planning, (ii) permitting (which is similar to the EPR requirements discussed below), and (iii) statutory nuisance.

Local authorities regulate and approve planning applications for all premises and, for those that may generate odours, may impose planning conditions to help control emissions. Planning authorities may ask for evidence of the extent of the process's odour impact when considering applications, but this is far from universal and may take the form of a comparison with an existing process or a dispersion model based on new or existing odour emission rates. Planning controls are generally less robust on smaller businesses, such as food take-aways and restaurants, than larger concerns such as intensive livestock farms. There is considerable variation in the levels of control exercised by different authorities in different council areas.

Where odour modelling is used, the local authority planning departments may assess the predicted impact of the process against benchmark criteria that have been agreed upon or previously used in other planning cases. An example of the potential criteria, based on the Environment Agency (EA) H4 Horizontal Guidance [48], is shown in Table 5 below.

Table 5. Odour impact criteria based on Environment Agency (EA) guidance [48].

Offensiveness Scale	OIC	Example of Odour Sources
Most offensive odours	$1.5\ ou_E \cdot m^{-3}$	Decaying animal or fish remains, septic effluent or sludge, biological landfill odours
Moderately offensive odours	$3\ ou_E \cdot m^{-3}$	Intensive livestock rearing, fat frying (food processing), sugar beet processing, well-aerated green waste composting
Less offensive odours	$6\ ou_E \cdot m^{-3}$	Brewery, confectionery, coffee roasting, bakery

These benchmarks may be used to support evidence submitted with a planning application or with a permit application. They are based on the annual 98th percentile of hourly average concentrations of odour modelled over 3 to 5 years at sensitive receptor locations.

These benchmarks are considered when determining setback distances from existing operations and levels of abatement that may be required for existing operations. Some councils sometimes impose lower odour concentration values.

2.5.1. Assessing Odour Impacts for Planning Purposes

The Institute of Air Quality Management has issued a guidance on the assessment of odour for planning and a guidance on interpreting dispersion modelling, which also includes methodologies for field odour studies and desk-study risk-based assessments [49]. The risk assessment methodology includes consideration of the source odour potential, pathway effectiveness, predicted odour exposure, and receptor sensitivity to qualitatively determine the magnitude of the odour effects at the specific receptor location odour effects, ranging from negligible impact, through slight adverse impact and moderate adverse impact, and up to substantial adverse impact. This methodology is quite dependent on the judgment/discretion of the assessor.

2.5.2. How Odours Are Assessed to Be Qualified as a Nuisance

If there are complaints of odours from non-EPR premises, the local authority environmental health department officers have a statutory duty to investigate the complaints [50]. To determine if odours constitute a statutory nuisance, local authorities can consider one or more of the following: where the odour is coming from, the character of the area, the number of people affected nearby, if the odour interferes with the quality of life of people nearby (for example, if they avoid using their gardens), how often the odour is present, and the characteristics of the odour [51]. Councils usually use at least two officers to confirm a nuisance.

For the odour to be determined a statutory nuisance, it must do one of the following: unreasonably and substantially interfere with the use or enjoyment of a home or other premises, injure health, or be likely to injure health.

The operator or premises have the potential to demonstrate that they are using best practicable means (BPM) as a defence [52]. The enforcement officers (usually designated Environmental Health Officers) should be objective and thorough in their investigation [47]. If the odourous process operator does not apply BPM to abate the odourous emissions, then an "Abatement Notice" is issued for the enforcement of the statutory nuisance, and legal proceedings will then apply. If the operator has used BPM to stop or reduce the odour, they may be able to use this as one of the grounds for appeal against the abatement notice or as a defence. If no appeal is made, then the Abatement Notice stands, and the operator can be prosecuted for not complying with the abatement notice in the criminal courts, although the BPM defence is also available in these circumstances.

Statutory nuisance laws do not apply to odours arising from residential properties, but they do apply to odours from business premises affecting residential properties.

2.5.3. Processes and Premises Regulated by the Environment Agencies under the EPR

The bodies responsible for regulating the industrial and farming activities not covered by the local authorities are Natural Resources Wales (NRW), the Scottish Environmental Protection Agency (SEPA) for Scotland, the Northern Ireland Environment Agency (NIEA) for Northern Ireland, and the Environment Agency (EA) for England [53–57].

The environment agencies use a permitting system to regulate the impact of the emissions. Concerning processes likely to causes odours, there is usually a condition within the permit setting out a requirement such as "the activities shall be free from odour at levels likely to cause pollution outside the site, as perceived by an authorised officer of the Environment Agency, unless the operator has used appropriate measures, including, but not limited to, those specified in any approved odour management plan, to prevent or where that is not practicable to minimise the odour. The operator shall submit to the Environment Agency for approval an odour management plan, which identifies and minimises the risks of pollution from odour; and they shall implement the approved odour management plan." [55].

Appropriate measures are normally assumed to include best available techniques (BAT) with BAT based on factors including best practice in the industry sector and relevant guidance, including European BREF [52] guidance for specific industry sectors. If an operator fails to comply with the terms of the permit, and in particular with their odour management plan, then a series of actions is taken by the environment agencies, ultimately leading to a withdrawal of the permit and prosecution.

In summary, the environment agencies set out their approach with the result in the following scenarios:

1. Where no odour is detectable or likely to be detectable, there will be no pollution beyond the boundary of the site concerning odour pollution.
2. Where odour is detectable, it may or may not cause offence, and the agency response will depend upon the degree of pollution and the cost and practicability of any remedial measures.
3. Where all appropriate measures are being used but are not completely preventing odour pollution, a level of residual odour will have to be accepted.
4. Where the odour is serious, even if all efforts have been made to apply BAT/appropriate measures, it may be necessary to suspend or revoke the permit in full or in part.

Normally, the process of enforcement leading ultimately to permit suspension or revocation will involve the regulator (the EA, SEPA, NRW, or NIEA) serving the operator with improvement or enforcement notices with the objectives of improving odour control management. Similar benchmarks to those used by local authorities are considered when determining setback distances from existing operations and levels of odour mitigation or abatement that may be required for existing operations and proposed new installations.

2.6. Spain

Spain also has an overall odour regulation based on the IED for any activity included in this regulation. The Law 5/2013 [58] and the Royal Decree 8/15/2013 [59] made the transposition of the European IED. The competences of the IED lie in the autonomous communities (AC). As a general approach, the procedure is to set ambient air odour limits for industrial activities, which are based on the following steps:

*   The facility/activity (new or existing) applies to obtain a permit.
*   The environmental administration evaluates if there is an odour concern and, if necessary, an odour assessment is requested.
*   There is no guideline for decision making on odour assessments results. The outcome completely depends on the environmental officer assigned to the case.
*   Upon completion of an odour assessment (if performed), the individual OIC are set by the environmental officer, which are typically based on the assessment results.

In June 2005, the region of Catalonia's AC presented the draft bill "Against Odourous Pollution" [60]. This draft was inspired by the first H4 Horizontal Guideline of the UK [48]. This draft received considerable pressure from the pig farming sector in Catalonia. This region is the main pork producer of Spain. Additionally, some political changes occurred in that region, having the consequence that the administrative procedure to approve the draft was finally interrupted. This draft was taken as a reference by many odour consultants in Spain.

In March 2019, the Canary Islands region sent to public inquiry the first regulation in Spain that sets odour limits. Again, political changes in the government prevented this regulation from being published.

In Spain, some small municipalities did regulate odours in the regions of Catalonia (Lliçà de Vall [61], Banyoles [62], Riudellots de la Selva [63], Sarrià de Ter [64], Valencia (Raspeig) [65], Murcia (Alcantarilla, San Pedro del Pinatar) [66,67], and the Canary Islands (Las Palmas) [68]. In the small town of Alcantarilla, the local odour regulation defines the areas with an "odour" saturation. This way, they limit the areas where an industrial facility that can potentially cause annoyance cannot be located or where the urban expansion has to be halted to avoid an odour impact. The OIC are set as 5 $ou_E \cdot m^{-3}$ (98th Percentile). The municipality of San Pedro del Pinatar set the OIC levels of the Catalonian draft.

The municipality of Las Palmas de Gran Canaria, similarly to San Vicente del Raspeig, has developed an "odour perception index" (IP) in its regulation. The equation to calculate the odour perception index is the following:

$$IP = \log 10 \, (C) \times FC \times FD \times FI \times FP \times FV \tag{1}$$

where C is odour concentration, FC is the hedonic tone factor, FD is the emission factor's duration, FI is the intermittency factor of the processes, FP is the emission period factor (varies from 1.0 to 1.2 depending on the time of the day/week; the lower value is used for the working day hours (7:00–22:00, M–F), and the higher value is used for the night period (22:00–7:00)), and FV is the wind direction factor. The OIC, in this case, are set as an odour perception index of 0.04.

2.7. The Netherlands

The Netherlands has an overall odour regulation based on the Industrial Emission Directive (IED) for any activity in this regulation. There is specific odour legislation only for livestock farming. For all activities except livestock, the protection against odour nuisance is regulated in the Activities Decree [69]. The premise here is to prevent or reduce the odour to an acceptable level by applying BAT. Additionally, odour regulations may be included in a customised decision or a permit. The local government may decide what levels are acceptable or not, but there is no clear national-level odour evaluation framework to do so. The competent authority may set a local odour policy to help determine the acceptable odour nuisance level [70]. The majority of the Dutch provinces have done so, while cities usually do not have an odour policy of their own but make use of the provincial one.

The local odour policies are either based on percentile values already in use in the last century or on an odour's hedonic tone. Common standards in use over many years are the calculated 98th or 95th percentile values of 0.5, 1.5, and 5 $ou_E{\cdot}m^{-3}$, representing different levels of protection. Popular limit values in local policies are also those based on the hedonic tone of the odour. To do so, measurements of hedonic tone are carried out by an olfactometry laboratory, according to the Dutch standard for hedonic tone [71]. The hedonic tone is expressed on a scale from -4 (very unpleasant) to $+4$ (very pleasant). In general, it is assumed that an odour nuisance can occur at odour concentrations higher than the odour concentration corresponding to the hedonic value of -0.5 as the 98th percentile. At concentrations above the concentration corresponding to a hedonic value of $H = -2$, severe odour nuisance and odour complaints are likely to occur. While developing an odour framework, the odour concentration at which a certain scale value for hedonic tone is reached (for example, $H = -2$) is taken as a guide value for the 98th percentile. Differences in acceptable nuisance levels are made between existing and new situations and between the residential areas and "scattered" houses.

Examples of provinces with local odour evaluation frameworks are Flevoland, Gelderland, Groningen, North Brabant, Overijssel, South Holland, and Zeeland. The local evaluation frameworks vary from one province to another. While setting local evaluation criteria, provinces or municipalities can base themselves on the following documents and considerations:

The Dutch Emission Guidelines (NeR) aimed to harmonise the environmental permits' emission requirements in the Netherlands. It contained guidelines for air emissions from industrial processes, including odour evaluation criteria, varying from, for example, 0.5 $ou_E{\cdot}m^{-3}$ at the 98th percentile for sewage treatment plants to 5 $ou_E{\cdot}m^{-3}$ at the 98th percentile for bread bakeries [72]. At the beginning of 2016, the NeR was cancelled, but the normative part was included in the Activities Decree. As the odour evaluation criteria were not normative, they were not included in this decree. However, they are still used as guidelines when drawing up local odour evaluation frameworks.

The letter from the Ministry of Housing, Spatial Planning and the Environment states that in most cases, a serious odour nuisance can be avoided when emission concentrations are below 5 $ou_E{\cdot}m^{-3}$ as the 98th percentile (continuous emission sources). For sources with short emission durations, the 98th percentile concentrations do not reflect the expected odour nuisance. For these sources, the use of a higher percentile is more appropriate.

Generally, at 98th percentile odour concentrations with a hedonic tone less than H = -2, a serious odour nuisance will occur (continuous emission sources). At odour concentrations below 0.5 $ou_E \cdot m^{-3}$ as the 98th percentile (and a hedonic tone not less than H = -0.5), no nuisance is expected (continuous emission sources).

The Law of 5 October 2006 on Livestock Odour Control of the Ministry of Housing, Spatial Planning and the Environment [73] of the Netherlands regulates odour nuisances caused by animal accommodation used in livestock farming. It contains some OIC for dwellings in the surrounding livestock farms. Limit values differ between 2 $ou_E \cdot m^{-3}$ as the 98th percentile for residential areas up to 8 or 14 $ou_E \cdot m^{-3}$ as the 98th percentile for rural areas. Those limit values were based on dose–response relationships which followed large investigations on odour emissions and odour nuisance around livestock farms in 2001 [74].

This legislation [75] contains the determination of odour emission factors, minimum distances for fur-bearing animals, the method of calculating odour intensity, and the method of determining the distance. This regulation is reviewed every year to add necessary changes.

The regulation appears to be based on the rationale that not all livestock operations are equal; i.e., it gives some odour emission factors depending on the animal type. Further, it regulates the separation distance stating that the setback distance between a livestock farm and an odour-sensitive receptor must be at least 100 (if the odour-sensitive receptor is situated in the built-up area) or 50 m (outside the built-up area). If animals are also kept in an animal category with no determined odour emission factor, a minimum distance of 50 to 100 metres must also be observed from the facilities where the animals are kept.

The present Law on Livestock Odour Control [73] is being reviewed at the moment. Health research over recent years [76] has shown unexpected high levels of nuisance (and lung diseases) in dense livestock areas. The possible revision of its content tends to lower acceptable intensity levels and enlarging setback distances. New investigations on health effects around livestock farms, including odour nuisance, as well as on possible measures, (BAT) are carried out to fundamentally review the law on this subject. An example of such investigation is the research reporting a strong relation between modelled odour exposure from livestock farming and odour annoyance among neighbours [77]. Public health authorities nowadays advise to use limit values of 2 $ou_E \cdot m^{-3}$ as the 98th percentile for residential areas, and no more than 5 $ou_E \cdot m^{-3}$ as the 98th percentile for rural areas [78].

2.8. Italy

Italy does not have a national-level regulation regarding odour. However, some efforts have been made at a regional level. The first regional regulation mentioning odour was the 2003 guideline of the region of Lombardy which covered the construction and operation of compost production facilities [79]. This guideline fixed some limit values for atmospheric emissions, thereby including a limit of 300 $ou_E \cdot m^{-3}$ relevant to odour emissions. Despite the old approach of giving an odour concentration value and the fact that this guideline is now obsolete (no longer valid), it is worth mentioning due to its historical significance.

Almost ten years later, in 2012, the region of Lombardy again acted as a pioneer in Italy by publishing a regional guideline specifically on odour emissions ("General determinations regarding the characterization of atmospheric emissions from activities with a high odour impact" [80]). This regional guideline was inspired by other regulations in Europe and adopted a more modern approach, based on odour dispersion modelling. The guideline does not have explicit acceptability criteria. However, it specifies that any plant with an odour impact shall evaluate the extent of this impact by drawing up impact maps indicating annual peak odour concentration values at the 98th percentile, thereby drawing the 98th percentile iso-concentration lines corresponding to the odour concentration values: 1, 3, and 5 $ou_E \cdot m^{-3}$, as resulting from atmospheric emission dispersion simulations (see example in Figure 2).

Figure 2. Example of a map resulting from atmospheric dispersion simulations representing the iso-concentration lines corresponding to the 98th percentile of the odour concentration values of 1, 3, and 5 $ou_E \cdot m^{-3}$.

Even though this approach is not particularly original concerning other European regulations, the guidelines present some innovative aspects that are worth highlighting. Annexe 1 fixes the requirements of the odour impact studies by emission dispersion simulation. Besides establishing criteria for input data quality and presentation of results, some other indications are provided regarding the dispersion model to be used. An interesting observation is that Gaussian models are excluded from the suggested models. This is further highlighted by the specification of the necessity for the model to treat calm winds, which are typical of the Lombard territory.

Another interesting aspect of this guideline is the precise definition given in Annexe 2 of the sampling procedures for gathering odour emission data from different source types, i.e., measuring representative odour concentrations and then evaluating odour emission rates which are required as model inputs. This aspect is particularly innovative, especially compared to EN 13725:2003 [1], which is extremely lacking in details of sampling procedures.

One positive aspect of this guideline is that it is based on a rather simple and sequential approach, and the required economic investment for its application is quite contained. The approach has been successfully accepted both by local authorities and plant owners or managers. Besides having raised the awareness of authorities and the public towards environmental odour pollution management, in some situations, the adoption of the Lombard guideline has already led to the identification and solution of odour problems for existing plants, or to the proper modification of the projects of new plants in order to limit their predicted odour impact to an acceptable extent.

Even though the guideline mentioned above is a regional guideline, it is currently used as the regulatory reference for most other Italian regions. Indeed, the region of Piemonte and the autonomous province of Trento have very recently issued their odour guidelines, which are substantially a copy of the Lombard guideline [81]. The latter's main innovation is that it fixes acceptability criteria in terms of the 98th percentile peak odour concentration limits that vary in function of the receptor distance from the source (Table 6).

Table 6. Proposed odour impact criteria (OIC) for the Italian province of Trento (98th percentile) [81].

Receptors in Residential Areas	
OIC (98th Percentile)	**Distance from the Source**
1 $ou_E{\cdot}m^{-3}$	>500 m
2 $ou_E{\cdot}m^{-3}$	200–500 m
3 $ou_E{\cdot}m^{-3}$	<200 m
2 $ou_E{\cdot}m^{-3}$	>500 m
3 $ou_E{\cdot}m^{-3}$	200–500 m
4 $ou_E{\cdot}m^{-3}$	<200 m

A different approach was proposed in the region of Puglia, with the publication of the D.g.r. 16 April 2015 [82]. This regional regulation is mainly based on an analytical approach to measure the "limit concentration" of 40 different odourous chemical compounds, each according to a specific analytical technique. The guideline also fixed odour concentration limits in terms of 2 000 $ou_E{\cdot}m^{-3}$ for point sources and 300 $ou_E{\cdot}m^{-3}$ for diffuse sources. This regional regulation has been strongly criticised because of its excessively complex approach—implying high costs of debatable usefulness for extensive chemical analyses— and to the fact that its principles are unnecessarily different from those adopted elsewhere since chemical analyses have been abandoned almost everywhere as a reference method for odour emission measurement.

2.9. Belgium

Belgium has federal and regional legislation. For environmental matters, the federal government and the regions share responsibility for the implementation of environmental policies. The competence in the evaluation of the odour impact lies in the Flemish and Wallonia region. The two regions have their own specificities.

Odour control is mainly based on the field inspection—the plume method, according to EN 16841-2 [4]. Following the plume method, the global emission rate is determined using the odour concentration at the receptor level and a reverse modelling approach. Odour dispersion modelling is frequently performed with the ADMS model in Wallonia and with the IMPACT model in Flanders (for both: 1 h mean values, peak-to-mean factor of 1).

The emission rate entered into the dispersion model is adjusted until the simulated average isopleths for 1 $su{\cdot}m^{-3}$ (su: "sniffing" unit) at about 1.5 m height fit the observed perimeter and the maximum perception distance. For field measurement, the results have to be expressed in "su" and not in ou. The reason is to make a clear distinction between the concentration of odour, collected in bags, measured according to EN 13725 [1], and the concentration obtained by field inspection. A fundamental difference with the European odour unit is that sniffing units are based on recognition of odour, whereas European odour units are determined by detection and not necessarily a recognition of the odour type. One sniffing unit per cubic metre is defined as the odour concentration at the border of the plume. It is impossible to quantify higher concentrations (e.g., 5 $su{\cdot}m^{-3}$) by observation in the field. Typically, 1 $su{\cdot}m^{-3}$ corresponds to a concentration of 1 to 5 $ou_E{\cdot}m^{-3}$.

Typically, about ten campaigns have to be organised over at least five different days in order to take most variations into account. Finally, the mean emission rate is calculated and introduced into the same dispersion model for the region's normal reference year to calculate the percentiles.

2.9.1. Walloon Region

There is no general legislation concerning odours in the Walloon region. The approach has been to provide guidelines for different activities. For example, there are specific regulations dealing with odour management for composting plants and farms. In the case of composting plants, the Walloon Decree from 2009 (2009/204053) [83] states that the odour concentration must not be greater than 3 $ou_E{\cdot}m^{-3}$ at the 98th percentile to the closest neighbour. In the case of farms, it is based on the calculation of a minimum separation distance to prevent odour annoyance [84]. The requirements for farms depend on the sector's area plan where they are established and whether the farm is new or existing. These values are not yet fully validated and are not yet compiled in a Walloon decree. Table 7 below shows the different guidelines.

Table 7. Odour impact criteria for the farming sector in Wallonia [83].

Area of the Sector Plan	OIC (98th Percentile)	
	Existing Farm	New Farm
Habitat area	$3\ ou_E \cdot m^{-3}$	$1\ ou_E \cdot m^{-3}$
Recreation area	$3\ ou_E \cdot m^{-3}$	$1\ ou_E \cdot m^{-3}$
Public service area	$3\ ou_E \cdot m^{-3}$	$1\ ou_E \cdot m^{-3}$
Cultivated area	$10\ ou_E \cdot m^{-3}$	$6\ ou_E \cdot m^{-3}$
Other areas	$6\ ou_E \cdot m^{-3}$	$3\ ou_E \cdot m^{-3}$

The AWAC (Walloon Agency for Air and Climate) is still working towards updating the Walloon odour regulation.

Besides the composting decree and the farming guidelines, the general rule is to define the conditions in the operating permits delivered by the Department of Permits and Authorizations (DPA) or the communes. Each of the four existing DPA (Namur-Luxembourg, Liège, Charleroi, Mons) works differently. Each activity is case-specific and whether an environmental permit is granted depends on the type of odours and their impact on the neighbourhood. By default, an odour concentration of $1\ ou_E \cdot m^{-3}$ at the 98th percentile is imposed.

The method used to measure odours is not set by any regulation. However, the lab/agency that carries out the odour concentration measurement by dynamic olfactometry [1] requires the Wallonia agreement. The way to check the OIC values set in a permit/authorisation depends on the environmental consultancy agency or the lab that performs the control. The organism in charge of checking the OIC has to justify to the DPA why they used that methodology. For example, the field inspection method is usually performed in municipal solid waste landfills [85–87]). The Walloon Environmental Police Division (DPE) has competence in both environmental permitting and complaint management. A new trend is to promote the use of resident diaries ("watchmen"). This approach is considered relevant by the DPE and efficient in solving odour annoyance [88].

2.9.2. Flemish Region

As far as odour nuisance is concerned, there is no legal framework in the Flemish region either. The Flemish odour policy is based on the following basic rules [89]: (1) when there is a nuisance, BAT measures must be taken to reduce it, (2) when there is no nuisance, no measures must be taken, (3) a severe odour nuisance is never acceptable, and (4) zero-emissions are not realistic.

While performing odour assessment studies, one of the key concepts is the "acceptable nuisance level". This level is situated between the no-effect level or target value and the limit value. The no-effect level is defined as the nuisance level from which no further decline in annoyance is observed; the limit value is the nuisance level at which severe nuisance occurs (structural complaints). Both values are expressed as 98th percentile concentration values. The acceptability level is determined, taking into account some environmental, legal, social, economic, financial, technological, and contextual aspects.

The way the target level, the limit level, and the acceptable nuisance level are derived is case-specific and must be determined by the odour consultant or odour lab that performs the odour assessment study. Some of the guidelines for doing this are summarised below [89].

For slaughterhouses and wastewater treatment plants (WWTPs), no effect levels or and limit values were scientifically determined based on dose–response studies derived from elaborate studies at different companies ([90,91] see Table 8). For some other odour-emitting sectors, only the no-effect levels were determined (see Figure 3). Some of these levels are determined based on dose–response relationships; others are deduced based on the hedonic tone of the odour.

Table 8. Target values and limit values for slaughterhouses and wastewater treatment plants (WWTPs) [92].

Sources	Target Value (No-Effect Value) [su·m^{-3} as 98th Percentile]	Limit Value [su·m^{-3} as 98th Percentile]
Slaughterhouses	0.5	1.5
WWTPs	0.5	2.0

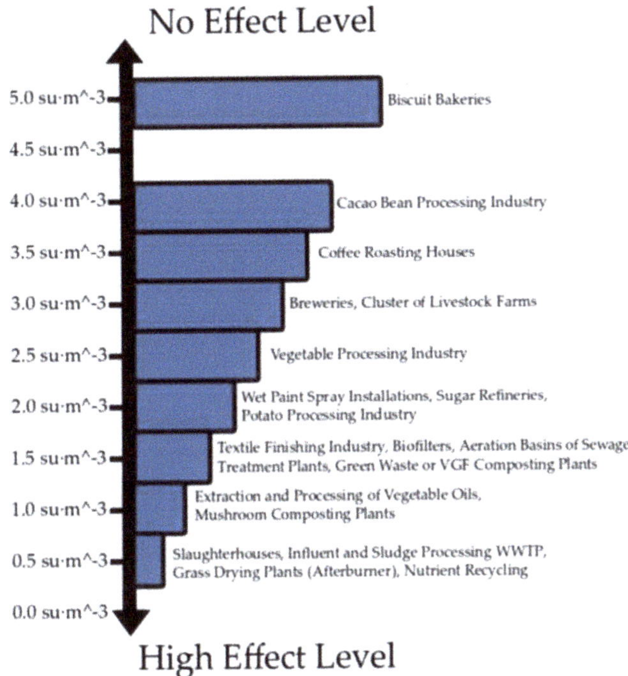

Figure 3. No-effect levels for different sectors [93].

Table 9 shows the no-effect levels for odours/sectors not mentioned in Figure 3:

Table 9. Target values not mentioned in Figure 3 [93].

Hedonic Tone	Target Value [su·m^{-3} as 98th Percentile]
strongly unpleasant	0.5
unpleasant	1.0–1.5
neutral	2.0
pleasant	2.5–3.0
strongly pleasant	3.5–5.0

The above-mentioned target values and limit values are used for odour impact assessment in highly sensitive places/areas (e.g., residential areas). If the limit value is exceeded, the odour impact is considered to be significantly negative. If the target value is exceeded, the odour impact is considered to be negative.

For moderate to low sensitive areas (e.g., industrial areas), the odour evaluation framework is less severe, and higher target and limit values are used. Table 10 shows the target and limit values for strongly unpleasant odours as a function of the odour sensitivity of the area. The odour evaluation framework based on these criteria is given in Table 11.

Table 10. Target and limit values as a function of the odour sensitivity of the area (strongly unpleasant odours) [93].

Odour Sensitivity of the Area	Target Value [su·m^{-3} as 98th Percentile]	Limit Value [su·m^{-3} as 98th Percentile]
Highly odour-sensitive locations	0.5	2.0
Moderate odour-sensitive locations	2.0	5.0
Low odour-sensitive locations	3.0	10

Table 11. Odour impact evaluation framework for strongly unpleasant odours [93].

98th Percentile-Concentration. [su·m^{-3}]	Low Odour-Sensitive Places	Moderate Odour-Sensitive Places	Highly Odour-Sensitive Places
>10	Significantly negative impact	Significantly negative impact	Significantly negative impact
5–10	N negative impact	Significantly negative impact	Significantly negative impact
3–5	Negative impact	Negative impact	Significantly negative impact
2–3	Negligible impact	Negative impact	Significantly negative impact
0.5–2	Negligible impact	Negligible impact	Negative impact
<0.5	Negligible impact	Negligible impact	Negligible impact

Similar odour evaluation frameworks can be derived for other types of odours.

In 2015 and 2018, sectoral Codes of Good Practice for prevention, assessment, and control of odour nuisance caused by asphalt plants and WWTPs were developed, including an odour evaluation framework [94,95]. For asphalt plants, the target and limit values (for highly sensitive areas) are fixed at 1 and 2.5 su·m^{-3} as the 98th percentile. As asphalt plants are non-continuous odour sources, also 99.99 percentile target and limit values are used. For asphalt plants producing 15 to 25% of the time, the target and limit values are fixed at 5 and 12.5 su·m^{-3} as the 99.99 percentile.

For WWTPs, a distinction is made between the sources that cause a very unpleasant odour (e.g., the primary treatment, sludge storage, and treatment) and sources that cause a neutral odour (such as the biological treatment). The impact of both odour types is determined in separate dispersion calculations. For the very unpleasant odours, the target and limit values given in Tables 10 and 11 are used. The target and limit values (for highly sensitive areas) are fixed at 1 and 2.5 su·m^{-3} as the 98th percentile. For the neutral odours, these values are, respectively, 1.5 and 3 su·m^{-3} as the 98th percentile. For less sensitive areas, higher target and limit values are used.

One last sector for which an odour impact evaluation framework was derived is livestock farming. In the environmental impact assessment guidebook for livestock farming [95], the following odour evaluation framework is included, which distinguishes between isolated livestock farms and livestock farms that belong to a cluster (Tables 12 and 13). The values in this framework are expressed in ou$_E$·m^{-3} (i.e., not in sniffing units). (Note: a livestock farm belongs to a cluster when one or more other farms are situated in the no-effect level contour (0.5 ou$_E$·m^{-3} as the 98th percentile) of the farm under investigation. Only livestock farms with odour emissions higher than 5% of the farm's odour emission under investigation should be taken into account).

Table 12. Odour impact evaluation framework for isolated livestock farms [96].

Concentration as 98% [$ou_E \cdot m^{-3}$]	Scattered Houses in Agricultural Area	Residential Area with Rural Character	Residential Area
>10	Significantly negative impact	Significantly negative impact	Significantly negative impact
3–10	Negative impact	Significantly negative impact	Significantly negative impact
1.5–3	Small negative impact	Negative impact	Significantly negative impact
1–1.5	Negligible impact	Small negative impact	Negative impact
0.5–1	Negligible impact	Negligible impact	Small negative impact
<0.5	Negligible impact	Negligible impact	Negligible impact

Table 13. Odour impact evaluation framework for livestock farms belonging to a cluster [96].

Concentration as 98th Percentile [$ou_E \cdot m^{-3}$]	Scattered Houses in Agricultural Area	Residential Area with Rural Character	Residential Area
>10	Significantly negative impact	Significantly negative impact	Significantly negative impact
5–10	Negative impact	Significantly negative impact	Significantly negative impact
3–5	Small negative impact	Negative impact	Significantly negative impact
<3	Negligible impact	Negligible impact	Negligible impact

3. Australia and New Zealand

Odours are the largest source of air pollution complaints in Australia (AU) and New Zealand (NZ). In AU and NZ, odour is managed and legislated in much the same way as other noxious pollutants such as SO_2 and NOx. Odour is controlled under the Protection of the Environment Operations Act 1997 in AU and the 1991 Resources Management Act of New Zealand, and the Resource Management Regulations of 2004 [97]. Strict odour assessment criteria exist in both countries.

Odour assessment criteria in AU and NZ are primarily used to compare odour concentrations from dispersion model outputs, in $ou \cdot m^{-3}$, to the respective country and state odour guideline values to determine whether objectionable or offensive effects are likely to occur, although there appears to be an increase towards a risk-based assessment approach, i.e., Western Australia. In general, the odour assessments in both countries take into account the following:

Odour guideline documents accompany each state in AU, with a single guideline document in NZ. It is emphasised that the guidelines and odour assessment criteria therein are not meant to be interpreted as a "pass or fail" test. The guidelines aim to provide a framework for effective project planning and a regulatory regime for odour-emitting activities. Other key points relating to odour assessment as per the AU and NZ guidelines are as follows.

Odour unit has the same meaning as that in the Australia and New Zealand Standard AS/NZS 4323.3, Stationary source emissions—Determination of odour concentration by dynamic olfactometry.

Peak-to-mean ratio. In New South Wales and Queensland, a "user-applied" conversion factor adjusts the mean dispersion model predictions to the peak concentrations perceived by the human nose. In New South Wales, the peak-to-mean value varies depending on whether the source is wake-free or wake-affected (due to structures), the source characteristics, the distance from the source, and atmospheric stability. In Queensland, the peak-to-mean value depends solely on whether a source is wake-free or not. In NZ and some states of AU, the peak-to-mean value has already been included in the odour assessment criteria. The peak-to-mean values can be applied to the emission rates or to the predicted odour concentrations.

Various percentile limits of 100%, 99.9%, and 99.5% are used throughout both countries. The percentiles allow for a small level of exceedances of the concentration predictions to account for the worst-case meteorological conditions, at which objectionable odours are unlikely to occur because the conditions occur infrequently or not at all.

Table 14 below presents the odour assessment criteria used throughout Australia and New Zealand, and Table 15 presents the peak-to-mean ratios applied in New South Wales, Australia, while Table 16 presents the peak-to-mean ratios applied in Queensland, Australia.

Table 14. Odour assessment criteria for New South Wales [98], Western Australia [99], Australian Capital Territory (ACT) [100], South Australia [101], Queensland [102], Victoria [103], and Tasmania [97] in Australia as well as New Zealand [104].

Odour Assessment Criteria	New South Wales Australia	Western Australia	ACTEW and South Australia	Queensland Australia	Victoria Australia	Tasmania Australia	New Zealand
Impact assessment criteria	2.0–7.0 ou Log scale based on population density	WA prefers a risk-based approach	2.0–7.0 ou ACT 2.0–10.0 SA Log scale based on population density	5 ou	Varies 5.0 ou Broiler farms 1.0 New Developments	2.0	1.0–10.0 ou Depends on the sensitivity of the receiving environment
Percentile value	99th or 100th Depends on the quality of Met and Emission Data	Dispersion modelling is no longer the first response	99.9	99.5	99.9	99.5 or 99.9 For an unknown and known mixture, respectively. Or, 100 if good-quality Met and Emissions	99.5 and 99.9
Averaging period	1 h but criteria are equivalent to 1 s	1 h	3 min	1 h	3 min	1 h	1 h
Peak-to-mean ratio	Peak-to-mean ratio applied by user to 1-h averaged conc. See Table 15.	Modelling is not used to compare against ou criteria	No peak-to-mean value applied but conc. must be scaled to 3 min using power law equation	Peak-to-mean ratio of 10:1 and 2:1 for wake-free and wake-affected + ground sources	No peak-to-mean applied but conc. must be scaled to 3 min using the power law equation	Peak-to-mean ratio is included in odour assessment criteria	Peak-to-mean ratio is included in odour assessment criteria

Table 15. Peak-to-mean values used in New South Wales, Australia [98].

Source Type	Stability Class (Unstable and Neutral) A, B, C, D	Stability Class (Stable) E, F
Area	2.5	2.3
Wake-affected point	2.3	2.3
Wake-free point	12	25
Volume	2.3	2.3

Table 16. Peak-to-mean values used in Queensland, Australia [102].

Source Type	Peak-to-Mean Value
Wake-affected point and all ground-based sources	2.0
Wake-free point	10.0

New South Wales further defines the peak-to-mean values as "near"-field or "far"-field. Near- and far-field distances are defined as "less than" and "greater than" ten times the largest source dimension.

For an unstable and neutral atmosphere, "near-field" area sources use a peak-to-mean value of 2.5, and for "far-field", a peak-to-mean value of 2.3. For a stable atmosphere, "near-field" area sources use a peak-to-mean ratio of 2.3, and for "far-field", a peak-to-mean value of 1.9.

In Queensland, "user-applied" peak-to-mean values are 2.0 for all wake-affected point sources and all ground-based sources. The peak-to-mean value is 10.0 for all wake-free point sources.

All peak-to-mean values in AU and NZ are based in some way on the original Katestone Scientific work [105,106] conducted in 1995 on behalf of the Environment Protection Authority of New South Wales.

Neither AU nor NZ provides different odour assessment criteria according to odour activity. This means that a broiler farm is assessed at the same odour rate as a piggery or a layer hen farm. However, odour assessment criteria can range depending on the size of the nearby potentially affected population. In several states in AU, namely, New South Wales, South Australia, and ACT (Canberra), a range of odour assessment criteria is applicable depending on the sensitivity of the population as determined by population numbers. For example, in New South Wales, a single residence is assessed at 7 ou, whilst for larger populations, where there will be a greater range of sensitivities to odour and a higher number of more sensitive individuals, the acceptable odour limit is defined as 2 ou. If an odour source is in an area with a rural residence to the north and a town of 500 people to the south, then the appropriate criterion would be 7 ou for the single residence and 3 ou for the town and adjoining houses.

In New Zealand, under the Resource Management Act, the environment's sensitivity must be taken into account and should be considered as part of any odour assessment. This is dictated by the district plan's provisions, which set out amenity expectations for each land use type. The sensitivity in a particular location is based on the characteristics of the land use, including the time of day and the reason people are at a particular location. For example, "people driving past a broiler farm may not find the odours offensive as their exposure is very brief. Similarly, odours from natural sources, such as mudflats or geothermal activity, are unlikely to be deemed offensive. However, people attending a wedding at a church may find odours from an anaerobic oxidation pond at a neighbouring wastewater treatment plant to be extremely offensive" [107].

In New Zealand, the sensitivity of the receiving environment is assessed according to three land use categories; highly sensitive, medium sensitivity, and low sensitivity. Hospitals, schools, childcare facilities, and residential areas are all assessed as highly sensitive, whilst a rural area that can carry just a handful of residences can be rated as having both high sensitivity and low sensitivity. The thinking goes that people "living in and visiting rural areas generally have a high tolerance for rural activities but they are still sensitive to other types of activities (e.g., industrial activities)" [107]. Along with this ambiguity of high and low sensitive land use activities, two different odour assessment criteria exist, where a highly sensitive area carries an odour assessment criteria of 2 ou and a low sensitive area is assessed at 10 ou.

Table 17 below provides the range of odour assessment criteria per population numbers for those states that include it. Table 18 includes the New Zealand odour assessment criteria as per sensitive land use types.

Western Australia has recently published its new June 2019 guideline document [99]. This new guideline does not recommend the comparison of the dispersion model output with the odour assessment criteria. The WA guideline has provided a range of emission sources, pathways, and receptor tools for analysing an odour that does not involve modelling. Emphasis is put on the characterisation of odour sources, field assessments, and analysis of the complaints register. Dispersion modelling is only recommended for "comparative" assessments. This refers to the comparison of two or more modelling scenarios without specific reference to air emission criteria.

Table 17. Odour assessment criteria range according to population numbers in South Australia, New South Wales, and ACT, and New Zealand [98,100,104].

South Australia (3-min Average 99.9 Percentile)		New South Wales (1-s *1 Average 99.9 Percentile)	
Number of People	**ou**	**Number of People**	**ou**
2000 or more	2	2000 or more	2
350 or more	4	Approx. 500	3
60 or more	6	Approx. 125	4
		Approx. 30	5
12 or more	8	Approx. 10	6
Single Residence	10	Single Residence	7
High Density	2		
300 or more	3		
50 or more	5	1 person or 2000 persons	1–10
10 or more	6		
Less than 10	7		

*1 Nose response time = 1 s averaging time.

Table 18. Details of the NZ sensitivities of the receiving environment [107].

Sensitivity of the Receiving Environment	Concentration	Percentile
High *1 (worst-case impacts during unstable to semi-unstable conditions)	1 ou·m^{-3}	0.1 and 0.5
High *1 (worst-case impacts during neutral to stable conditions)	2 ou·m^{-3}	0.1 and 0.5
Moderate *2 (all conditions)	5 ou·m^{-3}	0.1 and 0.5
Low *3 (all conditions)	5–10 ou·m^{-3}	0.5

*1 High sensitivity includes rural, rural residential, countryside living, commercial, and retail business. *2 Moderate sensitivity includes commercial, retail business, rural residential, countryside living, and light industry. *3 Low sensitivity includes rural, heavy industry, and public roads.

4. China

4.1. Background and Overview

The Chinese emission standard for odour pollutants [108] can be found at website of Ministry of Ecology and Environment of the People's Republic of China. The standard is only available in the Chinese language, with the title and keywords explained in English.

While odour legislation in Europe, America, and Australia is focused on minimising odour concentrations at receptors, with usually no specific requirements on odour emissions from the sources, the odour legislations in East Asian countries such as China and Japan have regulations both on disorganised odour emissions and on discharge limits from stacks. This is likely due to the higher population density in these areas, where odour pollution can be dense and complicated for tracking sources.

4.2. Odour Impact Assessment in the People's Republic of China (PRC)

In China, the emission standard for odour pollutants GB 14554-93 [108] is still valid even though it was legislated in 1994. An example of the validity of this standard is shown on the PRC environment website in a case dealing with odour pollutants (hydrogen sulfide and carbon disulfide) emitted from industry in 2007. Nevertheless, a revision of the standard GB 14554-93 is in progress, with the call for comments closed in March 2010. The consultation paper was released in December 2018, and the new version of this standard is therefore expected to be released soon (with more strict emission standards expected). GB 14554-93 [108] stated boundary odour concentrations and standard concentrations for disorganised odour emissions of eight odourants, as shown in Table 19. Industries such as livestock and poultry breeding have a specific pollutant discharge standard with an odour concentration limit of 70 [109]. Meanwhile, the GB 14554 standard [108] also legislated the discharge limit for the emissions of eight odourants and odour concentrations from stacks, as shown in Tables 20 and 21. Depending on stack height, various levels of emission rates standards (kg·h^{-1}) were given, with higher emission rates allowed under higher stack height. The odour concentration detection follows the "triangle odour bag method" [110], which is now also under revision. The "triangle odour bag method" requires six sniffing members each for sniffing three bags in which two bags are references with clean air inside. If the sniffing member can recognise the bag with an odour sample, the odour sample bag will then be diluted for the next level of sniffing until no recognition can be made among the three bags. The odour concentration can thus be estimated based on dilutions.

Table 19. Boundary standard values of odour pollutants (mg·m^{-3}) [111]. Names are formatted to the "IUPAC (common name)" convention.

Pollutant	Class 1 [*1]	Class 2 BER [a] [*2]	Existing	Class 3 BER [a] [*3]	Existing
Azane (ammonia)	1.0	1.5	2.0	4.0	5.0
N,N-dimethylmethanamine (trimethylamine)	0.05	0.08	0.15	0.45	0.8
Sulfane (hydrogen sulfide)	0.03	0.06	0.10	0.32	0.6
Methanethiol (methyl mercaptan)	0.004	0.007	0.010	0.020	0.035
Methylsulfanylmethane (dimethyl sulfide)	0.03	0.07	0.15	0.55	1.1
(Methyldisulfanyl)methane (dimethyl disulfide)	0.03	0.06	0.13	0.42	0.71
Carbon disulfide	2.0	3.0	5.0	8.0	19
Styrene (vinyl benzene)	3.0	5.0	7.0	14.0	19
Odour concentration [b]	10	20	30	60	70

[*1] Class 1—natural conservation areas, scenic areas, historical sites, and regions requiring special protection; [*2] Class 2—residential areas, areas of mixed activity (e.g., commercial and traffic, residential, cultural, industrial, and rural); [*3] Class 3—special industrial areas; [a] newly built, extended, or rebuilt (BER); [b] dimensionless.

The emission standards for odour pollutants of GB 14554-93 [108] have some drawbacks, partly because this is the first standard on odour in China, and it is now ~27 years old. First, the odour pollutants did not cover a wide representation from all industries with only the eight odourants.

Second, the three classes (Class 1, 2, and 3) based on which the standard boundary values were set were adopted from the Chinese Ambient Air Quality Standard GB3095-1996 [112]. However, this standard has been revised, and the new version considers only two classes of industrial area and non-industrial area for air quality. This standard was implemented in January 2016 [113], and thus should be revised accordingly.

Third, the different limits on the discharge of odourants as a function of the stack height are obstacles for applications of advanced odour reduction technologies since the industry has tried to avoid these new technologies by making a higher stack (and thus allowing a higher emission discharge) [114].

The local emission standards for odour pollutants [112] for the Shanghai area were implemented from 1 February 2017. In this standard, discharge limits were set for odour concentrations under various stack height levels for two classes of odour sources: industrial and other sources (Table 22).

Table 20. The discharge limits of emission rates for the 8 odourants and for the odour concentrations from stacks, where the IUPAC common name is provided in pararenthesis. O1—hydrogen sulfide (sulfane); O2—methyl mercaptan (methanethiol); O3—methylsulfanylmethane (dimethyl sulfide); O4—dimethyl disulfide ((methyldisulfanyl)methane); O5—carbon disulfide; O6—ammonia (azane); O7—trimethylamine (*N,N*-dimethylmethanamine); O8—vinyl benzene (styrene) [114].

Stack Height (m)	Discharge Limit of Emission Rate (kg·h^{-1})							
	O1	O2	O3	O4	O5	O6	O7	O8
15	0.33	0.04	0.33	0.43	1.5	4.9	0.54	6.5
20	0.58	0.08	0.58	0.77	2.7	8.7	0.97	12
25	0.90	0.12	0.90	1.2	4.2	14	1.5	18
30	1.3	0.17	1.3	1.7	6.1	20	2.2	26
35	1.8	0.24	1.8	2.4	8.3	27	3.0	35
40	2.3	0.31	2.3	3.1	11	35	3.9	46
60	5.2	0.69	5.2	7.0	24	75	8.7	104
80	9.3				43		15	
100	14				68		24	
120	21				97		35	

Table 21. The discharge limits of odour concentrations from stacks [114].

Stack Height (m)	Standard for Odour Concentration (Dilutions)
10	2000
20	6000
30	15,000
40	20,000
50	40,000
≥60	60,000

Table 22. The discharge limit for odour concentrations from stacks in the Shanghai area [115].

Pollutant	Stack Height (H; m)	Industrial Source	Non-Industrial Source
Odour Concentration	H < 15	500	800
	15 ≤ H < 30	1000	1000
	30 ≤ H < 50	1500	1500
	H ≥ 50	3000	3000
Odour Pollutants	H ≥ 15	See Table 21	

Compared to the national emission standard GB 14554-93 [108], this emission standard of odour concentration in Shanghai is much more stringent, with 500 (for the industrial area) or 800 (for the non-industrial area) compared to 2 000 under a stack height of 15 m or less. Further, 22 odourants were set for discharge limitations under a stack height of 15 m, both for the emitted concentration (mg·m^{-3}) and for the emission rate (kg·h^{-1}) (Table 23).

The standard on emitted concentrations was not included in the national standard of GB 14554-93 [108], while 14 more odour pollutants are newly included in the Shanghai standard. In addition, no difference was set for the Shanghai emission standard of odour pollutants under various stack heights, with apparently more stringent emission rate standards on single odour pollutants (e.g., for H$_2$S, 0.1 kg·h^{-1} in the Shanghai standard while ≥0.33 kg·h^{-1} in the national standard GB 14554-93). For fugitive odour emissions not emitted from specific stacks, limits of 20 and 10 dilutions were set for industrial areas and not-industrial areas, respectively. Additionally, further limits were set for 22 odourants for both typologies of land use (Table 24).

Table 23. Discharge limits for odour pollutants in the Shanghai area [115]. Names are formatted to the "IUPAC (common name)" convention.

Number	Pollutant	Maximum Acceptable Emission Concentration (mg·m^{-3})	Maximum Acceptable Emission Rate * (kg·h^{-1})
1	Azane (Ammonia)	30	1
2	Sulfane (Hydrogen sulfide)	5	0.1
3	Methanethiol (Methyl mercaptan)	0.5	0.01
4	Methylsulfanylmethane (Dimethyl sulfide)	5	0.1
5	(Methyldisulfanyl)methane (Dimethyl disulfide)	5	0.26
6	Carbon disulfide	5	1
7	Styrene (Vinyl benzene)	15	1
8	Ethylbenzene	40	1.5
9	Propanal (Propionic aldehyde) #	20	0.3
10	Butanal (Butyraldehyde) #	20	0.2
11	Pentanal (Valeraldehyde) #	20	0.2
12	Butan-2-one (Methyl ethyl ketone) #	50	5
13	4-methylpentan-2-one (Methyl isobutyl ketone) #	80	3
14	Prop-2-enoic acid (Acrylic acid) #	20	0.5
15	Methyl prop-2-enoate (Methyl acrylate) #	20	1
16	ethyl prop-2-enoate (Ethyl acrylate) #	20	1
17	Methyl 2-methylprop-2-enoate (Methyl methacrylate) #	20	0.6
18	Methanamine (Methylamine) #	5	0.11
19	N-methylmethanamine (Dimethylamine) #	5	0.15
20	N,N-dimethylmethanamine (Trimethylamine)	5	0.2
21	Ethyl acetate	50	1
22	Butyl acetate	50	1

*—if the efficiency of odour abatement technologies is higher than 95%, this criterion is by default fulfilled.
#—only implemented after the national standards of analytical methods were released.

Table 24. Boundary standard values for disorganised odour emissions in the Shanghai area [115]. Names are formatted to the "IUPAC (common name)" convention.

Number	Pollutant	Industry Area (mg·m^{-3})	Non-Industry Area (mg·m^{-3})
1	Azane (Ammonia)	1.0	0.2
2	Sulfane (Hydrogen sulfide)	0.06	0.03
3	Methanethiol (Methyl mercaptan)	0.004	0.002
4	Methylsulfanylmethane (Dimethyl sulfide)	0.06	0.02
5	(Methyldisulfanyl)methane (Dimethyl disulfide)	0.06	0.04
6	Carbon disulfide	2.0	0.3
7	Styrene (Vinyl benzene)	1.9	0.7

Table 24. *Cont.*

Number	Pollutant	Industry Area (mg·m^{-3})	Non-Industry Area (mg·m^{-3})
8	Ethylbenzene	0.6	0.4
9	Propanal (Propionic aldehyde)	0.26	0.08
10	Butanal (Butyraldehyde)	0.14	0.06
11	Pentanal (Valeraldehyde)	0.11	0.04
12	Butan-2-one (Methyl ethyl ketone)	2.0	1.0
13	4-methylpentan-2-one (Methyl isobutyl ketone)	1.2	0.7
14	Prop-2-enoic acid (Acrylic acid) #	0.6	0.11
15	Methyl prop-2-enoate (Methyl acrylate) #	0.7	0.4
16	ethyl prop-2-enoate (Ethyl acrylate) #	0.4	0.4
17	Methyl 2-methylprop-2-enoate (Methyl methacrylate) #	0.4	0.2
18	Methanamine (Methylamine) #	0.05	0.03
19	*N*-methylmethanamine (Dimethylamine) #	0.06	0.04
20	*N,N*-dimethylmethanamine (Trimethylamine)	0.07	0.05
21	Ethyl acetate	1.0	1.0
22	Butyl acetate #	0.9	0.4
23	Odour concentration	20 *	10 *

*—dimensionless; #—only implemented after the national standards of analytical methods were released.

The standard boundary limit values for the Shanghai standard generally show lower values than the national standard GB 14554-93 [108] for industrial and non-industrial areas. On the other hand, this standard includes 14 more odourants.

In addition, the "Technical specification on environmental monitoring of odour" [116], released on 29 December 2017 and taking effect on 1 March 2018, and the "Technical specification for olfactory laboratory construction" [117], released on 10 November 2017 and taking effect from that day, in China have been released after finishing the second round of comments in 2015.

The standard of the "Technical specification on environmental monitoring of odour" specifies the layout of sampling locations, odour sampling frequency, sampling methods, pre-treatment of collected odour samples, odour analysis methods, data processing and reporting, quality control and quality assurance, and so on. The odour sampling methods include sampling by vacuum bottles and sampling bags. The odour sampling and analysis should follow the standard method for odour concentration determination, GB/T 14675 [110], by applying the "triangle odour bag method".

The standard of the "Technical specifications for olfactory laboratory construction" also specifies the olfactory laboratory site selection and layout, interior design of the laboratory, etc. The olfactory laboratory should have at least three functioning areas, including a sampling preparation room, a sample mixing room, and an evaluation room, with two optional functioning areas of a buffer room and a restroom. The site selected for the olfactometric laboratory construction should have a maximum odour concentration of the ambient air lower than 10 ou.

4.3. Odour Impact Assessment in Hong Kong

Odour assessment criteria for Hong Kong can be found on the Hong Kong Environmental Protection Department's website. In Hong Kong, odour is assessed at a 5-s averaging period due to the shorter exposure period tolerable by human receptors. Conversion of model computed hourly average results to 5-s values is necessary to enable comparison against the recommended Hong Kong standard. The hourly concentration is first converted to a 3-min average value according to a power law relationship, which is stability-dependent due to the statistical nature of atmospheric turbulence. Another conversion factor (10 for unstable conditions and 5 for neutral to stable conditions) is then applied to convert the 3-min average to a 5-s average. In summary, to convert the hourly results to 5-s averages, the following factors listed in Table 25 need to be applied:

Table 25. Conversion factors to convert the 3-min odour concentrations to 5-s [118].

Stability Category	1-h to 5-s Conversion Factor
A and B	45
C	27
D	9

The values presented are similar to the peak-to-mean value approach applied in New South Wales.

Under "D" class stability, the 5-s concentration is approximately ten times the hourly average result. Note, however, that the combined use of such conversion factors together with the ISCST results may not be suitable for assessing the extreme close-up impacts of odour sources.

5. Japan

Japan has more than 40 years of odour legislation history at the national level. With industrial development and urbanisation in the 1960s, complaints against environmental pollution, including odours, drastically increased. To take measures against odour issues, the Offensive Odour Control Law (OOCL) [119] was enacted in 1971 and enforced in 1972. It regulates odours emitted from business activities and promotes preventive measures against odours to preserve the living environment and protect the health of the people [120]. Figure 4 depicts the framework of the OOCL.

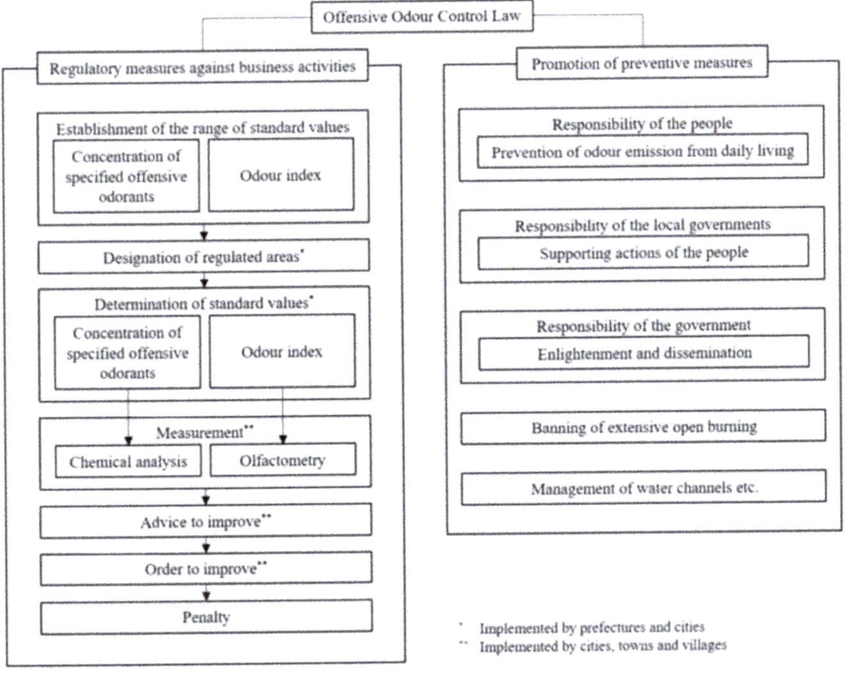

Figure 4. The framework of the Offensive Odour Control Law (OOCL) in Japan [119].

The OOCL provides three types of regulation standards on odours: (1) at the property line of the site, (2) discharged from stacks or other gas emission facilities, and (3) discharged from wastewaters.

Local authorities designate regulated areas in consideration of geographical and demographical conditions. Any kind of activity at factories or other businesses, including livestock farming within the regulated area, comes under odour legislation. Local authorities are entitled to demand reports and conduct on-site inspections at odour-emitting facilities, whereas they should carry out odour measurements by chemical analysis or olfactometry. If an odour-emitting facility in the regulated area does not meet the standard and simultaneously the living environment is impaired, the facility can be advised by the local authority to improve the operating conditions and take preventive measures. If the odour emission remains unchanged, the facility can be ordered to improve the situation. Penalties can be imposed on violators.

When the OOCL was enacted, odour regulations based on the concentrations of odourous compounds were introduced. Up to the present, twenty-two (22) substances shown in Table 26 have been designated as "specified offensive odourants". Local authorities determine the regulation standard values at the property line for each substance within a range established by the government (Table 26), considering the land use, geographical conditions, odour characteristics, and people's sensitivity to odours.

Table 26. Specified offensive odourants and the range of regulation standard values at the property line [119]. Names are formatted to the "IUPAC (common name)" convention.

Specified Offensive Odourant	Range of Standard Value at the Property Line (ppm)
Azane (Ammonia)	1–5
Methanethiol (Methyl mercaptan)	0.002–0.01
Sulfane (Hydrogen sulfide)	0.02–0.2
Methylsulfanylmethane (Dimethyl sulfide)	0.01–0.2
(Methyldisulfanyl) methane (Dimethyl disulfide)	0.009–0.1
N,*N*-dimethylmethanamine (Trimethylamine)	0.005–0.07
Acetaldehyde	0.05–0.5
Propanal (Propionaldehyde)	0.05–0.5
Butanal (Butyraldehyde)	0.009–0.08
2-methylpropanal (Isobutyraldehyde)	0.02–0.2
Pentanal (Valeraldehyde)	0.009–0.05
3-methylbutanal (Isovaleraldehyde)	0.003–0.01
2-methylpropan-1-ol (Isobutyl alcohol)	0.9–20
Ethyl acetate	3–20
4-methylpentan-2-one (Methyl isobutyl ketone)	1–6
Toluene	10–60
Styrene	0.4–2
Xylene	1–5
Propionic acid	0.03–0.2
Butanoic acid (Butyric acid)	0.001–0.006
Pentanoic acid (Valeric acid)	0.0009–0.004
3-methylbutanoic acid (Isovaleric acid)	0.001–0.01

However, these regulations are insufficient to deal with a considerable number of odour complaints caused by unregulated substances or complex odours since odour complaints have become more diversified. To improve this situation, the OOCL was amended in 1995, and odour regulations based on an "odour index", a sensory index of odour determined by the triangular odour bag method (TOBM), was introduced [121]. The TOBM is a static air dilution method by which the odour concentration or odour index is determined. In this method, an odour concentration is considered to be the dilution ratio when odourous air is diluted by odour-free air in an odour bag until the odour becomes unperceivable. The odour index is considered to be a logarithm of odour concentration, multiplied by ten.

The TOBM was first developed by the Tokyo metropolitan government in 1972 [122,123] and notified by the Japan Environment Agency in 1995. Since odour measurement is a crucial element of odour management and regulation, a quality control manual on the TOBM for laboratory use was published in 2002 to develop a reliable odour measurement method [124]. Local authorities determine the odour index standard values within a range from 10 to 21 established by the government. After the amendment of the OOCL, local authorities became entitled to choose either of the two regulations: (1) based on the concentrations of odourants, or (2) based on the odour index. According to the OOCL, the range of the regulation standards of both the concentrations of odourants and the odour index at the site's property line is equivalent to an odour intensity, which ranges from 2.5 to 3.5 on the six-point odour intensity scale shown in Table 27.

Table 27. Six-point odour intensity scale [119].

Scale	Odour Intensity
0	No odour
1	Barely perceivable (Detection threshold)
2	Faint but identifiable (Recognition threshold)
3	Easily perceivable
4	Strong
5	Extremely strong

Odours discharged from stacks and other gas emission facilities are regulated based on the standards at the site's property line. Table 28 summarises three types of odour regulation standards in Japan. Regulation standard values for odours discharged from smokestacks or other gas emission facilities are determined by dispersion modelling. The odour index of the wastewater is determined by the triangular odour flask method (TOFM) [121]. Odour emission facilities should meet all types of regulatory standards.

In addition to odour legislation at the national level, various investigations on chemical analysis and sensory measurement of odours have been carried out by local authorities since the 1960s. Up to the present, more than thirty local authorities have adopted their own odour legislation system as ordinances or guidelines. Moreover, several industrial cities have an agreement with local factories to preserve items such as air, water, noise, vibration, odour, hazardous substances, waste, and greenhouse gases. Based on the agreement, the factories voluntarily meet the desired odour index values, which are more rigid than the regulation standards at odour-emitting facilities. Some cities have also been conducting on-site inspections of potential odour-emitting facilities. These measurement results are released electronically on the city website [125].

Table 28. Summary of three types of odour regulation standards in Japan [121,122].

Regulation Type	Regulation Standard of the Concentration of Specified Offensive Odourants	Regulation Standard of Odour Index
Odours at the property line of the site	(Enforced in 1972) Determined by the local authority within a range shown in Table 26.	(Enforced in 1996) Determined by the local authority within a range from 10 to 21.
(2) Odours discharged from smokestacks or other gas emission facilities	(Enforced in 1972) Given as a flow rate calculated by the following equation: $q = 0.108\,H_e^2\,C_m$ where q: flow rate of specified offensive odourant (Nm^3/h), H_e: effective stack height (m), C_m: standard regulation value of specified offensive odourant at the property line (ppm). Applicable to the following 13 specified offensive odourants. Ammonia Hydrogen sulfide Trimethylamine Propionaldehyde Butyraldehyde Isobutyraldehyde Valeraldehyde Isovaleraldehyde Isobutyl alcohol Ethyl acetate Methyl isobutyl ketone Toluene Xylene	(Enforced in 1999) In the case the stack height (H_o) is 15 m or more Given as an odour emission rate (OER) calculated by the following equation: $q_t = (60 \times 10^A)/F_{max}$ $A = (L/10) - 0.2255$ where q_t: OER of discharged gas (Nm$^3 \cdot$min^{-1}), F_{max}: calculated value using the dispersion modelling in consideration of the building height in the vicinity (s$\cdot$Nm^{-3}), L: standard regulation value of odour index at the property line. (2) In the case H_o is less than 15 m Given as an odour index calculated by the following equation: $I = 10 \log C$ $C = K\,H_b^2 \times 10^B$ $B = L/10$ where I: odour index of discharged gas, K: coefficient determined depending on the stack diameter, H_b: maximum building height in the vicinity (m).
(3) Odours included in wastewater	(Enforced in 1995) Given as a concentration in the wastewater calculated by the following equation: $C_{Lm} = k\,C_m$ where C_{Lm}: concentration of specified offensive odourant in wastewater (mg$\cdot$L^{-1}), k: coefficient shown in Table 29 (mg$\cdot$L^{-1}). Applicable to the following four specified offensive odourants. Methyl mercaptan Hydrogen sulfide Dimethyl sulfide Dimethyl disulfide	(Enforced in 2001) Given as an odour index calculated by the following equation: $I_W = L + 16$ where I_W: odour index of wastewater.

Table 29. Coefficients used to calculate the standard values of specified offensive odourants in wastewater (mg$\cdot$L^{-1}) [121]. Names are formatted to the "IUPAC (common name)" convention.

Flow Rate of Wastewater: Q (m$^3 \cdot$s^{-1})	$Q \leq 0.001$	$0.001 < Q \leq 0.1$	$Q > 0.1$
Methanethiol (Methyl mercaptan)	16	3.4	0.71
Sulfane (Hydrogen sulfide)	5.6	1.2	0.26
Methylsulfanylmethane (Dimethyl sulfide)	32	6.9	1.4
(Methyldisulfanyl)methane (Dimethyl disulfide)	63	14	2.9

6. United States of America

In the United States, the Environmental Protection Agency (EPA) does not regulate odour as a pollutant; therefore, states and local jurisdictions have attempted to regulate odours. For the individual states, statutes approved by the legislature provide the legal framework for addressing odour emissions, while the corresponding state departments (e.g., Department of Environmental Quality, Department of Natural Resources) are responsible for enforcement of the odour rules or regulations. In the absence of "odour laws" or odour regulations, citizens and communities often find remedies and relief in basic "common law" nuisance lawsuits. However, exclusions and exemptions, such as "right-to-farm" laws and vague definitions of "nuisance", can sometimes make nuisance actions difficult and expensive to prosecute.

The National Air Pollution Control Administration of the U.S. Public Health Service commissioned the Copley International Corporation in 1970 to conduct a "National Survey of the Odor Problem". The Copley study's technical phase found the "dilution-to-threshold (D/T) ambient odour measurement method", embodied in the Scentometer device, to be a utilitarian and effective tool for investigation of odour and that odour judgment panels provide a definitive description of the odour emission [126].

Historically, the D/T values are based on the dilution ratio of the carbon-filtered air volume to the odourous air volume. This is different from laboratory olfactometry, where the dilution ratio is the total volume of air to the odourous air sample volume. The units of D/T are commonly used to specify the threshold value as being determined by field olfactometry and not laboratory olfactometry ($ou_E \cdot m^{-3}$). The difference in dilution ratio calculation provides a relationship of [threshold value in D/T] + 1 = (value in $ou_E \cdot m^{-3}$ or in ou). Examples include 7 D/T being the same as 8 $ou_E \cdot m^{-3}$ (8 ou) or 60 D/T being the same as 61 $ou_E \cdot m^{-3}$ (61 ou). The difference is negligible at relatively low threshold values.

The U.S. EPA commissioned a second Copley study in 1971, "Social & Economic Impacts of Odors", in the United States. The second Copley study found the Scentometer (D/T method) to be an effective and sensitive device and found odour judgment panels were a logistical challenge for responding to all complaints [127].

The U.S. EPA commissioned a third Copley study in 1972 for the "Development and Evaluation of a Model Odor Control Ordinance". The third Copley study recommended that odour regulation and enforcement be relegated to states and local jurisdiction using scientific approaches with trained inspectors using the Scentometer D/T method as well as source odour sampling [128]. This set the course in the U.S. for the EPA to pass jurisdiction of odours to individual states and municipalities.

Prior to these studies, in 1958, 1959, and 1960, the U.S. Public Health Service sponsored the development of an instrument and procedure for field olfactometry (ambient odour strength measurement) through official project grants [129]. The first field olfactometer device, called the Scentometer, was manufactured by the Barnebey-Cheney Company and subsequently manufactured by the Barnebey Sutcliffe Corporation. The only other field olfactometer, recognised by states as equivalent to the Scentometer, is the Nasal Ranger introduced by St. Croix Sensory in 2002.

6.1. State Regulations

As of 2018, field olfactometry still stands as the most commonly utilised method for odour regulation. Ten states currently utilise a D/T field olfactometry limit for their odour regulation: (1) Colorado [130], (2) Connecticut [131], (3) Delaware [132], (4) Illinois [133], (5) Kentucky [134], (6) Missouri [135], (7) Nevada [136], (8) North Dakota [137], (9) West Virginia [138], and (10) Wyoming [139]. Figure 5 displays the U.S. states with odour regulations. The ten field olfactometer states are displayed in red. Based on the original field olfactometer studies, most of these states have an odour limit of 7 D/T. As an example, Regulation 2 (5 CCR 1001-4) from the State of Colorado: " ... areas predominantly for residential or commercial purposes, it is a violation if odours are detected after the odourous air has been diluted with seven (7) or more volumes of odour-free air (7-D/T)" [130]. The Colorado regulation also designates a higher limit (15-D/T) for other land use areas, i.e., industrial. However, Colorado limits ambient odour to only 2 D/T at the receptor near large pig facilities. Once an enforcement agency within the state, such as the city of Denver, receives citizen complaints, enforcement personnel respond to the complaint location(s) and measure the D/T with field olfactometry every 10 min for 1 h. A violation exists if the enforcement agent twice measures the odour at 7 D/T or higher, with these measurements separated by at least 15 min, i.e., there is an odour above the limit with a duration/frequency.

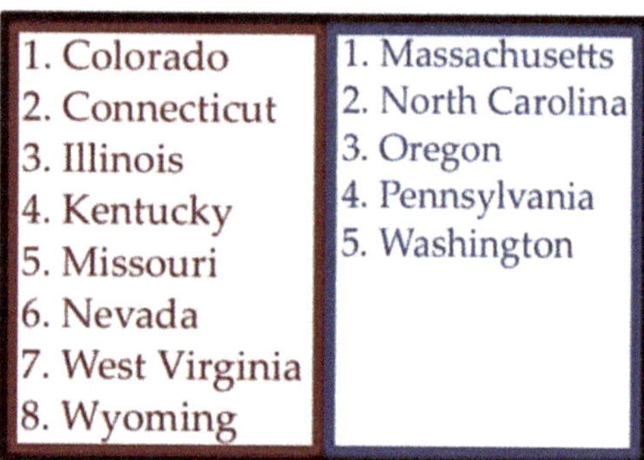

Figure 5. States of the U.S. where odour regulations exist based on dilution-to-threshold (D/T), indicated in red. The blue states have regulations with some reference to odour impacts [130–141].

Figure 5 also displays five states in blue, which have odour nuisance regulations with specific reference to odour properties, but without specific criteria for odour measurement or determination of nuisance. In 2013, and later updated in 2014, the Oregon Department of Environmental Quality (DEQ) published a document titled "Nuisance Odor Strategy" [140], which defines actions by the state for facilities under scrutiny for violating the Oregon nuisance code: " … may not generate odours that cause an unreasonable interference with another's enjoyment of their property" [141]. When complaints are issued to the DEQ, a facility is reviewed and prioritised based on a two-part nuisance score. One part is rated based on the frequency and duration of the odours, and a second value is based on the strength and offensiveness. An "Odor Intensity Referencing Scale" [8] is suggested for determining odour strength but not required by the nuisance law. Evidence is provided to a nuisance panel. If the panel issues notice of a nuisance, then a facility will be required to enter into a best work practices agreement and a complete a nuisance abatement proposal. While this is a detailed process, the determination of odour strength and offensiveness remains up to the subjective decision of an inspector.

6.2. Municipalities

Numerous municipalities in the U.S. have chosen to regulate odours when their state has not. One example of a U.S. municipal odour ordinance is from the city of Independence, Louisiana. Five stipulations of the Independence odour ordinance are as follows: (1) unlawful to cause emissions of an odour nuisance or odourous air contaminant, (2) odour that is unreasonably unpleasant, distasteful, disturbing, nauseating, or harmful to a person of ordinary sensibilities and which is detectable after it is diluted with seven volumes of odour-free air by a field olfactometer, 7 D/T, (3) the city may issue a citation for the violation, (4) any person may file a complaint, and the city will investigate the complaint, and (5) USD 500 penalty on conviction; penalty does not preclude further actions to abate violations. The use of the phrase " … to a person of ordinary sensibilities … " is commonly used in municipal codes for defining a nuisance; however, this remains arguable without a measurable parameter.

The second example of a U.S. municipal odour ordinance is from the city of Des Moines, Iowa. Des Moines code enforcement officers respond to citizen complaints as part of their normal code enforcement duties, i.e., restaurant inspections [142]. The city declares an "Odor Alert" when they receive ten complaints in a 24-h time period. An inspector responds, measures the ambient odour, identifies the probable source, and serves a notice of violation. A facility that receives three notices of violations in a 90-d period is designated by the city as a "significant odor generator" and is required to submit an "odor management plan" that may include air stack testing and air dispersion modelling. The designated "significant odor source" may appeal to a citizen "Odor Board", then the city council, and then the municipal court. Implemented in 1991, the city of Des Moines' citizen Odor Board is a unique, novel, and effective approach to addressing local odour nuisances.

When citizens find themselves in a position where the federal government, their state government, and their local municipality (or county, parish, district, or similar organisation) have not enacted an odour nuisance ordinance, there is the final option of bringing a "common law" legal suit against the facility. A judge and jury then determine the nuisance based on the evidence presented by the plaintiff and defendant.

6.3. Odour and Agriculture in the USA

Odour and agriculture are one of those "hot button" issues that exist worldwide, yet they have a very specific status in the USA. The Council for Agricultural Science and Technology [143] has published a white paper prepared by scientists from six U.S. universities summarising "Air Issues Associated with Animal Agriculture: A North American Perspective". In that study, odour emissions were discussed in the larger scope of gaseous and particulate emissions from pig, poultry, beef, and dairy production. While odour is mainly a local issue, hazardous gases (e.g., NH_3 and H_2S, some volatile organic compounds; VOCs) are regional and national concerns.

The CAST [143] argues for a "common sense" approach to regulating gaseous (including odour) emissions that recognises regulatory needs and market forces. All of this process needs to involve the public, regulatory agencies, and the livestock industry. Both the positive (economic development) and negative (e.g., lower real-estate values in the vicinity) aspects of animal production need to be reconciled for the greater good of rural communities.

The livestock industry in the U.S. funds air quality research aimed at baseline emission inventories (e.g., the National Air Emissions Monitoring Study, 2007–2009) [144] and research aimed at developing and field testing promising mitigation technologies. An important part of mitigation research is its practicality in the U.S. socio-economic climate. Economic analysis of tested technologies is a typical requirement for farm testing and possible future adoption.

The livestock industry funds educational tools for farmers, regulatory agencies, and scientists. For example, the Air Management Practices Assessment Tool [145] is an online resource to "provide an objective overview of mitigation practices best suited to address odour, emissions and dust at your livestock operation so that livestock and poultry producers may compare and narrow their options of mitigation techniques". Most recently, a scientific database was added to AMPAT. The scientific database summarises 265 papers reporting on the performance of technologies to mitigate emissions of odour and other gases from animal production operations [146,147]. There is growing evidence that some mitigation technologies offset benefits from regulating one pollutant by increasing emissions of another (e.g., NH_3 and N_2O).

Odour emissions are mainly generated by manure handling, storage, treatment, and land application. These processes are highly site-specific. The complexity of odour emissions is confounded by many factors at the nexus of species, local climate, geography, size and type of the facility, animal diet, manure management system, ventilation system, regulations, and human factors.

Many animal production facilities operate in areas that have a non-attainment status for regulated air pollutants, and thus face larger scrutiny because of the non-attainment of the air quality standard of the whole area (e.g., St. Joaquin Valley in California). Some regulated air pollutants associated with animal agriculture (e.g., PM-10, PM-2.5 [148]) are regulated by the National Ambient Air Quality Standards. Others (e.g., NH_3) are of concern due to the formation of secondary fine PM-2.5 and eutrophication. Many odourous gases are classified as VOCs, some reactive, and thus are of interest to ozone and NOx management.

The last thorough review of odour regulations focusing on U.S. agriculture was published by Redwine and Lacey (2000) [149]. Most states define a regulatory approach to confined animal feeding operations (CAFOs) based on the number of animal units (AU). Animal units are typically defined as 500 kg of live weight. Redwine and Lacey (2000) [149] summarised odour regulations in all U.S. states and grouped them according to the following criteria:

- Odour—Is there direct regulation of odour emissions?
- Setbacks—Are there setbacks (i.e., mandatory distances to neighbours)?
- Permits—Are permits required?
- Public—Is there public involvement in the permitting process?
- Training—Is some form of training required?
- LA—Are there land application [of manure] restrictions?
- Other—Any other approach to regulating odour from CAFOs or related
- information.

To date, the Redwine and Lacey report (2000) [149] is still the most comprehensive resource on odour regulations and animal agriculture in the US. They have summarised the following main points: 10 U.S. states regulate odour directly and 34 U.S. states have some rules or regulations designed to curtail odour emissions without explicit limitation (e.g., distance setback, manure management plan, permitting, land application regulations, manure application training).

As the U.S. Environmental Protection Agency (EPA) and state regulatory agencies are increasing monitoring of animal production operations, the emissions of odour and technologies to comprehensively mitigate them will become more important.

The U.S. EPA is also considering the application and possible implication of the existing CERLA/EPCRA ruling limiting emissions of any substance to air at or above 100 lbs (~45 kg) per day, per site.

7. Canada

Canadian federal legislation does not cover any odour regulations from industrial or agricultural facilities. Individual provinces and territories have an obligation for odour regulations [150,151]. In legislation, odour can be defined in different ways, such as a pollutant, contaminant, type of substance, nuisance, or an odourous substance and odourous contaminant. An odour may also be defined by its effects, which include being a contaminant that causes an adverse effect. The following is a brief summary of how odour is regulated by individual provinces in Canada [152].

7.1. Alberta

The Alberta Environmental Protection and Enhancement Act (EPEA) describes that an "adverse effect" means impairment of or damage to the environment, human health and safety, or property. In addition, the EPEA describes a "substance" as any matter that is capable of becoming dispersed in the environment or is capable of becoming transformed in the environment into matter. There is no specific mention of odour in the EPEA as a substance, which causes an adverse effect. However, odour might be a dispensed substance in the environment and, therefore, could be a prohibited contaminant.

7.2. British Columbia

In British Columbia, the Environmental Management Act does not mention odour; however, odour can be treated as an air contaminant that interferes with normal conduct of business or causes physical discomfort to a person. Odours attributed to any agricultural operations or activities on a farm in accordance with the Agricultural Waste Control Regulation, Code of Agricultural Practice for Waste Management are not prohibited.

7.3. Newfoundland and Labrador, Northwest Territories, Prince Edward Island

In Newfoundland and Labrador, the Northwest Territories, and Prince Edward Island, there are no standards for odours; however, odour is a prohibited contaminant [153]. In Newfoundland and Labrador, the Air Pollution Control Regulation 39/04, under Air Quality Standards-Schedule A, includes prescribed air quality standards which are relevant to agricultural operations for NH_3 (100 g·m^{-3} as 24-h average), H_2S (15 g·m^{-3} as 1-h standard, 5 g·m^{-3} as 24-h standard), and reduced sulfur compounds (30 g·m^{-3} as 1-h standard), expressed as an equivalent amount of H_2S.

In Prince Edward Island, under the Environmental Protection Act, odour is a contaminant; however, there is no standard for odour [154].

7.4. Nova Scotia and Saskatchewan

Under the Environmental Act, there is no odour standard in these provinces, but odour can be a contaminant [155].

7.5. Manitoba

In Manitoba, under the Environmental Act, odour is a pollutant, and there are some guidelines for odour concentrations in the ambient air with a maximum of two odour units for a residential area, a maximum of seven odour units for an industrial area, and one odour unit for all areas. The guideline states that to determine these concentrations, duplicate odour measurements should be taken not less than 15 min apart and not more than 60 min apart. The measurements are based on the ambient level. There are also some criteria for a maximum ammonia concentration of 1.4 µg·m^{-3}, and a maximum H_2S concentration of 15 µg·m^{-3} averaged over 1 h or 5 µg·m^{-3} averaged over 24 h [156].

7.6. Ontario

In Ontario, under the Environmental Protection Act (EPA), odour is a contaminant. Odour is a contaminant to the degree that it may cause discomfort, loss of enjoyment of normal use of the property, or interfere with the normal conduct of business. Another section of the act prescribes the maximum point of impingement concentrations for a variety of compounds [157]. A number of these are based on the odour potential of these compounds. Dispersion models are included in the regulation for calculating the maximum point of impingement concentrations from emission rate data [158]. Odour issues are routinely addressed by the Environmental Compliance Approval (ECA). Requirements for odour emission tests are often included as conditions for industrial sources, which are judged by the Ontario Ministry of the Environment, Conservation and Parks (MECP) to have a potential for odour impact. Emission test results are used with regulatory dispersion models to estimate the maximum point of impingement odour levels. A guideline of 1 ou odour concentration is based on the model's prediction when a 10-min averaging time is used [159].

However, the EPA does not apply to animal wastes disposed of in accordance with both normal farming practices and the regulations made under the Nutrient Management Act, 2002 [160].

The Ontario Municipal Act 2001 [161] allows municipalities to control odours within their jurisdiction. If the ministry receives an odour complaint, it is the role of the district office to follow-up the complaint, to verify the information provided, and to make an assessment on whether further action is required by the ministry. If the odour is deemed to be causing any adverse effects, steps will be taken to identify the source of the odour, address any adverse effects caused by the odour, and ensure that the responsible party takes all reasonable steps to mitigate the odour.

7.7. Quebec

In Quebec, odour is a contaminant under the Environmental Quality Act [162]. There are some odour standards for specific facilities, such as standards for odours discharged by a fried food plant or coffee roasting plant.

There is also an ambient air quality standard for H_2S—14 $\mu g \cdot m^{-3}$ averaged over 1 h. However, there is no standard for NH_3.

7.8. Examples on How to Deal with Odour Complaints

As an example, in Ontario, when odours are detected, complaints may occur. The complainants can contact the MECP, or they can complain directly to the facility "in question". The environmental officer investigates the odour episode and very often conducts a site visit to the area. During regular business hours, the local district office can be contacted directly; outside normal business hours, calls are directed to the ministry's Spills Action Centre toll-free line [163].

If the odour complaints are persistent for the area, the MECP can order the facility to perform an odour assessment, including odour testing, and, if the odour limit is exceeded, the facility is required to provide a plan of controlling measures.

7.9. Odour Assessments

There is no standard method across Canada for odour assessments; however, the most common approach is source odour measurements with dispersion modelling to predict off-site odour concentrations at any sensitive receptor [164]. Odour assessments are generally performed for the following reasons: to verify and investigate odour complaints; to determine the off-site odour impact from existing, expanding, or new operations; to assess long-term odour levels in an area; to determine compliance with odour legislation, or to rank potential odour sources for mitigation purposes.

In addition to source odour testing, an ambient odour assessment can be performed, which in most cases includes ambient testing at the affected or complained areas using a standard procedure and dynamic olfactometry evaluations using screened panellists. Further, ambient odour assessments may include community/resident odour surveys, odour observations, and observation forms for residents. Ontario is the strictest province in terms of odour regulations.

Odour testing in Ontario is performed according to Ontario's MECP Source Testing Code, Method ON-6: Determination of Odour Emissions from Stationary Sources [165]. The odour analyses follow the same method, which is similar to EN13725 with some exceptions, such as odour analysis only being conducted once and with eight panellists.

8. South America

There is a wide diversity of legislation related to odour management in the countries of Latin America. In this section, some countries that have developed legislation in terms of odour management are detailed.

8.1. Chile

Despite many disastrous socio-environmental conflicts triggered by odour episodes, Chile has not yet developed an odour regulation. The air pollution legislation has almost no specific standard for odours or compounds related to them, except the standard of total reduced sulfur odour generators associated with the manufacture of sulfated pulp [166].

The second body of law is the Law on General Environmental Framework [167] whose main instruments include environmental quality standards, emission standards, and the system of environmental impact assessment (SEIA). Regarding the existing environmental quality standards, there is no specific standard for odours in ambient air. There are ordinances in some municipalities, which establish restrictions on the generation of odours that may be a health risk or be annoying to the community [168].

As for the legal tools available to manage odours in the country, there is the Sanitary Code, which gives jurisdiction to the Health Authority (formed by the Ministry of Health and its Regional Ministerial Secretariats of Health) to issue general, or specific, provisions for the proper performance of the code, conferring the duty to monitor odour emissions and use sanctions such as fines, closures, cancellation of operating licenses or permits, or even closing facilities depending on the number of infractions. There is the use of an offensive odour indicator parameter, which is the number of complaints or allegations made by the community to the Health Authority or other agencies (Seremi, municipalities, etc.) which are channeled through the Health Authority.

In 2014, after a conflict caused by pig production, the Ministry of Environment (MMA) began developing a Strategy for Odour Management in Chile [169]. This aim was to strengthen the regulatory framework through short-, medium-, and long-term measures in order to help to quantify, control, and prevent odour generation. In this regard, the Ministry of the Environment is developing a "Regulation on Odour Prevention and Control" which may help some industrial sectors potentially generating odours to adopt improvements or technologies and practices to control odour.

The standardisation of odour measurement methodologies was also needed. To date, the standards homologated in Chile by the National Institute of Normalization (INN) are:

NCh3387:2015: Air Quality Assessment of Odour Annoyance Survey [170];

NCh3386:2015: Air Quality—Static sampling for olfactometry [171]; reference to German standard VDI 3883 Part 1:2015 [172];

NCh3190:2010: Air Quality—Determination of odour concentration by dynamic olfactometry [173]; reference to German standard VDI 3880:2011 [2] and European standard EN 13725:2003 [1].

In regulatory terms, as part of the Strategy for Odour Management, MMA establishes a prioritisation of these potential odour-generating activities based on the following criteria:

(1)   Activities with a greater number of complaints.
(2)   Activities with a greater number of facilities.
(3)   Activities involved in socio-environmental conflicts due to odours.

In November 2018, the draft standard for the emission of pollutants in pig farms was initiated and was expected to enter into force in 2020.

In 2019, a draft applying to the fishing industry was started, and it will be in effect in 2021.

It follows the beginning of the emission standard for wastewater treatment plants to end this stage with the cellulose industry and landfills.

8.2. Colombia

In Colombia, in recent years, there have been some interesting movements in odour regulation. The toolset for odour management has exponentially grown to a scale similar to that of many advanced European countries.

Although, in 1994, the regulation of the protection and control of air quality already established some restrictions and prohibitions related to emissions and the places that generate offensive odours, there has been a relevant advancement in the development of odour regulation in the last 6 years.

The first step was given in 2011. The Colombian Technical Norm NTC 5880 [174], "Air Quality. Determination of Odour Concentration by Dynamic Olfactometry", was published in December 2011. This norm defines a method for the objective determination of an odour concentration of a gas sample through dynamic olfactometry. EN 13725 was used as a reference document for this norm.

Some other norms related to odour measurement were published in 2013, such as the Colombian Technical Standard NTC 6011 [175], "Static Sampling for Dynamic Olfactometry"; the standard NTC 6012-1 [176] about the "Effects and Assessment of Odours. Psychometric Assessment of Odour Annoyance. Questionnaires"; or the standard NTC

6012-2 [177] about the "Effects and Assessment of Odours. Determination of Annoyance Parameters by Questioning; Repeated Brief Questioning of Neighbour Panellists".

In 2013, the Colombian Ministry of Environment and Sustainable Development approved Resolution 1541 [178] that sets acceptable levels for odours and some odourants in ambient air. This resolution also developed the procedures for activities of potential producers of odour complaints. It constitutes, in the first instance, an instrument for the promotion of the inclusion of good environmental practices considering that a process or a part of a process or activity causes an odour emission.

In t2014, further standards were published, such as the Colombian Standard NTC 6049-1 [179], "Measurement of Odour Impact by Field Inspection. Grid Measurement", the standard NTC 6049-2 [180], "Measurement of Odour Impact by Field Inspection. Plume Measurement", the NTC 6049-3 [181], "Measurement of Odour Impact by Field Inspection. Determination of Odour Intensity and Hedonic Tone", and, finally, the NTC 6049-4 [182], "Determination of the Hedonic Odour Tone Polarity Profiles".

The implementation of the regulatory scheme is focused on improving the process and activities' environmental performance, understanding that one of the main effects of the use of good environmental practices is the prevention, mitigation, and control of environmental impacts. In that sense, this resolution includes the plan for the reduction in the offensive odours impact—PRIO (because of its acronym in Spanish)—by which the activity or process proposes and puts under assessment and approval of the environmental authority the measures considered suitable for the management of their odour emissions.

Once the environmental authority assesses and approves the PRIO, the activity or productive process must fulfil the goals of this plan in the limited time for doing so and there is continuous surveillance from the environmental authority during the whole time of the activity or process. The implementation of the regulation is described as follows (Figure 6):

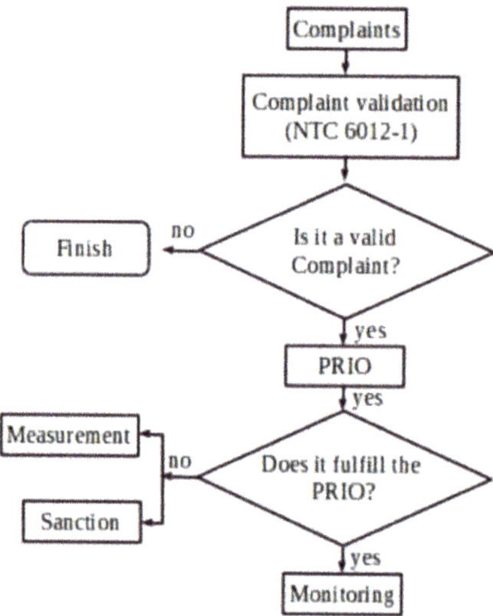

Figure 6. Implementation of Resolution 1541 [178].

The resolution also sets maximum acceptable limits of air quality for odourants substances such as H_2S, total reduced sulfur (TRS), and NH_3, applicable to those cases where these substances are the main ones responsible for odour issues. In Colombia, there are daily limits set for the odourants H_2S, TRS, and NH_3 of 7, 7, and 91 $\mu g \cdot m^{-3}$, respectively, and hourly limits of 30, 40, and 1400 $\mu g \cdot m^{-3}$, respectively.

On the other hand, Resolution 1541 [178] also establishes maximum acceptable limits for air quality in European odour units, so all the activities responsible for odour generation are under the fulfilment of one or another standard (for substances or odours). The following Table 30 shows the different limits in Colombia:

Table 30. Admissible concentration limits for odours in the air [178].

Activity	Admissible Level
Meat, fish, mollusc, and crustacean processing and preservation Oil refinery processes Paper pulp, paper, and cardboard manufacture Leathery and tanning of skins Nonhazardous waste collection, transport, transference, processing, or disposal WWTP Activities that collect water from water bodies receptors of wastewater discharges Manufacture of substances and basic chemical products Thermal destruction of animal by-products	$3\ \mathrm{ou_E \cdot m^{-3}}$
Farms Manufacture of vegetable oils and fats	$5\ \mathrm{ou_E \cdot m^{-3}}$
Decaffeination, roasting and grinding of coffee Other activities	$7\ \mathrm{ou_E \cdot m^{-3}}$

These limits are expressed as the 98th percentile of the hourly mean (equivalent to 175 exceedances per year). The method used to measure odours is detailed in the Colombian norm NTC 5880 [174], "Air Quality. Determination of Odour Concentration by Dynamic Olfactometry". The dispersion models allowed for odour are AERMOD and CALPUFF.

As mentioned above, the main purpose of the offensive odour regulation is to encourage sound environmental practices in activities or processes so the environmental impact can be managed properly in a comprehensive way, while also increasing the environmental and productive competitiveness.

Resolution 1541 [178] was further developed by Resolution 2087 [183], "Protocol for Monitoring, Control, and Surveillance of Offensive Odours". This resolution defines the methods, criteria, and suitable specifications for measurement of odourant substances or odours as well as the development of the PRIO.

9. Discussion

Odour is based on perception, the chemosensory response to odourants in the air. We experience odours throughout our days around the home and in our communities. The degree of an odour impact is based on five main factors, including the offensiveness and intensity/concentration of the odours, the frequency, and duration that the odours are present, and the location or context of the experience, all together commonly referred to as FIDOR (frequency, intensity, duration, offensiveness, and "receptor", which is also labelled as "location" in the alternative FIDOL). The personal experience and biases of the affected citizens have historically complicated the assessment by enforcement officials; however, standardised laboratory and field-based odour assessment protocols have provided the means to quantify a largely subjective experience objectively.

The complex interaction of the five FIDOR elements of odour makes it challenging to "regulate" odours at a country-level. Different philosophies of control, as well as different regulatory systems, hinder the development of one common approach to policy. However, utilising various quantification methods, several countries and provinces/states have adopted approaches that are suitable or politically feasible to legislate and enforce community odours.

The regulatory approaches outlined throughout this paper provide a foundation for highlighting important elements of regulation. Below is a list of questions that may be used for a discussion involving the formulation of odour regulation. This list is not complete, but it is an outline that can be useful.

Planning:

I. How do the existing local planning and zoning policies impact the proposed regulation and its implementation?
II. Who should be the stakeholders involved in drafting an odour regulation?
III. What are the costs of regulation (to the facility and the community/agency)?
IV. What are the costs of no regulation (to the facility and the community/agency)?
V. Choice of regulatory criteria:
VI. In which cases is an air quality regulation suggested, and in which cases is an emission regulation better?
VII. Why are only some industries regulated and not necessarily all types of emissions in a region or country?

Continuous improvement:

I. Which level of graduality has been reached by countries with a history of odour regulations, and what were the results?
II. Metrics:
III. What are the indicators of a successful odour regulation?
IV. How have various methods of current and past regulation been successful?
V. Is there a link between regulation and accreditation (operating permit, obligatory periodic audit)?

Recommendations:

I. Is there a list of common recommendations to countries/stakeholders that are considering an odour regulation?
II. Is there a need for a "clearinghouse" of best practices that document country-level experiences?
III. It is a challenge to answer these questions, and could the answers be different depending on the local/state situation.

For most of them, there is not one univocal answer. This paper describes approaches to the different regulations adopted by selected countries and regions within the countries. Table 31 summarises approaches categorised by methods, countries where they are adopted, and related pros/cons. Note that the identification of countries is based on the existence of regulatory enforcement. In some cases, an approach may still exist in a specific country based on specific facility permits. For example, while countries such as the USA or Spain may not regulate an odour concentration source emission measurement, a facility permit may be used to instill specific enforcement on one facility.

Monitoring emissions or rates of emission at the source, either perceived odours or chemical odourants, is a relatively simple approach, but it has the limitation that it does not account for the people's exposure and perception downwind.

Chemical analysis for the measurement of odourant concentrations has a lower uncertainty, but it is not always possible to relate chemical composition to odour perception. More research is needed to link specific chemicals with their influence on the overall odour. Chemical analysis alone can miss the impact of strong odourants that are present at low concentrations. Here, the use of an odour activity value (OAV) could be useful, but more data are needed on detection threshold values for important odourants.

Separation distances can be effective in preventing odour problems. However, more research is needed to improve models and/or adopt industrial models for odour regulations.

Table 31. Examples of approaches to odour regulations in selected countries [1,173,174,184–187].

General Approach	Methods	Country	Pros	Cons
(1) Emission measurement	(a) Measurement of odour concentration at the source of emissions	Japan (Measurement of odour index), China, Colombia, Canada (Quebec), Germany, UK	Standardised methodology (1)	No direct relationship with odour perception by citizens
	(b) Measurement of odour emission rate (at the source of emissions)	Japan, Canada, Germany, UK	Standardised methodology (1) for point sources and active area sources; More related to odour perception than just odour concentration measurement	Not standardised for passive area sources (except for Germany) Hardly achievable in the case of diffuse sources Not applicable to sources with variable emissions over time No direct relationship with odour perception by citizens (meteorological conditions and distance to receptors not considered)
	(c) Measurement of the concentration of specific odourants (chemical concentrations, mass/volume, volumetric mixing ratios)	USA (e.g., H_2S), Spain, Canada, Australia, New Zealand	High confidence level in the technique	Not representative of the odour of mixture. No direct relationship with odour perception by citizens
	(d) Measurement of the emission rate of specific odourants (chemical mass/time)	Japan, Canada, China	Standardised methodology	Not representative of the odour of mixture. No direct relationship with odour perception by citizens
(2) Fenceline measurement	(a) Measurement of odour index at the property line	Japan, China	Standardised methodology (Japan Environment Agency Notification No.63: 1995) Direct relationship with odour perception by citizens	
	(b) Measurement of the concentration of specific odourants at the property line	Japan, Canada, China	Standardised methodology (2)	Not representative of the odour of mixture. No direct relationship with odour perception by citizens
(3) Limitation of Impact	(a) Separation distances defined based on dispersion modelling			Only applicable to new installations. No direct relationship with odour perception by citizens
	(b) Separation distances defined based on empirical equations	USA, Canada (animal agriculture) Australia and New Zealand separation distances are defined by modelling and empirical equations	Ease of application (Less complex than dispersion models)	Only applicable to new installations. No direct relationship with odour perception by citizens
(4) Exposure assessment (OIC and complementary approaches (e.g., FIDOR factors))	(a) Dispersion modelling	Italy (Lombardy, Piemonte, Trento), Canada (Ontario), France (applicable for solvent industries), Germany	Applicability for predictive purposes	No standardisation. Different models and settings can be used leading to different results Hardly applicable to complex sources (diffuse or variable over time)

Table 31. *Cont.*

General Approach	Methods	Country	Pros	Cons
	(b) Field inspection	Germany (growing in AU and NZ), UK	Standardised methodology (European standard EN16841) Direct relationship with odour perception by humans	Long duration, limitations in extreme weather conditions, in not accessible areas, unsafe spots
	(c) Field olfactometry	USA (States and Municipalities)		
	(d) Citizen science		In general, less expensive than other techniques (no sophisticated equipment nor trained assessors needed). Not standardised. Bias can be reduced, and the technique can be very effective if relying on a large number of citizens and if observations are validated	Risk of bias due to the prejudice of involved citizens Might be ineffective in very conflictual situations (e.g., lawsuits) Challenging to verify each specific complaint
	(e) Collection of complaints (free-form or structured)	USA (municipalities), Colombia, New Zealand, Australia, UK	Easy to implement	Risk of bias due to the prejudice of involved citizens Might be ineffective in very conflictual situations (e.g., lawsuits) Challenging to verify each specific complaint
	(f) Regulator determination following complaints.	UK, Colombia	No measurements are needed. Regulators need to show permit or consent conditions are not being met	It can end in a court judgment
	(g) IOMS (instrumental odour monitoring systems)	France	Continuous measurement Possibility to discriminate odour/odourant sources	Not standardised technique It should be connected to odour measurements

The most common approaches to odour regulation are those entailing the use of dispersion modelling and field inspections for determining citizens' exposure to odours and compare it with odour impact criteria (OIC). There are two groups of OIC used in various jurisdictions. The first group is common in the Anglo-American countries with high threshold/low exceedance probability; the second group with low threshold/high exceedance is based on investigations in Germany. A more detailed discussion about OIC and their application in different countries in the form of Table S1 is provided in the Supplementary Material. A more comprehensive review of OIC and the manner in which they are applied is summarised by Brancher et al. (2017) [188].

Dispersion models have the advantage that they usually are less time-intensive and cost-intensive than field inspections. On the other hand, field inspections account for the real impact in the community. Field inspections are now regulated at a European level by EN 16841 [3,4].

Another possible approach to be considered for assessing odour impacts and regulating odours is advanced psychometry based on citizen science. Citizen science relies on observations from a large number of citizens. The methodology developed to do so is complex and involves engagement approaches and other aspects such as data plausibility checks and complex meteorological checks. Once this approach is made, there is no risk of personal biases from individual observations as each observation is validated, taking into account different factors. A recent review of assessment techniques in the context of malodour impact on communities was published by Hayes et al. (2014) [189]. Recent work by Braithwaite (2019) [190] proposes the use of the odour profile method for complex or unresolved cases and their sources, which involves an odour wheel, panel assessment of the intensity of the odour, and location information.

Instrumental odour monitoring systems (IOMS) have been developed with a wide range of technologies available. Results from various systems are not easily comparable, making it a challenge to use them for regulation while keeping an open market to allow for all technologies. Efforts have been made to regulate environmental odour monitoring with IOMS, but this is a very challenging and heavily debated task. The only regulation concerning IOMSs is in France. In this country, a plant may decrease the frequency of periodical measurements performed by olfactometry if it has an IOMS.

10. Conclusions

While many countries and regions regulate odours with different approaches, there can be agreement among all involved that the regulation of odours can be an immense problem. Odour regulation is a place where science, policy, economics, and public relations are interconnected.

These odours may be quantified based on odourant concentrations as well as human perception. Objective measurements of the odour experience include laboratory and field assessments with olfactometer devices and by direct observations. Air dispersion models and other computer algorithms, such as setback models, further analyse and quantify odour exposure.

More and more countries and communities are regulating odours, and the trend is bound to continue. There is an overall trend towards the measurement of odours instead of chemical odourants, while efforts to standardise odour concentration measurement and field assessments continue around the world.

There is expected to be an increase in approaches based on citizen science. Technology advancements will continue to make it easier to collect data efficiently and analyse the inputs more rapidly.

There are also promising advancements occurring with the standardisation of electronic noses and the development of more effective measurement tools. Multiple consensus working groups are currently discussing methods for testing and validating chemical sensing technologies.

In the end, integrated approaches are often needed to obtain the broadest vision of odour problems. Methods that can take into account all elements of the FIDOR model will go farthest to balance the interests of key stakeholders. Continual review of the various methods in use will provide lessons for countries and regions, creating new or modifying existing regulations.

Supplementary Materials: The following are available online at https://www.mdpi.com/2073-4433/12/2/206/s1, Table S1: Odour impact criteria (OIC) of various jurisdictions defined by the odour concentration threshold CT* (ou$_E$·m^{-3}) for the corresponding integration time of the ambient concentration and the exceedance probability p$_T$ (in %). The ambient odour concentration is determined either by the integration time or the peak-to-mean factor F.

Author Contributions: Writing—original draft preparation: all co-authors; writing—review and editing: all co-authors. All authors have read and agreed to the published version of the manuscript.

Funding: Partial support (for J.A.K.) came from the Iowa Agriculture and Home Economics Experiment Station: project number IOW05556 (Future Challenges in Animal Production Systems: Seeking Solutions through Focused Facilitation, sponsored by Hatch Act and State of Iowa funds). This project has received funding from the European Union's HORIZON 2020 research and innovation programme under grant agreement No 789315.

Institutional Review Board Statement: Not applicable.

Informed Consent Statement: Not applicable.

Data Availability Statement: No new data were created or analysed in this study. Data sharing is not applicable to this article.

Conflicts of Interest: The authors declare no conflict of interest.

Abbreviations

ADMS	Atmospheric Dispersion Modelling System developed by Cambridge Environmental Research Consultants.
AERMOD	American Meteorological Society/Environmental Protection Agency Regulatory Model
AMPAT	Air Management Practices Assessment Tool
AUSTAL2000G	Lagrangian odour dispersion model used in Germany
AWAC	Walloon Agency for Air and Climate
BAT	best available technology
BER	built, extended, rebuilt
BPM	best practicable means
CAFOs	confined animal feeding operations
CALPUFF	atmospheric pollution dispersion modelling system
DEFRA	Department for Environment, Food and Rural Affairs (United Kingdom)
DEQ	Department of Environmental Quality (Oregon)
D/T	detection/threshold
EA	Environment Agency (England)
ELV	emission limit value
EN	European Standard identified by a unique reference code "EN"
EPA	Environmental Protection Agency
EPEA	Environmental Protection and Enhancement Act (Alberta)
GRAL	Lagrangian dispersion model.
IMPACT	an odour dispersion model used in Flanders
IOMS	instrumental odour monitoring systems
INN	National Institute of Normalization
LASAT	Lagrangian dispersion model.
MMA	Ministry of the Environment (Chile)
MECP	Ministry of the Environment, Conservation and Parks (Ontario)
NAEMS	National Air Emissions Monitoring Study
NeR	Dutch Emission Guidelines
NIEA	Northern Ireland Environment Agency
NRW	Natural Resource Wales
NTC	Colombian Technical Norm (Colombian Institute of Technical Standards and Certification)
OAV	odour activity value
OIC	odour impact criteria
OOCL	Offensive Odour Control Law (Japan)
ou	American and Australian odour unit
$ou_E \cdot m^{-3}$	odour concentration, European odour unit per cubic metre
PRIO	Plan for the reduction of the offensive odours (Spanish)
SEPA	Scottish Environmental Protection Agency
Su	sniffing unit
TOBM	triangle odour bag method
TOFM	triangle odour flask method
TRS	total reduced sulfur
VDI	Verein Deutscher Ingenieure (VDI) (English: Association of German Engineers)
VOC	volatile organic compound
WWTPS	wastewater treatment plants

References

1. European Committee for Standardization CEN. *Air Quality—Determination of Odour Concentration by Dynamic Olfactometry*; EN 13725:2003; CEN: Brussels, Belgium, 2003.
2. The Association of German Engineers. *VDI. Olfactometry—Static Sampling*; VDI 3880; Beuth Verlag GmbH: Berlin, Germany, 2011.
3. European Committee for Standardization CEN. *Ambient Air—Determination of Odour in Ambient Air by Using Field Inspection—Part 1: Grid Method*; EN 16841–1:2016; CEN: Brussels, Belgium, 2016.
4. European Committee for Standardization CEN. *Ambient Air—Determination of Odour in Ambient Air by Using Field Inspection—Part 2: Plume Method*; EN 16841–2:2016I; CEN: Brussels, Belgium, 2016.
5. Watts, P.J.; Sweeten, J.M. Toward a better regulatory model for odour. In Ch. 15. In Proceedings of the Feedlot Waste Management Conference, Toowoomba, Queensland, Australia, 3–9 May 1995; Queensland Department of Primary Industries: Toowoomba, Queensland, Australia, 1995.
6. Good Practice Guide for Assessing and Managing Odour. Ministry for the Environment, New Zealand. 2016. Available online: https://www.mfe.govt.nz/sites/default/files/media/Air/good-practice-guide-odour.pdf (accessed on 6 October 2020).
7. The Association of German Engineers. *VDI. Olfactometry—Determination of Odour Intensity Part 1*; VDI 3882; Beuth Verlag GmbH: Berlin, Germany, 1992.
8. ASTM International. *Standard Practice for Referencing Suprathreshold Odor Intensity*; ASTM E544–10; ASTM International: West Conshohocken, PA, USA, 2010.
9. The Association of German Engineers. *VDI. Olfactometry: Determination of Hedonic Odour Tone Part 2*; VDI 3882; Beuth Verlag GmbH: Berlin, Germany, 1994.
10. Loriato, A.G.; Salvador, N.; Santos, J.M.; Moreira, D.M.; Reis, N.C., Jr. Odour—A vision on the existing regulation. *Chem. Eng. Trans.* **2012**, *30*, 25–30.
11. Giner, S.G.; Georgitzikis, K.; Scarlet, B.M.; Montobbio, P.; Roudier, S.; Delgado, L.S. Best available techniques (BAT) reference document for the intensive rearing of poultry and pigs. In *Industrial Emissions Directive 2010/75/EU (Integrated Pollution Prevention and Control)*; Publications Office of the European Union: Luxembourg, 2017; Available online: https://ec.europa.eu/jrc/en/publication/eur-scientific-and-technical-research-reports/best-available-techniques-bat-reference-document-intensive-rearing-poultry-or-pigs (accessed on 6 October 2020). [CrossRef]
12. European Environmental Bureau. EEB Position on the Relationship Between BAT AELs and ELVs. Available online: http://www.eeb.org/?LinkServID=F69A6726--9BBD-097A-F6825498E2D0618F&showMeta=0 (accessed on 26 July 2016).
13. Order of 12 February 2003 relating to the requirements applicable to classified installations subject to authorization under section 2730 treatment of by-products of animal origin, including debris, issues and corpses, excluding activities covered by other sections of the nomenclature, diagnostic, research and teaching establishments (Arrêté du 12 février 2003 relatif aux prescriptions applicables aux installations classées soumises à autorisation sous la rubrique 2730 traitement de sous-produits d'origine animale, y compris débris, issues et cadavres, à l'exclusion des activités visées par d'autres rubriques de la nomenclature, des établissements de diagnostic, de recherche et d'enseignement). *JORF* **2003**, *89*, 6654.
14. Order of 22 April 2008 setting the technical rules that must be met by composting facilities subject to authorization in application of Title I of Book V of the Environment Code (Arrêté du 22 avril 2008 fixant les règles techniques auxquelles doivent satisfaire les installations de compostage soumises à autorisation en application du titre Ier du livre V du code de l'environnement). *JORF* **2008**, *0114*, 8058.
15. Order of 20 April 2012 relating to the general requirements applicable to classified composting installations subject to registration under the heading n° 2780 (Arrêté du 20 avril 2012 relatif aux prescriptions générales applicables aux installations classées de compostage soumises à enregistrement sous la rubrique n° 2780). *JORF* **2012**, *0104*, 7794.
16. Order of 25 May 2012 relating to the general requirements applicable to classified installations subject to declaration under heading No. 2250 (production by distillation of alcoholic beverages of agricultural origin) (Arrêté du 25 mai 2012 relatif aux prescriptions générales applicables aux installations classées soumises à déclaration sous la rubrique no 2250 (production par distillation d'alcools de bouche d'origine agricole)). *JORF* **2012**, *0155*, 11055.
17. Order of 26 November 2012 relating to the general requirements applicable to installations falling under the registration regime under heading No. 2251 (preparation, packaging of wines) of the nomenclature of installations classified for environmental protection(Arrêté du 26 novembre 2012 relatif aux prescriptions générales applicables aux installations relevant du régime de l'enregistrement au titre de la rubrique n 2251 (préparation, conditionnement de vins) de la nomenclature des installations classées pour la protection de l'environnement). *JORF* **2012**, *0277*, 18660.
18. Order of 18 December 2014 relating to the general requirements applicable to installations covered by the declaration regime under heading No 2253 of the nomenclature of installations classified for the protection of the nomenclature of installations classified for environmental protection(Arrêté du 18 décembre 2014 relatif aux prescriptions générales applicables aux installations relevant du régime de la déclaration au titre de la rubrique n 2253 de la nomenclature des installations classées pour la protection de of the nomenclature of installations classified for environmental protection). JORF 2014. Available online: https://www.legifrance.gouv.fr/loda/id/JORFTEXT000030426541/ (accessed on 29 January 2021).
19. Federal Immission Control Act (Act on the Prevention of Harmful Effects on the Environment Caused by Air Pollution, Noise, Vibration and Similar Phenomena). Federal Ministry for Environment, Nature Conservation and Reactor Safety BImSchG. 2015. Available online: https://germanlawarchive.iuscomp.org/?p=315 (accessed on 28 December 2020).

20. Administrative regulation. Determination and Evaluation of Odour Immissions—Guideline on Odour in Ambient Air GOAA (1994/1999/2004/2008). Länderausschuss für Immissionsschutz GOAA 2008. Available online: https://ec.europa.eu/research/participants/documents/downloadPublic?documentIds=080166e5c7d98d4b&appId=PPGMS (accessed on 29 January 2021).
21. Both, R.; Sucker, K.; Winneke, G.; Koch, E. Odour intensity and hedonic tone—important parameters to describe odour annoyance of residents. *Water Sci. Technol.* **2004**, *50*, 83–92. [CrossRef]
22. The Association of German Engineers. *VDI. Measurement of Odour Impact by Field Inspection—Measurement of the Impact Frequency of Recognizable Odours—Grid Measurement*; VDI 3940 Part 1; Beuth Verlag GmbH: Berlin, Germany, 2006.
23. Janicke, L.; Janicke, U.; Ahrens, D.; Hartmann, U.; M&üller, W. *Development of the Odour Dispersion Model AUSTAL2000-G in Germany*; VDI-Berichte: Cologne, Germany, 2004; pp. 411–417.
24. The Association of German Engineers. *VDI. Olfactometry—Determination of Odour Concentration by Dynamic Olfactometry—Supplementary Instructions for Application of DIN EN 13725*; VDI 3884 Part 1; Beuth Verlag GmbH: Berlin, Germany, 2015.
25. Both, R. Directive on odour in ambient air: An established system of odour measurement and odour regulation in Germany. *Water Sci. Technol.* **2001**, *44*, 119–129. [CrossRef]
26. Sucker, K.; Both, R.; Bischoff, M.; Guski, R.; Winneke, G. Odor frequency and odor annoyance. Part I: Assessment of frequency, intensity and hedonic tone of environmental odors in the field. *Int. Arch. Occup. Environ. Health* **2008**, 671–682. [CrossRef]
27. Sucker, K.; Both, R.; Biscoff, M.; Guski, R.; Winneke, G. Odor frequency and odor annoyance Part II: Dose-response associations and their modification by hedonic tone. *Int. Arch. Occup. Environ. Health.* **2008**, 683–694. [CrossRef]
28. Swanson, J.R.; McGinley, C. Odour regulation in Germany—A new directive on odour in ambient air. Proceedings of an International Speciality Conference Sponsored by the A&WMA, Bloomington, MN, USA, 13–15 September 1995.
29. The Association of German Engineers. *VDI. Determination of the Hedonic Odour Tone—Polarity Profiles*; VDI 3940 Part 4; Beuth Verlag GmbH: Berlin, Germany, 2010.
30. State Agency for the Environment of Baden Württemberg LUBW. Creation of Polarity Profiles for the Concept of Smell and Scent for the Animal Species Fattening Bulls, Horses and Dairy Cows (In German). 2017. Available online: https://pudi.lubw.de/detailseite/-/publication/72525-Bericht.pdf (accessed on 24 August 2020).
31. The Association of German Engineers. *VDI. Measurement of Odour Impact by Field Inspection—Measurement of the Impact Frequency of Recognizable Odours Plume Measurement*; VDI 3940 Part 2; Beuth Verlag GmbH: Berlin, Germany, 2006.
32. State Agency for the Environment of Baden Württemberg LUBW. Creation of Polarity Profiles for the Concept of Smell and Scent for the Animal Species Goats and Sheep (In German). 2019. Available online: https://pudi.lubw.de/detailseite/-/publication/10083-Polarit%C3%A4tenprofile_Ziegen_Schafe.pdf (accessed on 24 August 2020).
33. TA Luft. *Technical Instruction on Air Quality Control (Erste Allgemeine Verwaltungsvorschrift zum Bundes-Immissionsschutzgesetz—Technische Anleitung zur Reinhaltung der Luft)*; Federal Ministry for Environment, Nature Conservation and Reactor Safety: Bonn, Germany, 2002.
34. Baumann, R.; Brandstätter, M.; Heimburger, G.; Kranabetter, A.; Moshammer, H.; Oitzl, S.; Rau, G.; Schauberger, G.; Schauer, U.; Scheicher, E.; et al. Guideline for recording and evaluating air quality in health resorts (Richtlinie zur erfassung und bewertung der luftqualität in kurorten). In *Commission for Climate and Air Quality (Kommission für Klima und Luftqualität)*; ÖAW: Vienna, Austria, 2013; p. 63.
35. ÖAW. *Umweltwissenschaftliche Grundlagen und Zielsetzungen im Rahmen des Nationalen Umweltplans für die Bereiche Klima, Luft, Geruch und Lärm*; Schriftenreihe der Sektion I. Band, 17; Kommissionfür Reinhaltung der Luft der Österreichische Akademie der Wissenschaften, Ed.; Bundesministeriums für Umwelt, Jugend und Familie: Vienna, Austria, 1994; Volume 17.
36. Oettl, D.; Moshammer, H.; Mandl, M.; Weitensfelder, L. *Guideline for the Assessment of Odor Immissions (Richtlinie zur Beurteilung von Geruchsimmissionen)*; Amt der Steiermärkischen Landesregierung Abteilung 15 Energie Wohnbau Technik: Graz, Austria, 2018.
37. Mainland, J.; Sobel, N. The sniff is part of the olfactory percept. *Chem. Senses* **2006**, *31*, 181–196. [CrossRef] [PubMed]
38. Schauberger, G.; Piringer, M.; Petz, E. Diurnal and annual variation of the sensation distance of odour emitted by livestock buildings calculated by the Austrian odour dispersion model (AODM). *Atmos. Environ.* **2000**, *34*, 4839–4851. [CrossRef]
39. Schauberger, G.; Piringer, M.; Petz, E. Corrigendum to "Diurnal and annual variation of the sensation distance of odour emitted by livestock buildings calculated by the Austrian odour dispersion model (AODM)". *Atmos. Environ.* **2013**, *67*, 459–462. [CrossRef]
40. Schauberger, G.; Piringer, M.; Schmitzer, R.; Kamp, M.; Sowa, A.; Koch, R.; Eckhof, W.; Grimm, E.; Kypke, J.; Hartung, E. Concept to assess the human perception of odour by estimating short-time peak concentrations from one-hour mean values. *Atmos. Environ.* **2012**, *54*, 624–628. [CrossRef]
41. Oettl, D.; Ferrero, E. A simple model to assess odour hours for regulatory purposes. *Atmos. Env.* **2017**, *155*, 162–173. [CrossRef]
42. Brancher, M.; Hieden, A.; Baumann-Stanzer, K.; Schauberger, G.; Piringer, M. Performance evaluation of approaches to predict sub-hourly peak odour concentrations. *Atmos. Env.* **2020**, *7*, 100076. [CrossRef]
43. Ferrero, E.; Oettl, D. An evaluation of a Lagrangian stochastic model for the assessment of odours. *Atmos. Env.* **2019**, *206*, 237–246. [CrossRef]
44. Bundesministerium fur Landwirtschaft, Forst, Umwelt und Waser BMLFUW. *Guideline for the Evaluation of Ambient Odour Emitted by Livestock Buildings*; Bundesministerium fur Landwirtschaft, Forst, Umwelt und Waser (Federal Ministry for Agriculture, Forestry, Environment, and Water): Vienna, Austria, 2017.

45. Schauberger, G.; Piringer, M.; Jovanic, O.; Petz, E. A new empirical model to calculate separation distances between livestock buildings and residential areas applied to the austrian guideline to avoid odor nuisance. *Atmos. Environ.* **2012**, *47*, 341–1347. [CrossRef]

46. Cseh, M.; Nárai, K.F.; Barcs, E.; Szepesi, D.B.; Szepesi, D.J.; Dicke, J.L. Odor setback distance calculations around animal farms and solid waste landfills. *Idojaras* **2010**, *114*, 303–318.

47. DEFRA UK 2010, Ref: PB13554. Odour Guidance for Local Authorities. Available online: https://www.gov.uk/government/uploads/system/uploads/attachment_data/file/69305/pb13554-local-auth-guidance-100326.pdf (accessed on 29 July 2020).

48. Environment Agency UK 2011, Ref: LIT 5419. Environmental Permitting: H4 Odour Management. Available online: https://www.gov.uk/government/uploads/system/uploads/attachment_data/file/296737/geho0411btqm-e-e.pdf (accessed on 28 July 2020).

49. Bull, M.; McIntyre, A.; Hall, D.; Allison, G.; Redmore, J.; Pullen, J.; Caird, L.; Stoaling, M.; Fain, R. *IAQM Guidance on the Assessment of Odour for Planning—Version 1.1*; Institute of Air Quality Management: London, UK, 2018.

50. DEFRA. *Code of Practice on Odour Nuisance from Sewage Treatment Works*; Department for Environment, Food, and Rural Affairs, DEFRA: London, UK, 2006.

51. Northern Ireland Environment Agency. *Odour Impact Assessment Guidance for Permitted and Licensed Sites*; Northern Ireland Environment Agency: Belfast, UK, 2012.

52. Guidance for Operators on Odour Management at Intensive Livestock IPPC Installations Version 3. Pollution Prevention and Control Northern Ireland. 2009. Available online: http://www.sepa.org.uk/media/60931/ippc_srg6_02_odour-management-at-intensive-livestock-installations-may-2005.pdf (accessed on 29 July 2020).

53. SEPA. *Odour Guidance*; Scottish Environment Protection Agency, SEPA, Scottish Government: Glasgow, UK, 2010.

54. Town and Country Planning (Scotland) Act. 1997. Available online: https://www.legislation.gov.uk/ukpga/1997/8/contents (accessed on 26 August 2020).

55. Pollution Prevention and Control (Scotland) Regulations. 2012. Available online: https://www.legislation.gov.uk/ssi/2012/360/contents/made (accessed on 28 July 2020).

56. Environmental Protection Act. 1990. Available online: http://www.legislation.gov.uk/ukpga/1990/43/part/III/crossheading/statutory-nuisances-england-and-wales (accessed on 28 July 2020).

57. Standard rules SR2009 No2—Low Impact Part A Installation, Environment Agency. Available online: https://assets.publishing.service.gov.uk/government/uploads/system/uploads/attachment_data/file/790718/SR2009_No2_Low_Impact_Part_A_Installation.pdf (accessed on 18 August 2020).

58. Spanish State Government. *Law 5/2013, of 11 June, which Modifies Law 16/2002, of 1 July, on Prevention and Integrated Control of Pollution and Law 22/2011, of 28 July, on Waste and Soil Contaminated (Ley 5/2013, de 11 de Junio, por la que se Modifican la Ley 16/2002, de 1 de Julio, de Prevención y Control Integrados de la Contaminación y la Ley 22/2011, de 28 de Julio, de Residuos y Suelos Contaminados)*; Spanish Official Gazette: Madrid, Spain, 2013. Available online: https://www.boe.es/eli/es/l/2013/06/11/5/con (accessed on 11 December 2018).

59. Spanish Ministry of Agriculture, Food and Environment. *Royal Decree 8/15/2013, of October 18, which Approves the Regulation of Industrial Emissions and the Development of Law 16/2002, of 1 July, on the Prevention and Integrated Control of Pollution (Real Decreto 8/15/2013, de 18 de Octubre, por el que se Aprueba el Reglamento de Emisiones Industriales y de Desarrollo de la Ley 16/2002, de 1 de Julio, de Prevención y Control Integrados de la Contaminación)*; Spanish Official Gazette: Madrid, Spain, 2013. Available online: https://www.boe.es/eli/es/rd/2013/10/18/815 (accessed on 11 December 2018).

60. Government of Catalonia. Draft of the Law against Odor Pollution, General Directorate of Environmental Quality. Government of Catalonia, Department of the Environment and Housing. General Directorate of Environmental Quality 2011. Barcelona, Spain. Available online: https://www.olores.org/images/pdfs/borrador_anteproyecto_ley_contaminacion_odorifera.pdf (accessed on 21 September 2020).

61. City Council Plenary of Lliçà del Val. *Ordinance Regulating the Release of Odors into the Atmosphere. Lliçà de Vall Town Hall, n.192 (Ordenanza de Regulación de la Liberación de Olores a la Atmósfera. Ayuntamiento de Lliçà de Vall, n.192)*; Official Gazette of the Province of Barcelona: Barcelona, Spain, 2007; Available online: https://dibaaps.diba.cat/vnis/temp/CIDO_bopb_2007_08_20070811_BOPB_20070811_027_029.pdf (accessed on 11 November 2020).

62. City Council of Banyoles. *Odor Ordinance of Banyoles City Council. City Council of Banyoles (Girona) (Ordenança d'Olors de l'Ajuntament de Banyoles. City Council of Banyoles (Girona), n. 120)*; Official Gazette of the Province of Girona: Banyoles Girona, Spain, 2008; Available online: https://seu.banyoles.cat/Portals/0/ORDENANA_A_OLORS.pdf (accessed on 28 July 2019).

63. City Council of Riudellots de la Selva. *Municipal Ordinance of Olors of Riudellots de la Selva. City Council of Riudellots de la Selva, n. 26 (Ordenança Municipal d'Olors de Riudellots de la Selva. City Council of Riudellots de la Selva, n. 26)*; Official Gazette of the Province of Girona: Banyoles Girona, Spain, 2009; Available online: https://dibaaps.diba.cat/vnis/temp/CIDO_bopg_2009_02_20090209_BOPG_20090209_027_036.pdf (accessed on 28 July 2019).

64. City Council of Sarrià de Ter. *Sarrià de Ter City Council Odor Ordinance. City Council of Sarrià de Ter. (Girona), n. 152 (Ordenança d'olors de l'Ajuntament de Sarrià de Ter. City Council of Sarrià de Ter. (Girona), n. 152)*; Official Gazette of the Province of Girona: Sarria de Ter Girona, Spain, 2018; Available online: https://ssl4.ddgi.cat/bopV1/pdf/2018/152/201815207044.pdf (accessed on 28 July 2019).

65. City Council of Sant Vicent del Raspeig. *Ordinance for the Protection of the Atmosphere (Ordenanza de Protección de la Atmósfera)*; City Council of Sant Vicent del Raspeig: Alicante, Spain, 1994; Available online: https://www.raspeig.es/uploads/ficheros/arbolficheros/descargas/201105/descargas-proteccion-de-la-atmosfera-es.pdf (accessed on 28 July 2019).

66. City Council of Alcantarilla. *Ordinance Regulating the Emission of Odors into the Atmosphere, Alcantarilla City Council (Ordenanza de Regulación de Emisión de Olores a la Atmósfera, Ayuntamiento de Alcantarilla)*; Boletín Oficial de la Región de Murcia: Murcia, Spain, 2016; Available online: https://www.alcantarilla.es/wp-content/uploads/2015/10/Ordenanza-Municipal-de-Calidad-Odor%C3%ADfera-del-Aire-1.pdf (accessed on 11 November 2020).

67. City Council of San Pedro del Pinatar. *Municipal Ordinance of Odoriferous Air Quality, San Pedro del Pinatar City Council (Ordenanza Municipal de Calidad Odorífera del Aire, Ayuntamiento de San Pedro de Pinatar)*; Boletín Oficial de la Región de Murcia: Murcia, Spain, 2011; Available online: https://www.sanpedrodelpinatar.es/wp-content/uploads/2019/09/ORDENANZA-DE-REGULACION-DE-EMISION-DE-OLORES-A-LA-ATMOSFERA.pdf (accessed on 11 November 2020).

68. City Council of Las Palmas de Gran Canaria. *Municipal Ordinance for the Protection of the Atmosphere Against Pollution by Forms of Matter, Las Palmas de Gran Canaria, Department of the Environment (Ordenanza Municipal de Protección de la Atmósfera Frente a la Contaminación por Formas de la Materia, Las Palmas de Gran Canaria, Concejalía de Medio Ambiente)*; Boletín Oficial de Las Palmas: Las Palmas, Spain, 1999; Available online: https://www.laspalmasgc.es/export/sites/laspalmasgc/.galleries/documentos-medio-ambiente/ONF_CONTAMINACIONFORMASMATERIA_17.pdf (accessed on 11 November 2020).

69. Decree on General Rules for Environmental Management Establishments (Besluit Algemene Regels voor Inrichtingen Milieubeheer). 2007. Available online: https://wetten.overheid.nl/BWBR0022762/2010--07--04/ (accessed on 28 July 2020).

70. Lagas, P. Odour policy in the Netherlands and consequences for spatial planning. *Chem. Eng. Trans.* **2010**, *23*, 7–12. Available online: http://www.aidic.it/cet/10/23/002.pdf (accessed on 28 July 2020).

71. NEN Royal Netherlands Standardization Institute. *Odour Quality—Sensory Determination of the Hedonic Tone of an Odour Using and Olfactometer*; NVN 2818; NEN Netherlands Standards Institute: Delft, The Netherlands, 2019.

72. Archives of the Dutch Air Emissions Directive (NeR). Available online: https://www.infomil.nl/onderwerpen/lucht-water/lucht/ner-archief/ (accessed on 29 December 2020).

73. The Law of 5 of October 2006 on Livestock Odour Impact Control. Ministry of Housing, Spatial Planning and the Environment: The Hague, The Netherlands. 2006. Available online: https://wetten.overheid.nl/BWBR0020396/ (accessed on 1 February 2021).

74. Noordegraaf, D.; Bongers, M. *Relationship between Odour Exposure and Odour Nuisance in the Neighbourhood of Livestock Farms*; PRA Odournet: Amsterdam, The Netherlands, 2007.

75. Ministry of Housing, Spatial Planning and the Environment. *Regulation on the Determination of Odour Emission Factors, Minimum Distances for Fur-Bearing Animals, the Method of Calculating Odour Exposure and Determining Setback Distances*; Ministry of Housing, Spatial Planning and the Environment: The Hague, The Netherlands, 2006; Available online: https://wetten.overheid.nl/BWBR0020711/.

76. Nijdam, R.; Dusseldorp, A.; Elders-Meijerink, M.; Jacobs, P.; Maassen, C.B.M.; van der Lelie, S.; Pasnagel, M.; van Strien, R.; van de Waal, N.; van de Weerdt, R.; et al. *Guidelines on Livestock Farming and Health*; Publication 2020–0092; National Institute on Public Health and Environment RIVM: Bilthoven, The Netherlands, 2020.

77. Boers, D.; Geelen, L.; Erbrink, H.; Smit, L.A.M.; Heederik, D.; Hooiveld, M.; Yzermans, C.J.; Huijbregts, M.; Wouters, I.M. The relation between modeled odor exposure from livestock farming and odor annoyance among neighboring residents. *Int Arch. Occup Environ. Health* **2016**, *89*, 521–530. [CrossRef] [PubMed]

78. Geelen, L.; Boers, D.; Brunekreef, B.; Wouters, I.M. *Geurhinder van Veehouderij Nader Onderzocht: Meer Hinder dan Handreiking Wgv Doet Vermoeden?—Actualisatie Blootstellingresponsrelatie Tussen Gemodelleerde Cumulatieve Geurbelasting en Geurhinder in Noord-Brabant en Limburg-Noord. Odor Nuisance from Livestock Farming Further Studied: More Nuisance than the Wgv Guide Suggests?—Update of Exposure-Response Relationship between Modeled Cumulative Odor Exposure and Odor Nuisance in Noord-Brabant and Limburg-Noord*; Public Health Services (GGD) of Brabant and the Institute for Risk Assessment Sciences (IRAS): Utrecht, The Netherlands, 2015; Available online: http://www.academischewerkplaatsmmk.nl/ufc/file2/hgm_internet_sites/graskl/fb5d0198ad97606c2afcff6d278343b8/pu/Eindrapport_GEUR_Loes_Geelen_23_3_2015.pdf (accessed on 14 January 2021).

79. Guidelines for the Construction and Operation of Compost Production Plants. Revocation of the d.g.r 2003, n. 7/12764 (Linee Guida Relative alla Costruzione e All'esercizio Degli Impianti di Produzione di Compost. Revoca della d.g.r 2003, n. 7/12764). Regione Lombardia, Bollettino Ufficiale. Available online: https://www.territorioambiente.com/wp-content/uploads/2014/07/dgr_7_12764_esercizio_impianti_produzione_compost.pdf (accessed on 26 August 2020).

80. General Determinations Regarding the Characterization of Gaseous Emissions into the Atmosphere Deriving from Activities with a Strong Odoric Impact D.G.R. 15 (Determinazioni Generali in Merito alla Caratterizzazione delle Emissioni Gassose in Atmosfera Derivanti da Attività a Forte Impatto Odorigeno D.G.R. 15). 2012. n. IX/3018.. Available online: http://www.olfattometria.com/download/dgr-lomb.pdf (accessed on 28 July 2020).

81. Guidelines for the Characterization, Analysis and Definition of Technical and Managerial Criteria for the Mitigation of Emissions from Odor Impact Activities. Autonomous Province of Trento (Linee Guida per la Caratterizzazione, L'analisi e la Definizione dei Criteri Tecnici e Gestionali per la Mitigazione delle Emissioni delle Attività ad Impatto Odorigeno. Provincia autonoma di Trento). 2012. Available online: https://www.ufficiostampa.provincia.tn.it/content/download/38536/643559/file/Linee%20guida%20odori.pdf (accessed on 28 July 2020).

82. Bollettino Ufficiale della Regione Puglia. *Discipline of Company Odor Emissions. Emissions Deriving from Pomace Plants. Emissions in Areas at High Risk of Environmental Crisis (Disciplina delle Emissioni Odorifere Delle Aziende. Emissioni Derivanti da Sansifici. Emissioni Nelle Aree a Elevato Rischio di Crisi Ambientale)*; Bollettino Ufficiale della Regione Puglia: Bari, Italy, 2015.

83. Decree of the Walloon Government Determining the Sectoral Conditions Relating to Composting Facilities When the Quantityg of Material Stored is Greater than or Equal to 500 m3 (Arrêté du Gouvernement Wallon Déterminant les Conditions Sectorielles Relatives aux Installations de Compostage Lorsque la quantitég de Matière Entreposée est Supérieure ou Égale à 500 m3); AGW, Walloon Government: Namur, Belgium. 2009. Available online: http://environnement.wallonie.be/data/dechets/cet/20CBD/20_2Compostage.htm (accessed on 28 July 2020).

84. Nicolas, J.; Delva, J.; Cobut, P.; Romain, A.C. Development and validating procedure of a formula to calculate a minimum separation distance from piggeries and poultry facilities to sensitive receptors. *Atmos. Environ.* **2008**, *42*, 7087–7095. [CrossRef]

85. Collart, C.; Lebrun, V.; Fays, S.; Salpeteur, V.; Nicolas, J.; Romain, A.C. Air survey around MSW landfills in Wallonia: Feedback of 8 years field measurements. In Proceedings of the ORBIT 2008, 6th Biomass and Organic Waste as Sustainable, Wageningen, The Netherlands, 13–15 October 2008; Available online: https://orbi.uliege.be/handle/2268/12829 (accessed on 28 July 2020).

86. Romain, A.C. Landfill of solid waste. In *Odour Impact Assessment Handbook*; Belgiorno, V., Ed.; Wiley: Hoboken, NJ, USA, 2013; pp. 232–250.

87. Romain, A.C.; Molitor, N. *Rapport D'étude pour la Convention ISSEP-ULg Estimation des Nuisances Olfactives*; CET Champ de Beaumont: Charleroi, Belgium, 2015.

88. Nicolas, J.; Cors, M.; Romain, A.C.; Delva, J. Identification of odour sources in an industrial park from resident diaries statistics. *Atmos. Environ.* **2010**, *44*, 1623–1631. [CrossRef]

89. Vision Document "The Road to a Sustainable Odor Policy" LNE Memorandum (Visiedocument 'De weg naar een duurzaam geurbeleid' LNE nota). Directieraad van Het Departement Leefmilieu, Natuur en Energie. 2008. Available online: www.lne.be/visie_duurzaam_geurbeleid_v6.7.doc (accessed on 29 January 2021).

90. Van Broeck, G.; van Langenhove, H. Odour Standardization Research: Development of a Methodology for Setting Odour Standards for Homogeneous Industrial Sectors—Evaluation of the Applied Methodology. Ministry of the Flemish Community—Environment and Infrastructure Department—Environment, Nature, Land and Water Management Administration 2000. (Onderzoek Geurnormering: Ontwikkelen van een Methodologie voor het Opstellen van Geurnormering voor Homogene Sectoren—Evaluatie van de Toegepaste Methodologie. Ministerie van de Vlaamse Gemeenschap—Departement Leefmilieu en Infrastructuur—Administratie Milieu-, Natuur-, Land- en Waterbeheer 2000.). Available online: https://www.un.org/esa/agenda21/natlinfo/countr/belgium/natur.htm (accessed on 29 January 2021).

91. University of Ghent; PRG; PRA Odournet; Eco2. Drafting of a Method to Translate No-Effect Levels into Odour Standards and Application ro 5 Pilot Sectors—Part III: Proposal of Standards for 5 Sectors—Voorstellen van een Geschikte Methode om Nuleffect Niveaus van Geurhinder te Vertalen naar Normen en Toepassing op 5 Pilootsectoren—Deel III: Formulering Voorstel Voor de 5 Pilootsectoren Ministerie van de Vlaamse Gemeenschap—Departement Leefmilieu en Infrastructuur—Administratie Milieu-, Natuur-, Land- en Waterbeheer 2002. Available online: https://omgeving.vlaanderen.be/sites/default/files/atoms/files/richtlijnenboek%2520lucht.pdf (accessed on 29 January 2021).

92. Van Broeck, G.; van Langenhove, H.; Nieuwejaers, B. Recent odour regulation developments in Flanders: Ambient odour quality standards based on dose-response relationships. *Water Sci. Technol.* **2001**, *44*, 103–110. [CrossRef]

93. Dermaux, D.; Vervaet, C.; Arts, P.; Lefebre, F.; van Broeck, G.; Knight, D.; Wouters, J.; van Mierlo, T.; Wackenier, L. Richtlijnenboek Air—Updated Version January 2012 (Lucht—Geactualiseerde Versie Januari 2012). Antea Group & VITO i.s.m. LNE. 2012. Available online: https://omgeving.vlaanderen.be/sites/default/files/atoms/files/richtlijnenboek%20lucht.pdf (accessed on 28 July 2020).

94. Government of Flanders. *Sectoral Code of Good Odor Practice—Preventing, Assessing and Controlling Odor Nuisance Caused by an Asphalt Plant (Sectorale Code van Goede Geurpraktijk—Voorkomen, Beoordelen en Beheersen van Geurhinder Veroorzaakt Door een Asfaltcentrale)*; Department Omgeving: Brussels, Belgium, 2015; Available online: https://omgeving.vlaanderen.be/sites/default/files/atoms/files/cvggp-asfaltcentrales-v1.pdf (accessed on 2 February 2021).

95. Sectoral Code Of Good Odor Practice Preventing, Assessing And Controlling Odor Nuisance Caused By A Sewage Treatment Plant (Sectorale Code Van Goede Geurpraktijk Voorkomen, Beoordelen En Beheersen Van Geurhinder Veroorzaakt Door Een Rioolwaterzuiveringsinstallatie). Departement Omgeving LNE and Aquafin. 2018. Available online: https://omgeving.vlaanderen.be/sites/default/files/atoms/files/Sectorale_Code_van_Goede_geurpraktijk%20_RWZI_RWZI_versie_sept2018_0.pdf (accessed on 28 July 2020).

96. Willems, E.; Monseré, T.; Dierckx, J. *Updated Guidelines Book Environmental Impact Assessment "Basic Guidelines per Activity Group—Agricultural Animals". ABO i.c.w. (Geactualiseerd Richtlijnenboek Milieueffectrapportage 'Basisrichtlijnen per Activiteitengroep—Landbouwdieren'. ABO i.s.m)*; Departement Omgeving: Brussels, Belgium, 2019; Available online: https://omgeving.vlaanderen.be/sites/default/files/atoms/files/Richtlijnenboek%20landbouwdieren%20%28mei%202019%29.pdf (accessed on 28 July 2020).

97. Environment Protection Policy (Air Quality). Environment Division, Department of Tourism, Arts and Environment. 2004. Available online: http://epa.tas.gov.au/documents/epp_air_quality_2004.pdf (accessed on 28 July 2020).

98. New South Wales Environmental Protection Authority. *Technical Framework: Assessment and Management of Odour from Stationary Sources in NSW*; New South Wales Environmental Protection Authority: Parramatta, Australia, 2006.

99. Department of Water and Environmental Regulation, Western Australia. *Odour Emissions Guideline*; Department of Water and Environmental Regulation, Western Australia: Joondalup, Perth, Australia, 2019.
100. Methodology for Impact Assessment of Nuisance Odours in ACT. *ACTEW Odour Assessment Guidelines 2007*; ActAGL: Canberra, Australia, 2007. Available online: https://environment.des.qld.gov.au/__data/assets/pdf_file/0021/90246/guide-odour-impact-assess-developments.pdf (accessed on 29 January 2021).
101. South Australia EPA. *Odour Assessment Using Odour Source Modelling. South Australia EPA Guidelines*; EPA 373/07; South Australia EPA: Adelaide, Australia, 2007.
102. Department of Environment and Heritage Protection, Queensland Government. *Odour Impact Assessment from Developments*; Department of Environment and Heritage Protection, Queensland Government: Brisbane, Queensland, Australia, 2012. Available online: https://www.ehp.qld.gov.au/licences-permits/business-industry/pdf/guide-odour-impact-assess-developments.pdf (accessed on 27 July 2020).
103. Odour and EPA's Role. Available online: https://www.epa.vic.gov.au/for-community/environmental-information/odour/odour-epa-role (accessed on 10 November 2020).
104. New Zealand Ministry for the Environment. *Good Practice Guide for Assessing and Managing Odour*; New Zealand Ministry for the Environment: Wellington, New Zealand, 2016.
105. Katestone Scientific. Peak-to-mean ratios for odour assessments. In *Report to the Environment Protection Authority of New South Wales*; Katestone Scientific: Brisbane, Australia, 1995.
106. Katestone Scientific. Peak-to-mean ratios for odour assessments. In *Report to Environmental Protection Agency New South Wales*; Katestone Scientific: Brisbane, Australia, 1998.
107. New Zealand Ministry for the Environment. *Microbiological Water Quality Guidelines for Marine and Freshwater Recreational Areas*; New Zealand Ministry for the Environment: Wellington, New Zealand, 2003.
108. *Emission Standards for Odor Pollutants*; GB 14554–93; Beijing, China Environmental Protection Agency and China Technical Supervision Bureau, China Standard Press: Beijing, China, 1993.
109. *Discharge Standard of Pollutants for Livestock and Poultry Breeding*; GB 18596–2001; Beijing, China Environmental Protection Agency and China General Administration of Quality Supervision, Inspection and Quarantine, China Standard Press: Beijing, China, 2001.
110. *Air Quality-Determination of Odor-Triangle Odor Bag Method*; GB/T 14675–93; China Environmental Protection Agency, China Standard Press: Beijing, China, 1994.
111. Harrop, D.O. *Air Quality Assessment and Management A Practical Guide*; CRC Press: Boca Raton, FL, USA, 2002.
112. *Ambient Air Quality Standard*; GB3095–1996; Beijing, China Environmental Protection Agency and China Technical Supervision Bureau, China Standard Press: Beijing, China, 1996.
113. *Ambient Air Quality Standard*; GB3095–2012; Beijing, China Environmental Protection Agency and China Technical Supervision Bureau, China Standard Press: Beijing, China, 2012.
114. Wang, G.; Wang, Z.S.; Li, Y.M.; Geng, J. Odor standard progress in abroad. In Proceedings of the 4th National Symposium on Odor Pollution Monitoring and Control, Zibo, China, 10–12 October 2012.
115. Emission Standards for Odor Pollutants; DB31/1025–2016; Shanghai Environmental Protection Agency and Shanghai Quality and Technical Supervision Bureau, Shanghai, China. 2016. Available online: https://sthj.sh.gov.cn/ (accessed on 30 January 2021).
116. Technical Specification for Environmental Monitoring of Odor; HJ 905–2017; China Environmental Protection Agency: Beijing, China 2017. Available online: http://www.mep.gov.cn/ (accessed on 11 January 2018).
117. *Technical Specification for Olfactory Laboratory Construction*; HJ 865–2017; China Environmental Protection Agency, China Environmental Science Press: Beijing, China, 2017.
118. Long, Y.; Tin, K. Sewerage and Sewage Disposal Stage 2. Environmental Impact Assessment Ordinance. Hong Kong Drainage Services and Environmental Protection Departments 2001, Environmental Impact Assessment Study Brief No. ESB-082/2001. Available online: https://www.epd.gov.hk/eia/register/study/latest/esb-082.htm (accessed on 28 July 2020).
119. Japan Ministry of the Environment. *The Offensive Odor Control Law*; Japan Ministry of the Environment: Tokyo, Japan, 2003. Available online: http://www.env.go.jp/en/laws/air/offensive_odor/ (accessed on 1 November 2020).
120. Segawa, T. *Odor Regulation in Japan. East Asia Workshop on Odor Measurement and Control Review*; Japan Ministry of the Environment: Tokyo, Japan, 2004; pp. 36–41. Available online: http://www.env.go.jp/en/air/odor/eastasia_ws/2--1-3.pdf (accessed on 1 November 2020).
121. Japan Ministry of the Environment. *Odor Index Regulation and Triangular Odor Bag Method*; Japan Ministry of the Environment: Tokyo, Japan, 2003. Available online: http://www.env.go.jp/en/air/odor/regulation/ (accessed on 1 November 2020).
122. Iwasaki, Y.; Ishiguro, T.; Koyama, I.; Fukushima, H.; Kobayashi, A.; Ohira, T. On the new method of determination of odor unit. In Proceedings of the 13th Annual Meeting of the Japan Society of Air Pollution, Oita, Japan, 7–9 November 1972.
123. Iwasaki, Y.; Fukushima, H.; Nakaura, H.; Yajima, T.; Ishiguro, T. A new method for measuring odors by triangle odor bag method (I)—Measurement at the source. *J. Jpn. Soc. Air Pollut.* **1978**, *13*, 246–251.
124. Japan Ministry of the Environment. *Olfactory Measurement Method Quality Control Manual*; Japan Ministry of the Environment: Tokyo, Japan, 2002. Available online: http://www.env.go.jp/en/air/odor/olfactory_mm/02quality.pdf (accessed on 1 November 2020).
125. Higuchi, T.; Ukita, M.; Sekine, M.; Imai, T. A case study and recent improvements in odor management in Japan. *Water Pract.* **2007**, *1*, 1–7. [CrossRef]

126. Copley International Corporation. *National Survey of the Odor Problem—Study of Social and Economic Impact of Odors—Phase I*; U.S. Department of Commerce—National Technical Information Services: Alexandria, VA, USA, 1970.

127. Copley International Corporation. *National Survey of the Odor Problem—Study of Social and Economic Impact of Odors—Phase II*; U.S. Department of Commerce—National Technical Information Services: Alexandria, VA, USA, 1971.

128. Copley International Corporation. *National Survey of the Odor Problem—Study of Social and Economic Impact of Odors—Phase III: Development and Evaluation of a Model. Odor Control. Ordinance*; U.S. Department of Commerce—National Technical Information Services: Alexandria, VA, USA, 1973.

129. Huey, N.A.; Broering, L.C.; Jutze, G.A.; Gruber, C.W. Objective odor pollution control investigations. *Air Pollut. Control Assoc.* **1960**, *10*, 441–444. [CrossRef]

130. Department of Public Health and Environment, State of Colorado. *5 CCR 1001–4: Regulation No. 2 Odor Emissions*; 24-4-103 C.R.S; Department of Public Health and Environment, State of Colorado: Denver, CO, USA, 2013. Available online: https://www.colorado.gov/pacific/sites/default/files/5-CCR-1001--4.pdf (accessed on 27 July 2020).

131. Department of Energy & Environmental Protection, State of Connecticut. *Control of Odors*; Department of Energy & Environmental Protection, State of Connecticut: Hartford, CT, USA, 2022. Available online: http://eregulations.ct.gov/eRegsPortal/Browse/RCSA?id=Title%2022a\T1\textbar{}22a-174\T1\textbar{}22a-174\T1\textbar{}23\T1\textbar{}22a-174\T1\textbar{}23 (accessed on 27 July 2020).

132. Department of Natural Resources and Environmental Control. *Control of Odorous Air Contaminants*; Sections 1100 and 1119; Department of Natural Resources and Environmental Control, State of Delaware: Dover, DE, USA, 1982. Available online: http://regulations.delaware.gov/AdminCode/title7/1000/1100/1119.shtml (accessed on 27 July 2020).

133. Environmental Protection Agency of Illinois. *Title 35: Environmental Protection, Subtitle B: Air Pollution, Chapter I: Pollution Control Board, Subchapter 1: Air Quality Standards and Episodes, Part 245 Odors*; Environmental Protection Agency of Illinois: Springfield, IL, USA, 2019. Available online: http://air.ky.gov/Pages/DAQ-Regulations.aspx (accessed on 27 July 2020).

134. Department of Environmental Protection, Division of Air Quality. *Ambient Air Quality*; Department of Environmental Protection, Division of Air Quality, State of Kentucky: Lexington, KY, USA, 2007. Available online: http://air.ky.gov/Pages/DAQ-Regulations.aspx (accessed on 27 July 2020).

135. Department of Natural Resources State of Missouri Division 10—Air Conservation. *Air Quality Standards, Definitions, Sampling and Reference Methods and Air Pollution Control Regulations for the Entire State of Missouri*; Department of Natural Resources State of Missouri Division 10—Air Conservation Commission: Jefferson City, MO, USA, 2013. Available online: https://www.sos.mo.gov/adrules/csr/current/10csr/10csr (accessed on 27 July 2020).

136. Division of Environmental Protection, State of Nevada. *Odors*; Division of Environmental Protection, State of Nevada: Carson City, NV, USA, 2016; Available online: https://www.leg.state.nv.us/nac/NAC-445B.html#NAC445BSec22087 (accessed on 27 July 2020).

137. Department of Health, State of North Dakota. *Restriction of Odorous Air Contaminants*; Department of Health, State of North Dakota EPA Administrative Code: Bismarck, ND, USA, 2007; Chapters 33, 15, 16. Available online: http://www.legis.nd.gov/information/acdata/pdf/33--15--16.pdf (accessed on 27 July 2020).

138. Department of Environmental Protection. *To Prevent and Control the Discharge of Air Pollutants into the Open Air Which Causes or Contributes to an Objectionable Odor or Odors*; Department of Environmental Protection, State of West Virginia: Charleston, WV, USA, 1967; p. 45.

139. Department of Environmental Quality State of Wyoming. *Ambient Standards for Odors*; Department of Environmental Quality State of Wyoming: Cheyenne, WY, USA, 2016; Chapter 2, Section 11. Available online: https://eqc.wyo.gov/Public/ViewPublicDocument.aspx?DocumentId=12592 (accessed on 27 July 2020).

140. State of Oregon Department of Environmental Quality. *Nuisance Odor Strategy*; State of Oregon Department of Environmental Quality: Portland, OR, USA, 2014.

141. *Oregon Administrative Rule: Chapter 340: Environmental Quality, Division 208: Visible Emissions and Nuisance Requirements*; Chapter 340, Division 208; State of Oregon: Salem, OR, USA, 2016.

142. Chapter 42 Environment: Article V—Odor Control. In *City of Des Moines, Iowa—Code of Ordinances*; Sections 301–317. 2020. Available online: https://www.municode.com/library/ia/des_moines/codes/code_of_ordinances?nodeId=MUCO_CH42EN_ARTVODCO (accessed on 27 July 2020).

143. Jacobson, L.D.; Auverman, B.W.; Massey, R.; Mitloehner, F.M.; Sutton, A.L.; Xin, H. Air issues associated with animal agriculture: A North American perspective, CAST. *Issue Paper-Counc. Agric. Sci. Technol.* **2011**, *47*, 1–24. Available online: https://www.cast-science.org/publication/air-issues-associated-with-animal-agriculture-a-north-american-perspective/ (accessed on 31 August 2020).

144. U.S. Environmental Protection Agency. National Air Emissions Monitoring Study. Available online: https://www.epa.gov/afos-air/national-air-emissions-monitoring-study (accessed on 31 August 2020).

145. Air Management Practices Assessment Tool (AMPAT). Iowa State University Extension and Outreach. Available online: http://www.agronext.iastate.edu/ampat/ (accessed on 31 August 2020).

146. Koziel, J.A.; Andersen, D.; Harmon, J.; Hoff, S.; Maurer, D.; Rieck-Hinz, A. Improvement to air management practices assessment tool (AMPAT): Scientific literature database. In Proceedings of the ASABE Annual International Meeting, New Orleans, LA, USA, 29 July 2015. Paper No. 152190993.

147. Maurer, D.L.; Koziel, J.A.; Harmon, J.D.; Hoff, S.J.; Rieck-Hinz, A.M.; Andersen, D.S. Summary of performance data for technologies to control gaseous, odor, and particulate emissions from livestock operations: Air management practices assessment tool (AMPAT). *Data Brief* **2016**, *7*, 1413–1429. [CrossRef]

148. EPA national ambient air quality standards for particulate matter; Final rule. In *Fed. Regist.*; 2013; 78, pp. 3086–3287. Available online: https://www.federalregister.gov/documents/2013/01/15/2012-30946/national-ambient-air-quality-standards-for-particulate-matter (accessed on 30 January 2021).

149. Redwine, J.S.; Lacey, R.E. A summary of state odor regulation pertaining to confined animal feeding operations. In Proceedings of the Second International Conference, Des Moines, IA, USA, 9–11 October 2000; pp. 33–41.

150. Bokowa, A.H. Review of odour legislation. *Chem. Eng. Trans.* **2011**, *23*, 32. Available online: https://www.cetjournal.it/index.php/cet/issue/view/vol23 (accessed on 5 October 2020). [CrossRef]

151. Government of New Brunswick. *Clean Air Act*; Government of New Brunswick: Fredericton, NB, Canada, 2013; Available online: https://www2.gnb.ca/content/gnb/en/departments/elg/environment/content/air_quality/clean_air.html (accessed on 1 February 2021).

152. Furberg, M.; Preston, K.; Smith, B. *Final Report, Odour Management in British Columbia: Review and Recommendations*; RWDI AIR Inc.: Vancouver, BC, Canada, 2005; pp. 830–999.

153. Government of Newfoundland and Labrador. *Environmental Protection Act*; Government of Newfoundland and Labrador: St John's, NL, Canada, 2013; Chapter E-14.2; Available online: https://www.assembly.nl.ca/legislation/sr/statutes/e14-2.htm (accessed on 1 February 2021).

154. Government of Northwest Territories and Nunavut. *Environmental Rights Act RSNWT*; Government of Northwest Territories and Nunavut: Iqaluit, NT, Canada, 1988; Volume C83, pp. 1–8. Available online: http://www.canlii.org/en/nu/laws/stat/rsnwt-nu-1988-c-83-supp/latest/rsnwt-nu-1988-c-83-supp.html (accessed on 5 October 2020).

155. Government of Nova Scotia. *Environment Act Chapter 1 of the Acts of 1994–95*; Government of Nova Scotia: Halifax, NS, Canada, 2014; Available online: https://www.nslegislature.ca/sites/default/files/legc/statutes/environment.pdf (accessed on 1 February 2021).

156. Government of Manitoba. *The Environment Act. C.C.S.M. c. E125*; Government of Manitoba: Winnipeg, MB, Canada, 2014. Available online: https://web2.gov.mb.ca/laws/statutes/ccsm/e125e.php (accessed on 1 February 2021).

157. Standards Development Branch. *Ontario's Ambient Air Quality Criteria*; (MOE) Standards Development Branch: Toronto, ON, Canada, 2012.

158. Ontario Ministry of Agriculture, Food and Rural Affairs. *Minimum Distance Separation (MDS) Formulae: Implementation Guidelines*; Publication 707; Ontario Ministry of Agriculture, Food and Rural Affairs (OMAFRA): Guelph, ON, Canada, 2006.

159. Environmental Protection Act Ontario Regulation 419/05: Air—Local Air Quality. Government of Ontario. 2013. Available online: http://www.e-laws.gov.on.ca/html/regs/english/elaws_regs_050419 (accessed on 5 October 2020).

160. Nutrient Management Act. 2002. Available online: http://www.omafra.gov.on.ca/english/engineer/facts/06--073.htm (accessed on 5 October 2020).

161. Ontario Municipal Act. 2001. Available online: https://www.ontario.ca/laws/statute/01m25 (accessed on 5 October 2020).

162. Environmental Quality Act. Government of Quebec. Available online: http://legisquebec.gouv.qc.ca/en/ShowDoc/cs/Q-2 (accessed on 19 October 2020).

163. Odours in Our Environment. Ontario Ministry of the Environment, Conservation and Parks. Available online: https://www.ontario.ca/page/odours-our-environment (accessed on 5 October 2020).

164. Guo, H.; Yu, Z. Supporting Information for the Development of an Odour Guideline for Saskatchewan. Ph.D. Thesis, University of Saskatchewan, Saskatoon, SK, Canada, 2011. Available online: https://harvest.usask.ca/bitstream/handle/10388/8629/HUANG-DISSERTATION-2018.pdf?sequence=1&isAllowed=y (accessed on 5 October 2020).

165. Ontario Ministry of Environment (MOE). *Ontario Source Testing Code, Part G, Method ON-6: Determination of Odour Emissions from Stationary Sources*; Ontario Ministry of Environment (MOE): Toronto, ON, Canada, 2010.

166. Ministry of the Environment, Chile. *Law N°37 2012. Standard for the Emission of Total Reduced Sulfur Compounds, Odors Generators Associated with the Manufacturing of Sulfated Pulp*; Ministry of the Environment: Santiago, Chile, 2012. Available online: http://www.leychile.cl/Navegar?idNorma=1049596 (accessed on 29 July 2020).

167. Ministerio Secretaría General de la Presidencia. *Law N° 19.300: Law of General Basis on the Environment*; Ministerio Secretaría General de la Presidencia: Santiago, Chile, 1994. Available online: http://www.leychile.cl/Navegar?idNorma=30667 (accessed on 29 July 2020).

168. ECOTEC Engineering Ltd. *Background for Regulation of Odors in Chile*; ECOTEC Engineering Ltd.: Santiago, Chile, 2013. Available online: https://olores.mma.gob.cl/wp-content/uploads/2019/03/ECOTEC-Ingenieria.pdf (accessed on 2 February 2021).

169. Ministry of the Environment Chile. *Strategy for The Management of Odors in Chile*; Ministry of the Environment Chile: Santiago, Chile, 2017. Available online: https://olores.mma.gob.cl/wp-content/uploads/2019/03/Estrategia_Olores_Actualizacion2017.pdf (accessed on 2 February 2021).

170. Instituto Nacional de Normalización. *NCh3387:2015: Air Quality and Assessment of Odour Annoyance Survey*; Instituto Nacional de Normalización: Santiago, Chile, 2015.

171. Instituto Nacional de Normalización. *NCh3386:2015: Air Quality—Static Sampling for Olfactometry*; Instituto Nacional de Normalización: Santiago, Chile, 2015.

172. VDI 3883 Part 1: Effects and Assessments of Odours—Assessment of odor Annoyance. VDI 3883 Part 1. 2015. Available online: https://www.vdi.de/en/home/vdi-standards/details/vdi-3883-blatt-1-effects-and-assessment-of-odours-assessment-of-odour-annoyance-questionnaires (accessed on 29 January 2021).
173. Instituto Nacional de Normalización. *NCh3190:2010: Air Quality—Determination of Odour Concentration by Dynamic Olfactometry*; Instituto Nacional de Normalización: Santiago, Chile, 2010.
174. ICONTEC. *Calidad Del Aire. Determinación de la Concentración de Olor por Olfatometría Dinámica*; NTC 5880; ICONTEC: Bogota, Colombia, 2011.
175. ICONTEC. *Static Sampling Olfactometry*; NTC 6011; ICONTEC: Bogota, Colombia, 2013.
176. ICONTEC. *Effects and Evaluation of Odors Psychometric Evaluation of Odor Discomfort Questionnaires*; NTC 6012—1; ICONTEC: Bogota, Colombia, 2013.
177. ICONTEC. *Effects and Evaluation of Odors Determination of Nuisance Parameters by Repeated Short Questions, to Panelists from a Neighborhood*; NTC 6012—2; ICONTEC: Bogota, Colombia, 2013.
178. Ministerio de Ambiente y Desarrollo Sostenible. *Resolution 1541 by which the Permissible Levels of Air Quality or Immission are Established, the Procedure for the Evaluation of Activities that Generate Offensive Odors and other Provisions are Dictated(Resolución 1541 por la cual se Establecen los Niveles Permisibles de Calidad del Aire o de Inmisión, el Procedimiento para la Evaluación de Actividades que Generan Olores Ofensivos y se Dictan otras Disposiciones)*; Ministerio de Ambiente y Desarrollo Sostenible: Bogota, Colombia, 2013.
179. ICONTEC. *Measurement of Odor Impact by Field Inspection. Measurement of the Impact Frequency of Recognizable Odors. Mesh Measurement*; NTC 6049—1; ICONTEC: Bogota, Colombia, 2014.
180. ICONTEC. *Measurement of Odor Impact by Field Inspection. Measurement of the Impact Frequency of Recognizable Odors. Pen Measurement*; NTC 6049—2; ICONTEC: Bogota, Colombia, 2014.
181. ICONTEC. *Measurement of Odor Impact by Field Inspection. Determination of Odor Intensity and Hedonic Odor Tone*; NTC 6049—3; ICONTEC: Bogota, Colombia, 2014.
182. ICONTEC. *Measurement of Odor Impact by Determining Hedonic Tone of Odor Polarity Profiles*; NTC 6049—4; ICONTEC: Bogota, Colombia, 2014.
183. Ministerio de Ambiente y Desarrollo Sostenible. *Resolution 2087: Protocol for the Monitoring, Control and Surveillance of Offensive Odors(Resolución 2087: Protocolo para el Monitoreo, Control y Vigilancia de Olores Ofensivos)*; Ministerio Ambiente y Desarrollo Sostenible: Bogota, Colombia, 2014.
184. Japan Environment Agency. *Calculation Method of Odor Index and Odor Intensity*; No. 63; Japan Environment Agency: Tokyo, Japan, 1995. Available online: https://www.env.go.jp/en/laws/air/odor/cm.html (accessed on 12 October 2020).
185. Joint Standards Australia/Standards New Zealand Committee EV-007, Methods for Examination of Air. Stationary Source Emissions Part 3: Determination of Odour Concentration by Dynamic Olfactometry. *Aust. New Zealand Stand.* **2001**, *4323*, 1–56. Available online: https://www.saiglobal.com/PDFTemp/Previews/OSH/as/as4000/4300/43233.pdf (accessed on 12 October 2020).
186. ASTM International. *Standard Practice for Determination of Odor and Taste Thresholds by a Forced-Choice Ascending Concentration Series Method of Limits*; ASTM E679–19; ASTM International: West Conshohocken, PA, USA, 2019.
187. Japan Environment Agency. *Measurement Method of Specified Offensive Odor Substances*; Japan Environment Agency: Tokyo, Japan, 1972. Available online: https://www.env.go.jp/en/laws/air/odor/mm.html (accessed on 12 October 2020).
188. Brancher, M.; Griffiths, K.D.; Franco, D.; de Melo Lisboa, H. A review of odour impact criteria in selected countries around the world. *Chemosphere* **2017**, *168*, 1531–1570. [CrossRef] [PubMed]
189. Hayes, J.E.; Stevenson, R.J.; Stuetz, R.M. The impact of malodour on communities: A review of assessment techniques. *Sci. Total Environ.* **2014**, *500–501*, 395–407. [CrossRef] [PubMed]
190. Braithwaite, S. Sensory Analysis and Health Risk Assessment of Environmental Odors. Ph.D. Thesis, UCLA, Los Angeles, CA, USA, 2019. Available online: https://escholarship.org/uc/item/21g660d3 (accessed on 28 December 2020).

MDPI
St. Alban-Anlage 66
4052 Basel
Switzerland
Tel. +41 61 683 77 34
Fax +41 61 302 89 18
www.mdpi.com

Atmosphere Editorial Office
E-mail: atmosphere@mdpi.com
www.mdpi.com/journal/atmosphere